THE LOEB CLASSICAL LIBRARY
FOUNDED BY JAMES LOEB

EDITED BY
G. P. GOOLD

PREVIOUS EDITORS

T. E. PAGE   E. CAPPS
W. H. D. ROUSE   L. A. POST
E. H. WARMINGTON

FRONTINUS

LCL 174

FRONTINUS

THE STRATAGEMS
THE AQUEDUCTS OF ROME

WITH AN ENGLISH TRANSLATION BY

CHARLES E. BENNETT

EDITED BY MARY B. McELWAIN

TRANSLATION OF THE AQUEDUCTS
BEING A REVISION OF THAT OF
CLEMENS HERSCHEL

HARVARD UNIVERSITY PRESS
CAMBRIDGE, MASSACHUSETTS
LONDON, ENGLAND

First published 1925
Reprinted 1950, 1961, 1969, 1980, 1993, 1997

LOEB CLASSICAL LIBRARY® is a registered trademark
of the President and Fellows of Harvard College

ISBN 0-674-99192-3

*Printed in Great Britain by St Edmundsbury Press Ltd,
Bury St Edmunds, Suffolk, on acid-free paper.
Bound by Hunter & Foulis Ltd, Edinburgh, Scotland.*

CONTENTS

PREFACE	ix
LIFE AND WORKS OF FRONTINUS	xiii
THE MANUSCRIPTS	xxviii
BIBLIOGRAPHY	xxxv

THE STRATAGEMS

Book I	3
Book II	89
Book III	205
Book IV	267

THE AQUEDUCTS OF ROME — 329

Book I	339
Book II	389

INDEXES	469
ILLUSTRATIONS	485

ILLUSTRATIONS

FOLLOW PAGE 484

PORTA MAGGIORE

MAP OF AQUEDUCTS

RUINS OF AQUA CLAUDIA

THE SEVEN AQUEDUCTS AT THE PORTA MAGGIORE

AD SPEM VETEREM

RUINS OF AQUA CLAUDIA NEAR THE APPIAN WAY

MAP OF ROME AND VICINITY

TABLES SHOWING THE WATER SUPPLY OF THE
  CITY OF ROME

PREFACE

BEFORE his death in May 1921, Professor Bennett had finished the draft of his translation of the *Strategemata* and of his revision of Clemens Herschel's translation of the *De Aquis*. He had also, through various footnotes, indicated clearly his attitude toward the texts he had adopted as the basis for his translation. For the editorial revision of the versions, the introductory material, the index, many of the footnotes and the general matters of typography, the responsibility should rest with the undersigned. The references to the sources of the *Strategemata* have been selected for the most part from those cited in Gundermann's *conspectus locorum*.

The translation of the *Strategemata* is based upon Gundermann's text, Leipzig, 1888, with very few changes, which are indicated in the footnotes. The brackets indicating glosses or conjectures have been omitted for the sake of appearance. Professor Bennett's translation is the first English rendering of the *Strategemata* with any accuracy of interpretation, the only other English version, published in London in 1811 by Lieutenant Robert B. Scott, leaving much to be desired both in the matter of interpretation and the manner of expression. A French version, prepared under the direction of M. Nisard, is a careful piece of work, well annotated, and indicating knowledge of the sources as well as of the standard editions of the *Strategemata*.

PREFACE

The text of the *De Aquis* is that of Bücheler, Leipzig, 1858, with certain changes in spelling and punctuation and with the omission of his diacritical marks. Some variants in readings have been admitted, and where the text is unreadable, conjectures have been accepted and the translation bracketed. The translation is a revision of that of Herschel, 1899, since the credit for the first English version of this treatise must go not to a Latinist but to an hydraulic engineer. In the preface to his book, Mr. Herschel thus explains his undertaking of this translation: "Having had the study of Frontinus for my pastime and hobby for many years, it has seemed to me fitting that others should be enabled likewise to partake of the instruction and pleasure this has given me. . . . We laymen have been waiting for a long time for Latin scholars to do this thing for us, and they have not responded." Mr. Herschel was very familiar with the translations of the French architect, Rondelet, and the German builder, Dederich, and he resolved to add an English rendering to these translations by men whose interests were primarily scientific.

But though not a Latin scholar, Mr. Herschel achieved what few Latin scholars can hope to do. In his quest for first-hand information about his subject, he went to Rome, studied the aqueducts in person, conferred with Lanciani and others, and finally went to the monastery at Monte Cassino and succeeded in having this, the sole original manuscript of the *De Aquis*, photographed for publication in his book. This is an excellent facsimile, and a most valuable gift to the student of this work.

During the preparation of his translation, Mr.

PREFACE

Herschel found in Professor Bennett an interested and helpful adviser. In the conclusion of his preface he says: "Professor Charles E. Bennett gave me active aid, countenance and encouragement at every stage of the work, and ended off by reading the proofs and correcting the copy of the translation to an extent that makes the translation fairly his own." When, therefore, Professor Bennett was considering the preparation of this volume, Mr. Herschel generously turned over to him for revision his translation, which for twenty-five years has continued to be the only English version of the *De Aquis*. And within the past year this pioneer translator has given further proof of his continued interest in Frontinus and his spirit of co-operation with the students of the classics by patiently solving for the writer some of the technical difficulties encountered in proof reading. For this help grateful acknowledgment is here made.

MARY B. MCELWAIN.

Smith College, Northampton, Mass.

THE LIFE AND WORKS OF SEXTUS JULIUS FRONTINUS[1]

Of the details of the life of Frontinus we are but scantily informed. His personality, as will be shown, stands out in his works in no ambiguous fashion, but the events of his career, so far as we can glean them, are few, disjointed and indefinite. Even the year of his birth is not known, but since Tacitus[2] speaks of him as *praetor urbanus* in the year 70 A.D., we may infer that he was born not far from the year 35.

Of his family and of his birthplace we know as little. His family name, Julius, and the fact that he held the office of water commissioner, which, as he tells us,[3] was from olden times administered by the most eminent men of the State, would point to patrician descent. His writings on surveying,[4] so far as we have knowledge of them, betray the teachings of the Alexandrian school of mathematics, especially of Hero of Alexandria, and it is not unlikely that he was educated in that city.

He was three times elected consul, first in 73 or

---

[1] The biographical sketch here given is taken largely from Professor Bennett's article, "A Roman Waring" (*Atlantic Monthly*, March 1902), to which the reader is referred for a fuller and very sympathetic account of Frontinus as water commissioner, and from Herschel's *Life and Works of Frontinus*.
[2] *Hist.* iv. 39.   [3] *De Aquis*, 1, p. 331.   [4] *Cf.* p. xviii.

LIFE AND WORKS OF FRONTINUS

74,[1] again in 98,[2] and a third time in 100.[3] After his first incumbency of this office, he was dispatched to Britain as provincial governor.[4] In this post, as Tacitus[5] tells us, Frontinus fully sustained the traditions established by an able predecessor, Cerialis, and proved himself equal to the difficult emergencies with which he was called upon to cope. He subdued the Silures, a powerful and warlike tribe of Wales, and with the instinct for public improvements which dominated his whole career, at once began in the conquered district the construction of a highway, named from him the Via Julia, the course of which can still be made out, and some of whose ancient pavement, it is thought, may still be seen.[6]

From this provincial post he returned to Rome in 78, after which the next twenty years of his life are a blank. But to this period, from his forty-third to his sixty-second year, we attribute a large part of his writings. His treatise on the Art of War[7] may have been written immediately after his return from Britain in 78. His *Strategemata* is assigned by Gundermann to the years 84–96.[8] Within this period

---

[1] C. Nipperdey, *Opuscula* (Berlin, 1877), p. 520 ff., places the date of his first consulship in 73.

[2] *C.I.L.* iii. 2, p. 862 ; *Mart.* x. xlviii. 20.

[3] *C.I.L.* viii. 7066 ; vi. 2222.

[4] His exact tenure of office there is uncertain. *Cf.* E. Hübner, *Die römischen Legaten von Britannien.* Rhein. Mus. xii. (1857), p. 52 ; Nipperdey, *Opuscula, loc. cit.*

[5] Agricola, xvii : *sustinuitque molem Iulius Frontinus, vir magnus, quantum licebat, validumque et pugnacem Silurum gentem armis subegit, super virtutem hostium locorum quoque difficultates eluctatus.*

[6] *Cf.* Wm. Camden, *Britannia*, iii. p. 113 ; D. Williams, *History of Monmouthshire*, p. 36 ff.

[7] *Cf.* p. xviii.   [8] *Cf.* p. xx.

LIFE AND WORKS OF FRONTINUS

also his services as Augur doubtless began, an office in which the younger Pliny succeeded him at his death in 103 or 104.[1]

In 97 he was appointed to the post of water commissioner, the office whose management gives him probably his best title to eminence, and during the tenure of this he wrote the *De Aquis*. The office of water commissioner he held presumably until his death.

The *De Aquis* is primarily a valuable repository of information concerning the aqueducts of Rome. But it is much more than that. It gives us a picture of the faithful public servant, charged with immense responsibility, called suddenly to an office that had long been a sinecure and wretchedly mismanaged, confronted with abuses and corruption of long standing, and yet administering his charge with an eye only to the public service and an economical use of the public funds. It is this aspect of the *De Aquis* which lends it, despite its generally technical nature and its absolute lack of stylistic charm, a certain literary character. It depicts a man; it depicts motives and ideals, the springs of conduct.

The administration of which Frontinus was a part was essentially one of municipal reform. Nerva and Trajan alike aimed to correct the abuses and favouritism of the preceding régime. They not only chose able and devoted assistants in their new policy; they themselves set good examples for imitation.

In Frontinus they found a loyal and zealous champion of their reforms. Realizing the importance of his office, he proceeded to the study of its

[1] *Cf.* Pliny, *Epist.* IV. viii. 3 ; X. xiii.

LIFE AND WORKS OF FRONTINUS

details with the spirit of the true investigator, displaying at all times a scrupulous honesty and fidelity. Were one asked to point out, in all Roman history, another such example of civic virtue and conscientious performance of simple duty, it would be difficult to know where to find it. Men of genius, courage, patriotism are not lacking, but examples are few of men who laboured with such whole-souled devotion in the performance of homely duty, the reward for which could certainly not be large, and might possibly not exceed the approval of one's own conscience.

In Martial [1] we have a picture of Frontinus spending his leisure days in a delightful environment. Pliny [2] writes of appealing to him as one well qualified to help to settle a legal dispute. In the preface to an essay on farming [3] which Frontinus wrote, it is stated that he was interrupted in his writing by being obliged to serve as a soldier, and it is thought that this may have been on the occasion of Trajan's expedition against the Dacians in 99; this, however, is pure conjecture.

Near Oppenheim in Germany has been found an inscription [4] dedicated by Julia Frontina, presumably

---

[1] x. lviii. 1–6:
> Anxuris aequorei placidos, Frontine, recessus,
>   et propius Baias, litoreamque domum,
> et quod inhumanae cancro fervente cicadae
>   non novere nemus, flumineosque lacus
> dum colui, doctas tecum celebrare vacabat
>   Pieridas; nunc nos maxima Roma terit.

[2] *Epist.* v. i. 5 : *adhibui in consilium duos quos tunc civitas nostra spectatissimos habuit, Corellium et Frontinum.*

[3] *Cf.* p. xviii.

[4] *Cf.* A. Dederich, *Zeitschr. für die Alterthums-Wissenschaft,* vi. (1839), p. 841.

LIFE AND WORKS OF FRONTINUS

the daughter of Frontinus; its date is supposed to be about 84. Another inscription near the ancient Vetera Castra[1] is dedicated to Jupiter, Juno and Minerva in recognition of the recovery from illness of Sextus Julius Frontinus; and there is also a lead pipe, said to have been found near the modern Via Tiburtina, inscribed SEXTIULIFRONTINI.

Pliny[2] has preserved for us a saying of Frontinus, "Remembrance will endure if the life shall have merited it," and the truth of the words is most aptly exemplified in the case of their author. Rich and valuable as is his treatise, the *De Aquis*, in facts relating to the administration of ancient aqueducts, it is the personality of the writer that one loves to contemplate, his sturdy honesty, his conscientious devotion to the duties of his office, his patient attention to details, his loyal attachment to the sovereign whom he delighted to serve, his willing labours in behalf of the people whose convenience, comfort and safety he aimed to promote. We sympathize with him in his proud boast[3] that by his reforms he has not only made the city cleaner, but the air purer, and has removed the causes of pestilence that had formerly given the city such a bad repute; and we can easily pardon the Roman Philistinism with which, after enumerating the lengths and courses of the several aqueducts, he inquires in a burst of enthusiasm,[4] "Who will venture to com-

---

[1] A Roman camp on the Rhine, now Birten or Xanthen.

[2] *Epist.* IX. xix. 1, 6: *addis etiam melius rectiusque Frontinum, quod vetuerit omnino monumentum sibi fieri. Vetuit exstrui monumentum; sed quibus verbis? "Impensa monumenti supervacua est; memoria nostri durabit, si vita meruimus.*

[3] Cf. *De Aquis*, 88, p. 417.

[4] Cf. *De Aquis*, 16, p. 357.

LIFE AND WORKS OF FRONTINUS

pare with these mighty works the pyramids or the useless though famous works of the Greeks?" A thorough Roman of the old school, he has surely by his life, as revealed in the *De Aquis*, abundantly merited the remembrance which posterity has accorded and will long continue to accord him.

The works of Frontinus are all of a technical nature, written, as he tells us,[1] partly for his own instruction, and partly for the advantage of others. The first of these was probably a treatise on the Art of Surveying, of which fragments are extant. It consisted originally of two books, and the excerpts, collected by Lachmann,[2] treat the following subjects: *de agrorum qualitate, de controversiis, de limitibus, de controversiis agrorum*. The work is known to us principally through the *codex Arcerianus* at Wolfenbüttel, dating probably from the sixth or not later than the seventh century, which appears to have been a book used by the Roman State employees and contains treatises on Roman law and land surveying, including some pages of Frontinus. Various citations in other authors from this work of Frontinus point to the latter as a pioneer in this practical work of the Roman surveyor, and to his writings as the standard authority for many years.

The composition by Frontinus of a military work of a theoretical nature is attested first by his own words in the preface to his *Strategemata*,[3] and also by

[1] *De Aquis*, Pref. 2, p. 333.

[2] *Die Schriften der röm. Feldmesser*, Berlin, 1848, 1852. *Cf.* also M. Cantor, *Die röm. Agrimensoren und ihre Stellung in der Gesch. der Feldmesskunst*, Leipzig, 1875.

[3] *Cum ad instruendam rei militaris scientiam unus ex numero studiosorum eius accesserim eique destinato, quantum cura nostra valuit, satisfecisse visus sim*, etc.

LIFE AND WORKS OF FRONTINUS

statements of Aelian,[1] a late contemporary, and of Vegetius,[2] who wrote on the Art of War some three centuries later. This treatise is wholly lost, except in so far as Vegetius may have incorporated it in his own work.

The *Strategemata*, presumably following the lost work on the Art of War, which it was designed to supplement, narrates varied instances of successful stratagems, which illustrate the rules of military science, and which may serve to foster in other generals the power of conceiving and executing like deeds.[3] As it has come down to us, this work consists of four books, three of them written by Frontinus, the fourth by an author of unknown identity.[4] These four books were still further increased by additional examples, interpolated here and there throughout the work.

Such is Gundermann's conclusion, resulting from his own investigations[5] added to those of

[1] *De Instruendis Aciebus*, Pref. : παρὰ Φροντίνῳ τῶν ἐπισήμων ὑπατικῶν ἡμέρας τινὰς διέτριψα, δόξαν ἀπενεγκαμένῳ περὶ τὴν ἐν τοῖς πολέμοις ἐμπειρίαν . . . εὗρον οὐκ ἐλάττονα σπουδὴν ἔχοντα εἰς τὴν παρὰ τοῖς Ἕλλησι τεθεωρημένην μάθησιν.

[2] *De Re Militari*, i. 8: *compulit evolutis auctoribus ea me in hoc opusculo fidelissime dicere, quae Cato ille Censorius de disciplina militari scripsit, quae Cornelius Celsus, quae Frontinus perstringenda duxerunt;* and ii. 3: *nam unius aetatis sunt, quae fortiter fiunt; quae vero pro utilitate rei publicae scribuntur, aeterna sunt. Idem fecerunt alii complures, sed praecipue Frontinus, divo Traiano ab eius modi comprobatus industria.*   [3] Cf. *Strat.* Pref. p. 3.

[4] E. Fritze, P. Esternaux and F. Kortz dissent from this view and claim the fourth book also for Frontinus. *Cf.* also note 6 on p. xxii.

[5] G. Gundermann, *Quaestiones de Iuli Frontini Strategematon Libris*, Fleckeisen Jahrb. Supplementbd. 16 (1888), p. 315.

LIFE AND WORKS OF FRONTINUS

Wachsmuth[1] and Wölfflin.[2] From internal evidence Gundermann places the composition of the first three books between 84 and 96, basing this inference upon references to Domitian, who is repeatedly called Germanicus,[3] a title not given to him until after his expedition against the Germans in 83, and who is nowhere called *divus*, as is Vespasian in the *Strategemata*,[4] and Nerva in the *De Aquis*,[5] so that the composition of the work evidently fell within the lifetime of Domitian. The dating of the fourth book is a matter of conjecture. Wachsmuth assigned it to the fourth or fifth century, believing it the work of a *ludi magister*, who compiled it when seeking examples suitable for *declamationes* or *controversiae*. Wölfflin saw no reason to dissent from this conclusion. Gundermann, while admitting that there is no argument to prove that it was not written then,—except that if this view is correct, the pseudo-Frontinus must have imitated the purer speech of Frontinus *summo studio*,—thinks that its composition belongs rather to the beginning of the second century, and that its author was a student of rhetoric who lived not long after Frontinus, a dull man who did not weigh the value of his sources in his compilation. Gundermann cites IV. iii. 14 to support his theory, but Wachsmuth would transpose this example to the second book as being applicable to Frontinus himself. Schanz[6] enters into the controversy and

[1] C. Wachsmuth, *Ueber die Unächtheit des vierten Buchs der Frontinschen Strategemata*, Rhein. Mus. xv. (1860), p. 574.
[2] E. Wölfflin, *Frontins Kriegslisten*, Hermes, ix. (1875), p. 72.
[3] *Cf.* I i. 8 ; II. iii. 23 ; II. xi. 7.
[4] *Cf.* II. i. 17.
[5] *Cf.* 102.
[6] M. Schanz, *Zu Frontins Kriegslisten*, Philol. xlviii. (1889), p. 674.

LIFE AND WORKS OF FRONTINUS

says that the language of the fourth book conclusively refutes Wachsmuth's view; he submits instead the theory that the author of this book was a contemporary of Frontinus, the officer to whom the Lingones[1] submitted in 70 A.D., who drew his examples from Frontinus and other sources, and that a third person joined the two works, wrote a preface for the fourth book, and added to the preface of the first book. This hypothesis, he thinks, removes the troublesome problem of duplicates,[2] which could easily creep in with a third reader somewhat superficially handling new material.

The points of dissimilarity between the first three books and the fourth are treated in detail by Wachsmuth, and even more exhaustively by Wölfflin. The two works differ first of all in the plan followed by their respective authors. Frontinus in his preface[3] outlines the arrangement which he proposes to follow in presenting examples: in the first book he will give illustrations of stratagems employed before the battle begins; in the second, those that refer to the battle itself and that tend to effect the complete subjugation of the enemy; the third will contain stratagems connected with sieges and the raising of sieges. To this arrangement the titles of the chapters in the first three books conform, whereas the headings of the chapters in the fourth book give no suggestion of historical stratagems, but belong rather, as Wachsmuth says, to a *militarisches Moralbüchlein*. Stewechius, for this reason, conjectured that this fourth book might be Frontinus's theoretical work, but its preface controverts this idea.

[1] *Cf.* IV. iii. 14.
[2] *Cf.* p. xxv.
[3] *Cf.* p. 5.

LIFE AND WORKS OF FRONTINUS

In his further proof of the spurious character of the fourth book, Wachsmuth points out that of the duplicate illustrations found throughout the entire work,[1] all but one occur in Book IV.; he notes also that the examples in this book are drawn much more largely from Valerius Maximus than are those of the earlier books, and that several of its titles correspond to titles employed by Valerius Maximus,[2] and he further proceeds to cite thirty-two passages,[3] which he claims are taken almost verbatim from that author. He contrasts the use of such words as *traditur, fertur, dicitur*,[4] which he claims are found in no genuine example in the first three books, with the *constat*[5] of the true Frontinus, who would regard illustrations of unsafe tradition as of little benefit to the generals whom he wished to instruct.[6]

Wachsmuth finds traces of the pseudo-Frontinus in the fourth paragraph of the preface to Book I., which are designed to pave the way for the fourth book, where the στρατηγικὰ outnumber the στρατηγή-

---

[1] *Cf.* p. xxv.

[2] *i.e.* chapters i., iii., iv., v.. vi. Cf. Val. Max. II. vii; IV. iii.; VI. v.; III. viii.; IV. i.

[3] *i.e.* IV: i. 1, 2, 13, 17, 18, 23, 26, 31, 32, 38, 39, 40, 42, 44, 46; iii. 1, 3, 12; iv. 1, 2; v. 4, 13, 14, 15, 16, 17, 20, 23; vi. 3; vii. 29, 36, 39.

[4] *Cf.* IV: i. 1; i. 3; ii. 1; iii. 1, 10, 11; v. 13, 14; vii. 4.

[5] *Cf.* II. i. 13; II. iii. 21.

[6] H. M. Connor, in an appendix to her thesis, *A Study of the Syntax of the Strategemata of Frontinus* (Ithaca, 1921), makes a comparative study of the syntactical uses of the first three books and the fourth, and concludes: "A thoughtful examination of the four books has revealed to me no compelling argument in respect to syntactical structure, diction or content, which establishes the existence of a pseudo-Frontinus."

ματα, and where the writer has a distinct preference for *dicta*.[1]

On these and other grounds Wachsmuth brands as spurious a number of examples in their entirety and parts of various others. In his decisions against the following twenty, Wölfflin and Gundermann concur: I. iii. 7; I. vii 4; I. vii. 7; I. xi. 15; II. iii. 11; II. iv. 14; II. iv. 19; II. viii. 5; II. viii. 9; II. xi. 6; III. iv. 2; III. iv. 4; III. vii. 5; III. xii. 3; III. xiii. 3–5; III. xv. 2; IV. iii. 10; IV. vii. 11.

Wölfflin agrees with Wachsmuth in his general conclusions and continues this line of investigation. He begins by comparing the preface to the *De Aquis* with what he considers the genuine preface to the *Strategemata*, and notes similarities of style and structure. He then goes on to compare the first three books of the *Strategemata* with the fourth in points of Latinity, arrangement or subject matter. He contrasts the two authors' methods of employing proper names,[2] notes the frequent recurrence in Frontinus of certain phrases[3] not found in the pseudo-Frontinus, observes that Frontinus customarily places the author of his stratagems at the beginning of the story, and follows certain subordinate principles of subdivision[4] within the general divisions of his

[1] *Cf.* IV. v. 12, 13; vi. 3; vii. 4. Also IV. i. 17; IV. vii. 1, 2, 3, 16.

[2] *e.g.* Frontinus sometimes speaks of the elder Scipio merely as Scipio, sometimes as Africanus; he mentions the younger Scipio only once, as Scipio. In Book IV. the younger Scipio is once called Africanus, once Aemilianus; the elder is spoken of merely as Scipio.

[3] i.e. *ob hoc, ob id, ideoque,* as against *ob eam causam* in Book IV.

[4] *i.e.* that of nationality, followed by Val. Max.; or of locality, *e.g.* I. iv. 1–7, operations on land; 8–10, on rivers; 11–14, on sea.

LIFE AND WORKS OF FRONTINUS

work, neither of which usages characterizes the pseudo-Frontinus, and adds examples of other variations in Latinity and subject matter. Of Wachsmuth's thirty-two examples,[1] Wölfflin recognizes twenty as surely and directly taken from Valerius Maximus, and he adds to the list I. xi. 11–13, not mentioned by Wachsmuth. He considers the relation of the real and the pseudo-Frontinus to other authors from whom they drew their material, and finds a difference in their attitude toward Sallust, Caesar and Vegetius; and in general he discerns in the true Frontinus a truthfulness toward the facts given in his sources, whereas the pseudo-Frontinus, while exhibiting at times a slavish dependence on form, has no conscience about changing the facts. He believes that it was not by accident but by design that the fourth book was united to the other three, that the author of this book wished to be considered Frontinus and took certain definite measures to achieve that end, attempting to imitate the style of Frontinus in the use of certain phrases,[2] keeping all his stories within the period which would be known to Frontinus, and in the preface to Book IV. virtually claiming the authorship of the first three books. Wölfflin rejects also, as belonging to the pseudo-Frontinus, the third paragraph of the preface to Book I., as well as the fourth (rejected by Wachsmuth and Gundermann), since he finds in it a rhetorical exaggeration,[3] not characteristic of the true Frontinus, and a lack of consistency between the apology for incompleteness

---

[1] *Cf.* p. xxii.

[2] e.g. *quodam deinde tempore, tum cum maxime.*

[3] i.e. *quis enim ad percensenda omnia monumenta, quae utraque lingua tradita sunt, sufficiat?*

LIFE AND WORKS OF FRONTINUS

here expressed [1] and Frontinus's avowed intention of citing only as occasion shall demand.[2] But his strongest reason for suspecting the genuineness of these two paragraphs lies in the fact that their insertion here interferes with an arrangement exhibited elsewhere by Frontinus of annexing the summary of succeeding chapters directly to some such statement as *quibus deinceps generibus suas species attribui*.[3]

Gundermann reviews the arguments of Wachsmuth and Wölfflin, accepts many of their conclusions and adds to the evidence. He disagrees with Wölfflin as to the ungenuineness of the third paragraph of the preface, and defends the authenticity of several examples. Of the duplicates, the critics agree that IV. v. 8, 9, 10, 11, and IV. vii. 6 are interpolations from Book I., and that II. iv. 15, 16 are interpolations from Book IV. Wölfflin and Gundermann regard I. i. 11 as transposed from chapter v.; Wachsmuth thinks it originated in chapter i.

Besides these duplicates, there are several cases in which the same story has apparently been drawn from different sources and is, therefore, told differently in two places; *i.e.* I. iv. 9 and I. iv. 9*a*; I. v. 10 and III. ix. 9; I. v. 24 and II. xii. 4; II. viii. 11 and IV. i. 29; III. xvi. 1 and IV. vii. 36; III. ix. 6 and III. xi. 3; IV. ii. 5 and IV. ii. 7.

In addition to the stories suspected as a whole, various other portions of the text are regarded as interpolated, *i.e.* parts of I. ii. 6; I. xi. 13; II. iii. 7;

---

[1] *ne me pro incurioso reprehendat, qui praeteritum aliquod a nobis reppererit exemplum.*

[2] *quemadmodum res poscet.*

[3] *Cf.* Books II. and III., and the *De Aquis*.

LIFE AND WORKS OF FRONTINUS

II. v. 31; II. v. 34; II. ix. 2; IV. v. 14, which are condemned on grounds of Latinity or other lack of agreement with the genuine or even the pseudo-Frontinus. In all sections of the book are found errors in names and in facts, and many changes in order have been suggested. Wachsmuth would put II. ix. 3, 5 in III. viii., and IV. iii. 14 after II. xi. 7. Gundermann thinks II. viii. 5 should follow II. viii. 3, and II. viii. 9 follow II. viii. 10. For the transposition of a whole leaf of the manuscript, see p. xxxi. The errors in general Gundermann thinks should be attributed in small part to copyists, in larger part to the carelessness or the error of the author, but in largest part to the sources from which the material is drawn, many of which no longer exist.

In his preface to the *De Aquis*[1] Frontinus himself tells us how it came to be written. Having been invested with the duties of water commissioner, he deemed it of the greatest importance to familiarize himself with the business he had undertaken, considering nothing so disgraceful as for a decent man to conduct an office delegated to him according to the instructions of assistants. He therefore gathered together scattered facts bearing on his subject, primarily to serve for his own guidance and instruction, though not unmindful of the fact that his efforts might be found useful by his successor.

Animated by this spirit and purpose, he wrote his little manual, faithfully carrying out the programme which he had laid down for himself at the outset of the work. He tells us the names of the aqueducts existing in his day, when and by whom each was constructed, at what points each had its

[1] *Cf.* p. 331.

LIFE AND WORKS OF FRONTINUS

source, how far they were carried underground and how far on arches, the height and size of each, the number of taps and the distributions made from them, the amount of water supplied to public reservoirs, public amusements, State purposes and private persons, and finally what laws regulated the construction and maintenance of aqueducts, and what penalties enforced these laws, whether established by resolutions of the Senate or by edicts of the Emperors. And what he records is based not on hearsay, but on personal examination of all details, supplemented by the study of plans and charts which he had made.

The work is a simple and truthful narration of facts, containing a mass of technical detail essential to a complete understanding of the system described. As an honest and thorough-going exposition of that system, the *De Aquis* will always remain the starting-point for any investigation pertaining to the water supply of ancient Rome.

THE MANUSCRIPTS

I. OF THE *STRATEGEMATA*

THE codices of the *Strategemata*[1] are of two classes. Of the better class, three manuscripts survive; of the second, inferior class, there are many representatives.

The three manuscripts of the first class are the *codex Harleianus* 2666 (*H*), which contains the entire work, but is very carelessly written; the *codex Gothanus* I. 101 (*G*), and the *codex Cusanus* C. 14 (*C*), each of which contains excerpts only. These three have been taken from a copy which, though not free from errors and lacunae, was still most carefully written, and must be thought to have preserved with the greatest fidelity even mutilated words.

The second class is derived from a codex which the copyist had corrected in many places according to his pleasure, and the manuscripts of this class never bring any help to readings where the first class is in error. The best representative of the second class is *codex Parisinus* 7240 (*P*), which far surpasses all the rest in authority. Both classes come from the same archetype in which a leaf had been transposed.[2] A copyist of *P* about the thir-

---

[1] For a full account of all the manuscripts, *cf.* G. Gundermann, *De Iuli Frontini Strategematon libro quo fertur quarto*, Commentationes Philologae Jenenses (Leipzig, 1881), p. 83.

[2] *Cf.* p. xxxi.

THE MANUSCRIPTS

teenth century, noticing this transposition, attempted a re-arrangement which appears in some of the later codices and editions.

The archetype had many lacunae, glosses and duplicates. It seems to have contained, besides the *Strategemata*, a breviarium of Eutropius, since these two works are found combined in most of the oldest codices of both classes. After the thirteenth century, copyists put Vegetius and Frontinus in one codex.

The *codex Harleianus*, a parchment of the ninth or tenth century, is now in the British Museum. It contains many errors and omissions. Some of these were corrected by the copyist himself, others by another hand of the same period. The latter (designated by h) became the model for the second class of manuscripts, the readings coinciding in many instances with those of P. As the corrector frequently used the eraser, for readings lacking in G and C we have only the testimony of the second class. The readings of h are sometimes happy amendments, sometimes atrocious corruptions. The writer omitted, copied incorrectly, amended and changed the spelling, assimilating most prepositions.

The *codex Gothanus*, a parchment of the ninth century, contains a breviarium of Eutropius, a breviarium of Festus and excerpts from Frontinus.[1] It was written by two hands. The copyist of Frontinus frequently separated the words wrongly, and misunderstood signs of abbreviation.

---

[1] *i.e.* all of Book IV., followed by II. ix. 7—II. xii. 2 (from *quarum metu illi* through *secundum consuetudinem*); pref. to Book I.; I. i. 1-2; pref. to Book II.; II. i. 1-3; pref. to Book III.; III. i. 1-3; III. iii. 1-7; III. vii. 1-6.

THE MANUSCRIPTS

The *codex Cusanus* is a parchment of the twelfth century.[1] It is written in two columns and contains many other things besides excerpts[2] from Frontinus. Few of the latter are intact, most of them being contracted or otherwise changed; but this codex contains some excellent notes.

The *codex Parisinus* is a parchment dating from the end of the tenth or the beginning of the eleventh century. It contains the *Strategemata* and a breviarium of Eutropius. The order of examples found in the early codices is retained,[3] and the passage II. ix. 7—II. xii. 2 (*quarum metu illi—secundum consuetudinem*) stands at the end of Book IV., but after the words *persecuti aciem in fossas deciderunt et in eo modo victi sunt* (the concluding words of II. xii. 2 in our present arrangement, the conclusion of II. ix. 7 as it then stood), the copyist wrote *duo capitula sunt requirenda*.

d is the designation given by Gundermann to all the codices of the second class excluding *P*. These date mostly from the fourteenth or fifteenth century. The oldest and best among them are *Harleianus* 2729, *Oxoniensis-Lincolniensis* 100, *Parisinus* 5802, *Gudianus* 16. Here and there these codices show an amendment which is an improvement, but almost every-

---

[1] *Cf.* Joseph Klein, *Ueber eine Handschrift des Nicolaus von Cues.*, Berlin, 1866, p. 6.

[2] *i.e.* pref. to Book I. (*si qui erunt—hostis sit, cf.* p. 6), followed by I: i. 12, 13, 4; ii. 7; v. 1; vi. 4; vii. 2, 3; viii. 8; **x**. 4; xi. 5, 8, 19; xii. 1, 5, 12; II: i. 1, 2, 13, 17; ii. 2; v. 41; vi. 3; viii. 6, 7, 11; ix. 6, 7; xiii. 2, 3, 8; III: vii. 6; ix. 6; IV: i. 3, 5, 17, 29, 35, 36, 38, 42, 45; iii. 1, 2, 3, 9, 10, 12; iv. 1, 2; v. 12, 13; vii. 1, 4, 10, 14, 15, 37; II. xi. 2.

[3] *Cf.* p. **xxxi**.

THE MANUSCRIPTS

where are signs of corruptions which have been afterwards corrected.

In the manuscript from which the archetype of our present manuscripts was derived, a leaf (or leaves) at some time became transposed which contained the following: IV. vii. 42 (beginning with *continere vellet*), IV. vii. 43 (now II. ix. 8), 44 (II. ix. 9), 45 (II. ix. 10), II. x. 1, 2, II. xi. 1–7, II. xii. 1, 2 (through *secundum consuetudinem*). This leaf had stood originally after the leaf ending with the words *quarum metu illi con-* in II. ix. 7, and before that beginning with the words *adventaret recepit aciem* in II. xii. 2, and when transposed stood between the leaf which ends with the words *cum in agmine milites*, in IV. vii. 42, and the leaf beginning with the words *-nere vellent pronuntiavit* in the same chapter.

To F. Haase[1] and E. Hedicke[2] belongs the credit of the present arrangement of the examples. Haase, having chanced on a manuscript of Frontinus in a book-shop, bought it fairly cheaply, being attracted by a note after II. ix. 7, which in all the better manuscripts ended in this way: *quarum metu illi cum adventarent, recepit aciem; persecuti aciem in fossas deciderunt et eo modo victi sunt.* In this copy there followed in red ink the words: *nota hic defectum magnum*. Haase was led by this to investigate more fully and he decided, as a result of his researches, that the passages found in the codices at the end of Book IV. should not be placed *after* II. ix. 7, but *within* it, since the conclusion of II. ix. 7 was quite unsuitable to its beginning. Hedicke later followed up Haase's work, and by the discovery of like in-

---

[1] Cf. *Rhein. Mus.* iii. (1845), p. 312.
[2] Cf. *Hermes*, vi. (1872), p. 156.

consistencies in the two parts of II. xii. 2, and IV. vii. 42, as they stood, and by the use of slight emendations, restored the order and consistency of the arrangement as it stands to-day.

II. OF THE *DE AQUIS*

As early as 1425 Poggio Bracciolini had learned that at Monte Cassino there was a copy of the *De Aquis* of Frontinus, but it was not until he visited the monastery in 1429 that the manuscript, *C*, was actually found. It was carried off to Rome, copied and returned to Monte Cassino, where it still remains. It is an original manuscript in the sense that at the time of its discovery no other manuscript of the work was known, nor has any since come to light, excepting those derived from copies made by Poggio at that time.

Eight of these copies are described by Bücheler in his preface, only two of them being of any significance. According to the judgment of Poleni, in which Bücheler concurs, these two were written a little after the middle of the fifteenth century. The *codex Urbinus,* or *Vaticanus* 1345, agrees closely with the original at Monte Cassino; the *codex Vaticanus* 4498, which contains many errors and seems to have come from an inferior copy, was used by Pomponius Laetus and Sulpicius in bringing out their first edition in 1484–1492. Jocundus may have had access to the original copy, since he agrees with it in certain readings not found in earlier editions. Poleni used a copy of *C* made by Gattola [1] and both Vatican manuscripts. Bücheler

[1] *qui satis ut illa aetate religiose Poleno librum descripsit.* B, xxxii

THE MANUSCRIPTS

consulted both Vaticans and a copy of *C* which Kellermann[1] had made for Schultze.

The original manuscript, *C* 361, is of parchment and contains, besides the *De Aquis*, the *De Re Militari* of Vegetius and a part of Varro's *De Lingua Latina*. It had seen hard usage before it came to the monastery, some leaves being torn and some chapters much mutilated. Poleni places its date at the end of the thirteenth or the beginning of the fourteenth century, while the catalogue of the library of Monte Cassino assigns it to the end of the eleventh or the beginning of the twelfth. Bücheler thinks it belongs to the thirteenth rather than the eleventh. It is written in minuscules which were growing dim even when copied by Gattola; to-day some parts of the manuscript are so obscure as to be difficult to read. Various portions were copied in red ink;[2] the punctuation is erratic; sentences sometimes begin with capitals, again with small letters; no intervals are left between words except where the intention is to show that something is missing from the text; but in places where lacunae are found, the spaces left do not always seem proportioned to the words to be supplied.

There are traces of emendation by some hand of a later century; dots, originally omitted, have been placed over the letter *i*; marks of abbreviation have

[1] *quo qui epigraphica studia attigerunt, auctorem sciunt nullum esse certiorem.* B.

[2] *e.g.* the words at the beginning of the preface: *Incipit prologus iuliifrontini in libro deaqueductu urbis;* at the end of chapter iii.: *Explicit prologus;* after chapter lxiii.: *liber primus explicit. liber secundus incipit;* the names of the aqueducts at the beginning of chapters lxv–lxxiii.

THE MANUSCRIPTS

been mischievously doubled and tripled in places,[1] words have been changed[2] and lacunae filled in.

In the mathematical part of the work, the numbers are often quite obviously incorrect, although it is hard to tell whether the inaccuracies are to be attributed to the author, whose interests were not primarily in the field of mathematics, and who may have used approximate figures at times, or to the copyist, who did not understand the signs or who may have erred in the substitution of the names of the figures for their symbols.

The title of the work given in *C* is *de aqueductu urbis Romae*; that of the early editions, *de acquaeductibus*; for these infelicities Bücheler has substituted the title suggested by Heinrich, *de aquis urbis Romae*.

The division into two books is indicated in *C* and in the *codex Urbinus*. The numbers of the chapters were added by Poleni. For a new collation of *C*, cf. Petschenig, Wien. Stud. vi. (1884), p. 249. For a facsimile of the manuscript, cf. Clemens Herschel, *The Two Books on the Water Supply of the City of Rome*.

---

[1] *Cf.* 112.   [2] e.g. *quarto* to *quinto* in 11.

BIBLIOGRAPHY

The *editio princeps* of the *Strategemata* was printed at Rome in 1487 by Eucharius Silber in the same book with Aelian, Vegetius and Modestus; this was reprinted in 1494 and 1497. In 1495 an edition appeared at Bologna containing the same works, and in the sixteenth century no fewer than ten editions were printed at Antwerp, Basel, Cologne, Leyden and Paris, in which Frontinus was either combined with Aelian, Vegetius and Modestus or was one of the authors represented in larger collections of examples of military achievements. In the seventeenth century six or seven editions appeared, including that of Tennulius (Leyden and Amsterdam, 1675), with notes and emendations, and that of Scriverius (Leyden, 1607), who was the first to unite the works of Frontinus in one volume. In his collection of treatises on military matters, which included the *Strategemata*, he incorporated also the *De Aquis*, *De Re Agraria*, *De Limitibus* and *De Coloniis*. He included also in this edition notes and comments of Stewechius and Modius. The edition of Keuchen (Amsterdam, 1661) also contains all these works; the Bipontine edition of 1788 and that of Dederich (Leipzig, 1855) contain the *Strategemata* and the *De Aquis*.

Oudendorp's edition (Leyden, 1731 and 1779) included notes and emendations of Modius, Stewechius, Tennulius, Casaubon, Salmasius, Gronovius

BIBLIOGRAPHY

and Scriverius. He consulted eleven codices in the preparation of his book, as well as the earlier editions, added copious notes of his own, incorporated in the volume Poleni's life of Frontinus,[1] and provided the book with an index. The edition of Schwebel (Leipzig, 1772) contained Poleni's life of Frontinus and selected notes from previous works with some additions of his own.

The authoritative critical edition is that of Gundermann, (Leipzig, 1888). This contains, in addition to a full critical apparatus, a *conspectus locorum*, a fairly exhaustive list of references to the sources of the *Strategemata*.

Among the books and monographs dealing with various matters connected with the *Strategemata*, the following may be mentioned:

G. Gundermann, *Quaestiones de Juli Frontini Strategematon Libris*, Fleckeis. Jahrb. Supplementbd. 16 (1888), p. 315.

M. Petschenig, *Sprachliches zu Frontins Strategemata*, Philol. Supplementbd. 6 (1892), p. 399.

A. Eussner, *Zu Frontins Strategemata*, Blätter für das bayr. Gymn. vii. (1871), p. 84.

J. Zechmeister, *De Iuli Frontini Strategematon Libris*, Wien. Stud. v. (1883), p. 224.

G. Hartel, *Analecta*, Wien. Stud. vi. (1884), p. 98.

P. Esternaux, *Die Komposition von Frontins Strategemata*, Progr. Berlin, 1899.

F. Kortz, *Quaest. Grammaticae de Julii Frontini Operibus Institutae*, Diss. Münster, 1893.

H. Düntzer, *Domitian in Frontins Strategemata*, Bonner Jahrbücher, H 96/97 (1896), p. 172.

---

[1] *Cf.* p. xxxviii.

BIBLIOGRAPHY

G. Grasso, *Una questione di topogr. stor. ed un errore di Frontino tra le imprese di Filippo II di Macedonia*, Rendiconti d. R. Istituto lombardo Ser. 2, vol. 31 (1898), p. 976.

A. Dederich, *Bruchstücke aus dem Leben des Sextus Julius Frontinus*, Zeitschr. für die Alterthumsw. 1839, pp. 834, 1077.

E. Fritze, *De Iuli Frontini Strategematon Libro* iv., Diss. Halle, 1888.

C. Wachsmuth, *Rhein. Mus.* xv. (1860), p. 574.

E. Wölfflin, *Hermes*, ix. (1875), p. 72.

F. Haase, *Rhein. Mus* iii. (1845), p. 312.

E. Hedicke, *Hermes*, vi. (1872), p. 156.

G. Gundermann, *Comment. philol. Jen.* 1 (1881), p. 83.

H. M. Connor, *A Study of the Syntax of the Strategemata of Frontinus*, Diss. Ithaca, 1921.

M. Schanz, *Philol.* xlviii. (1889), p. 674.

The *editio princeps* of the *De Aquis* is that of Pomponius Laetus and Sulpicius, brought out at Rome 1484–1492 in connection with Vitruvius, with whom this work of Frontinus was combined in most of the earlier editions. Then followed editions printed in Venice and Florence, 1495–1497, several[1] editions in the sixteenth century, including that of Jocundus published at Florence in 1513, and again in 1522 and 1523, and three in the seventeenth century, one of which was the edition of Scriverius.[2]

In 1722 Poleni[3] brought out his elegant edition, which far surpassed all the preceding works. He used three manuscripts in its preparation, *C* and

---

[1] One at Basel in 1530, and at Strassburg in 1543 and 1550.

[2] Cf. p. xxxv.

[3] *vir de Frontino optime meritus.* B.

BIBLIOGRAPHY

both Vaticans, and consulted former editions, using Jocundus as his base, but deferring to the codices in case of disagreement, except where the greater probability of Jocundus's reading won him to that. He claims that as an editor he forbore to make conjectures, unless driven to this by the most urgent necessity, which led to his being criticized for leaving many places unamended. He included much in the way of explanatory and illustrative material, being especially fitted to do this as a mathematician and a man of letters, appended a collection of imperial edicts concerning aqueducts, incorporated conjectures and notes of Opsopoeus, Scriverius, Scaliger and Keuchen, pruned out corruptions of copyists and editors, added a life of Frontinus, a prolegomenon and an index of matters treated there and in the notes. In the mathematical part of the work, in order to correct manifest errors, he recklessly changed numbers, disregarding the codex when he failed to understand the symbols, and using the system adopted by Metius[1] in his calculations, whereas Frontinus must, with all antiquity, have used the system of Archimedes.[2]

Other editions of the eighteenth century are the Bipontine (1788) and the Adler, printed at Altona in 1792. Dederich's edition (Wesel, 1841) included notes of Heinrich and Schultze. It is characterized by careless mistakes, bad judgment in the choice of readings, and a lack of intelligence in interpretation.

---

[1] Adrien Metius (1571–1635) was celebrated for adopting the fraction 355/113 to represent the relation of the circumference of a circle to its diameter, this relationship having been previously indicated by mathematicians from the time of Archimedes by 22/7.

[2] *Cf.* tables at end of book.

BIBLIOGRAPHY

The authoritative critical edition is that of Bücheler (Leipzig, 1858). In preparing it he consulted *C* and both Vatican manuscripts, and aimed to retain the reading of the codex whenever possible. He has taken great pains to atone for the copyist's carelessness by indicating for each lacuna the exact number of letters to be supplied. He has also bracketed many conjectures which have crept into the text as glosses, and where he has admitted his own or others' conjectures, he has clearly marked them as such. His text differs in places from the reading of *C* as shown by the facsimile in Herschel's edition, where he has failed to note the divergence. This may be due to inaccuracies in his copy of *C*.

We now have *Les Aqueducs de la Ville de Rome*. Ed. and French translation by P. Grimsal, Paris. Budé, 1944.

In 1899 Clemens Herschel visited the monastery at Monte Cassino and succeeded in securing a facsimile of *C* which stands in the front of his work, *The Two Books on the Water Supply of the City of Rome*. This work includes not only the translation of the *De Aquis*, but also several chapters treating of the measuring and distribution of water, water rights in Rome, the building of aqueducts, etc.

Among the important works relating to the study of the Roman aqueduct system, the first to mention is, of course, the epoch-making treatise of Rodolfo Lanciani, *Topografia di Roma antica, i commentarii di Frontino intorno le aque e le aque dotti, silloge epigrafica aquaria*, Memorie delle classe di scienze morali, stor. e filol. della accad. dei Lincei bol. 4 (1881), p. 315. Other books and monographs are:

BIBLIOGRAPHY

A. Rocchi, *Il diverticolo Frontin. all' acqua Tepula*, Studi e docum. di storia e diritto 17 (1896), p. 125.

K. Merckel, *Die Ingenieurtechnik im Altertum*, Berlin, 1899.

L. Cantarelli, *La Serie dei curatores aquarum*, Bolletino della commissione arch. com. di Roma 29 (1901), p. 180.

M. H. Morgan, *Remarks on the Water Supply of Ancient Rome*, Transactions and Proceedings of the Am. Phil. Ass. xxxiii. (1902), p. 30.

Th. Ashby, *Die antiken Wasserleitungen der Stadt Rom*, Neue Jahrb. für das klass. Altertum 23 (1909), p. 246.

G. Gundermann, *Berl. philol. Wochenschr.* xxiii. (1903), p. 1450.

R. Lanciani, *Ruins and Excavations of Ancient Rome.*

J. H. Middleton, *The Remains of Ancient Rome.*

S. B. Platner, *Ancient Rome.*

C. E. Bennett, "A Roman Waring," *Atlantic Monthly*, 1902, p. 382.

Clemens Herschel, Lecture delivered before the Engineering Students of Cornell University, 1894.

SIGLA

$H =$ Codex Harleianus 2666.
$P =$ Codex Parisinus 7240.
$d =$ Codices (singly or collectively) of an inferior class.
$C =$ Monte Cassino Codex.
cod. Vat. = Codex Vaticanus 4498.
cod. Urb. = Codex Urbinus.
$B =$ Bücheler.

Bibliographical addendum (1980)

Strategemata

G. Bendz, *Die Echtheitsfrage des vierten Buches der frontinschen Strategemata*, Lund 1938 (as against the views expressed in this Loeb volume, pp. xixff, Bendz convincingly argues for the authenticity of Book 4)

G. Bendz, *Textkritische und interpr. Bemerkungen*, Lund 1943.

De aquis

Critical edition by F. Krohn, Leipzig 1922
Les aqueducs, edited by P. Grimal (Budé 1944–61)
T. Ashby, *Aqueducts of Ancient Rome*, Oxford 1935.

THE STRATAGEMS OF
SEXTUS JULIUS FRONTINUS

BOOK I

IULI FRONTINI
STRATEGEMATON

LIBER PRIMUS

Cum ad instruendam rei militaris scientiam unus ex numero studiosorum eius accesserim eique destinato, quantum cura nostra valuit, satisfecisse visus sim, deberi adhuc institutae arbitror operae, ut sollertia ducum facta, quae a Graecis una στρατηγηµάτων appellatione comprehensa sunt, expeditis amplectar commentariis. Ita enim consilii quoque et providentiae exemplis succincti duces erunt, unde illis excogitandi generandique similia facultas nutriatur; praeterea continget, ne de eventu trepidet inventionis suae, qui probatis eam experimentis comparabit.

Illud neque ignoro neque infitior, et rerum gestarum scriptores indagine operis sui hanc quoque partem esse complexos et ab auctoribus exemplorum quidquid insigne aliquo modo fuit traditum. Sed, ut opinor, occupatis velocitate consuli debet. Longum est enim, singula et sparsa per immensum corpus

---

[1] The praenomen appears at the beginning of Book II in one MS., *P*.
[2] Frontinus alludes to his work on the Art of War. This is now lost.

THE STRATAGEMS OF SEXTUS[1] JULIUS FRONTINUS

BOOK I

Since I alone of those interested in military science have undertaken to reduce its rules to system,[2] and since I seem to have fulfilled that purpose, so far as pains on my part could accomplish it, I still feel under obligation, in order to complete the task I have begun, to summarize in convenient sketches the adroit operations of generals, which the Greeks embrace under the one name *strategemata*. For in this way commanders will be furnished with specimens of wisdom and foresight, which will serve to foster their own power of conceiving and executing like deeds. There will result the added advantage that a general will not fear the issue of his own stratagem, if he compares it with experiments already successfully made.

I neither ignore nor deny the fact that historians have included in the compass of their works this feature also, nor that authors have already recorded in some fashion all famous examples. But I ought, I think, out of consideration for busy men, to have regard to brevity. For it is a tedious business to hunt out separate examples scattered over the vast

historiarum persequi; et hi, qui notabilia excerpserunt, ipso velut acervo rerum confuderunt legentem. Nostra sedulitas impendet operam, ut, quemadmodum res poscet, ipsum quod exigitur quasi ad interrogatum exhibeat; circumspectis enim generibus, praeparavi opportuna exemplorum veluti consilia. Quo magis autem discreta ad rerum varietatem apte conlocarentur, in tres libros ea diduximus: in primo erunt exempla, quae competant proelio nondum commisso; in secundo, quae ad proelium et confectam pacationem pertineant; tertius inferendae solvendaeque obsidioni habebit στρατηγήματα; quibus deinceps generibus suas species adtribui.

Huic labori non iniuste veniam paciscar, ne me pro incurioso reprehendat, qui praeteritum aliquod a nobis reppererit exemplum. Quis enim ad percensenda omnia monumenta, quae utraque lingua tradita sunt, sufficiat? At multa et transire mihi ipse permisi; quod me non sine causa fecisse scient, qui aliorum libros eadem promittentium legerint; verum facile erit sub quaque specie suggerere. Nam cum hoc opus, sicut cetera, usus potius aliorum quam meae commendationis causa adgressus sim, adiuvari me ab his qui aliquid illi adstruent, non argui credam.

STRATAGEMS, I

body of history; and those who have made selections of notable deeds have overwhelmed the reader by the very mass of material. My effort will be devoted to the task of setting forth, as if in response to questions, and as occasion shall demand, the illustration applicable to the case in point. For having examined the categories, I have in advance mapped out my campaign, so to speak, for the presentation of illustrative examples. Moreover, in order that these may be sifted and properly classified according to the variety of subject-matter, I have divided them into three books. In the first are illustrations of stratagems for use before the battle begins; in the second, those that relate to the battle itself and tend to effect the complete subjugation of the enemy; the third contains stratagems connected with sieges and the raising of sieges. Under these successive classes I have grouped the illustrations appropriate to each.

It is not without justice that I shall claim indulgence for this work, and I beg that no one will charge me with negligence, if he finds that I have passed over some illustration. For who could prove equal to the task of examining all the records which have come down to us in both languages! And so I have purposely allowed myself to skip many things. That I have not done this without reason, those will realize who read the books of others treating of the same subjects; but it will be easy for the reader to supply those examples under each category. For since this work, like my preceding ones, has been undertaken for the benefit of others, rather than for the sake of my own renown, I shall feel that I am being aided, rather than criticized, by those who will make additions to it.

SEXTUS JULIUS FRONTINUS

Si qui erunt, quibus volumina haec cordi sint, meminerint στρατηγικῶν et στρατηγηγημάτων perquam similem naturam discernere. Namque omnia, quae a duce provide, utiliter, magnifice, constanter fiunt, στρατηγικὰ habebuntur; si in specie eorum sunt, στρατηγήματα. Horum propria vis in arte sollertiaque posita proficit tam ubi cavendus quam opprimendus hostis sit. Qua in re cum verborum quoque inlustris exstiterit effectus, ut factorum ita dictorum exempla posuimus.[1]

Species eorum, quae instruant ducem in his, quae ante proelium gerenda sunt:

I. De occultandis consiliis.
II. De explorandis consiliis hostium.
III. De constituendo statu belli.
IIII. De transducendo exercitu per loca hosti infesta.
V. De evadendo ex locis difficillimis.
VI. De insidiis in itinere factis.
VII. Quemadmodum ea, quibus deficiemur, videantur non deesse aut usus eorum expleatur.
VIII. De distringendis hostibus.
VIIII. De seditione militum compescenda.
X. Quemadmodum intempestiva postulatio pugnae inhibeatur.
XI. Quemadmodum incitandus sit ad proelium exercitus.
XII. De dissolvendo metu, quem milites ex adversis conceperint ominibus.

---

[1] *This paragraph is regarded by modern critics as interpolated. See Introduction, p. xxiv.*

STRATAGEMS, I

If there prove to be any persons who take an interest in these books, let them remember to discriminate between "strategy" and "stratagems," which are by nature extremely similar. For everything achieved by a commander, be it characterized by foresight, advantage, enterprise, or resolution, will belong under the head of "strategy," while those things which fall under some special type of these will be "stratagems." The essential characteristic of the latter, resting, as it does, on skill and cleverness, is effective quite as much when the enemy is to be evaded as when he is to be crushed. Since in this field certain striking results have been produced by speeches, I have set down examples of these also, as well as of deeds.

Types of stratagems for the guidance of a commander in matters to be attended to before battle:

  I. On concealing one's plans.
 II. On finding out the enemy's plans.
III. On determining the character of the war.
 IV. On leading an army through places infested by the enemy.
  V. On escaping from difficult situations.
 VI. On laying and meeting ambushes while on the march.
VII. How to conceal the absence of the things we lack, or to supply substitutes for them.
VIII. On distracting the attention of the enemy.
 IX. On quelling a mutiny of soldiers.
  X. How to check an unseasonable demand for battle.
 XI. How to arouse an army's enthusiasm for battle.
XII. On dispelling the fears inspired in soldiers by adverse omens.

SEXTUS JULIUS FRONTINUS

I. DE OCCULTANDIS CONSILIIS

1. M. PORCIUS CATO devictas a se Hispaniae civitates existimabat in tempore rebellaturas fiducia murorum. Scripsit itaque singulis, ut diruerent munimenta, minatus bellum, nisi confestim obtemperassent, epistulasque universis civitatibus eodem die reddi iussit. Unaquaeque urbium sibi soli credidit imperatum; contumaces conspiratio potuit facere, si omnibus idem denuntiari notum fuisset.

2. Himilco dux Poenorum, ut in Siciliam inopinatus appelleret classem, non pronuntiavit, quo proficisceretur, sed tabellas, in quibus scriptum erat, quam partem peti vellet, universis gubernatoribus dedit signatas praecepitque, ne quis legeret nisi vi tempestatis a cursu praetoriae navis abductus.

3. C. Laelius, ad Syphacem profectus legatus, quosdam ex tribunis et centurionibus per speciem servitutis ac ministerii exploratores secum duxit. Ex quibus L. Statorium, qui saepius in isdem castris fuerat et quem quidam ex hostibus videbantur agnoscere, occultandae condicionis eius causa baculo ut servum castigavit.

4. Tarquinius Superbus pater, principes Gabinorum interficiendos arbitratus, quia hoc nemini volebat commissum, nihil nuntio respondit, qui ad eum a filio erat missus; tantum virga eminentia papaverum capita, cum forte in horto ambularet, decussit. Nuntius sine responso reversus renuntiavit adulescenti

---

[1] 195 B.C. *Cf.* Appian *Hisp.* 41.
[2] 396 B.C. *Cf.* Polyaenus v. x. 2.
[3] 203 B.C. *Cf.* Livy xxx. 4.
[4] The surname Superbus, here given to Tarquinius Priscus, the father, is usually applied only to his son, the last Roman king.

I. ON CONCEALING ONE'S PLANS

MARCUS PORCIUS CATO believed that, when opportunity offered, the Spanish cities which he had subdued would revolt, relying upon the protection of their walls. He therefore wrote to each of the cities, ordering them to destroy their fortifications, and threatening war unless they obeyed forthwith. He ordered these letters to be delivered to all cities on the same day. Each city supposed that it alone had received the commands; had they known that the same orders had been sent to all, they could have joined forces and refused obedience.[1]

Himilco, the Carthaginian general, desiring to land in Sicily by surprise, made no public announcement as to the destination of his voyage, but gave all the captains sealed letters, in which were instructions what port to make, with further directions that no one should read these, unless separated from the flag-ship by a violent storm.[2]

When Gaius Laelius went as envoy to Syphax, he took with him as spies certain tribunes and centurions whom he represented to be slaves and attendants. One of these, Lucius Statorius, who had been rather frequently in the same camp, and whom certain of the enemy seemed to recognize, Laelius caned as a slave, in order to conceal the man's rank.[3]

Tarquin the Proud,[4] having decided that the leading citizens of Gabii should be put to death, and not wishing to confide this purpose to anyone, gave no response to the messenger sent to him by his son, but merely cut off the tallest poppy heads with his cane, as he happened to walk about in the garden. The messenger, returning without an

SEXTUS JULIUS FRONTINUS

Tarquinio, quid agentem patrem vidisset. Ille intellexit, idem esse eminentibus faciendum.

5. C. Caesar, quod suspectam habebat Aegyptiorum fidem, per speciem securitatis inspectioni urbis atque operum ac simul licentioribus conviviis deditus, videri voluit captum se gratia locorum ad mores Alexandrinos vitamque deficere; atque inter eam dissimulationem praeparatis subsidiis occupavit Aegyptum.

6. Ventidius Parthico bello adversus Pacorum regem, non ignarus Pharnaeum quendam, natione Cyrrhestem, ex his qui socii videbantur, omnia quae apud ipsos agerentur nuntiare Parthis, perfidiam barbari ad utilitates suas convertit. Nam quae maxime fieri cupiebat, ea vereri se ne acciderent, quae timebat, ea ut evenirent optare simulabat. Sollicitus itaque, ne Parthi ante transirent Euphraten, quam sibi supervenirent legiones quas in Cappadocia trans Taurum habebat, studiose cum proditore egit, uti sollemni perfidia Parthis suaderet, per Zeugma traicerent exercitum, qua et brevissimum iter est et demisso alveo Euphrates decurrit; namque si illa venirent, adseverabat se opportunitate collium usurum ad eludendos sagittarios, omnia autem vereri, si

---

[1] *Cf.* Livy i. 54; Val. Max. VII. iv. 2. Herod. v. 92 tells the same story of Periander and Thrasybulus.
[2] Alexandria.
[3] 48 B.C. *Cf.* Appian *C.* ii. 89.

answer, reported to the young Tarquin what he had seen his father doing. The son thereupon understood that the same thing was to be done to the prominent citizens of Gabii.[1]

Gaius Caesar, distrusting the loyalty of the Egyptians, and wishing to give the appearance of indifference, indulged in riotous banqueting, while devoting himself to an inspection of the city[2] and its defences, pretending to be captivated by the charm of the place and to be succumbing to the customs and life of the Egyptians. Having made ready his reserves while he thus dissembled, he seized Egypt.[3]

When Ventidius was waging war against the Parthian king Pacorus, knowing that a certain Pharnaeus from the province of Cyrrhestica, one of those pretending to be allies, was revealing to the Parthians all the preparations of his own army, he turned the treachery of the barbarian to his own advantage; for he pretended to be afraid that those things would happen which he was particularly desirous should happen, and pretended to desire those things to happen which he really dreaded. And so, fearful that the Parthians would cross the Euphrates before he could be reinforced by the legions which were stationed beyond the Taurus Mountains in Cappadocia, he earnestly endeavoured to make this traitor, according to his usual perfidy, advise the Parthians to lead their army across through Zeugma, where the route is shortest, and where the Euphrates flows in a deep channel; for he declared that, if the Parthians came by that road, he could avail himself of the protection of the hills for eluding their archers; but that he

SEXTUS JULIUS FRONTINUS

se infra per patentis campos proiecissent. Inducti hac adfirmatione barbari inferiore itinere per circuitum adduxerunt exercitum dumque fusiores ripas et ob hoc operosiore ponte[1] iungunt instrumentaque moliuntur, quadraginta amplius dies impenderunt. Quo spatio Ventidius ad contrahendas usus est copias eisque triduo, antequam Parthus adveniret, receptis acie commissa vicit Pacorum et interfecit.

7. Mithridates, circumvallante Pompeio, fugam in proximum diem moliens, huius consilii obscurandi causa latius et usque ad applicitas hosti valles pabulatus, conloquia quoque cum pluribus avertendae suspicionis causa in posterum constituit, ignes etiam frequentiores per tota castra fieri iussit; secunda deinde vigilia praeter ipsa hostium castra agmen eduxit.

8. Imperator Caesar Domitianus Augustus Germanicus, cum Germanos, qui in armis erant, vellet opprimere nec ignoraret maiore bellum molitione inituros, si adventum tanti ducis praesensissent, profectioni suae census obtexuit Galliarum; sub quibus inopinato bello adfusus contusa immanium ferocia nationum provinciis consuluit.

---

[1] operosiore ponte *Salmasius*; operosiores pontes *MSS. and edd.*

---

[1] Since Ventidius really wished the Parthians to take the longer route, that his reinforcements might have time to arrive, he led Pharnaeus to believe that he hoped they would cross by the shorter route. For he knew that Pharnaeus would counsel the Parthians to take the course of action not desired by himself.

[2] The text is very uncertain at this point, though the general meaning is clear.

[3] 38 B.C. *Cf.* Dio xlix. 19.

feared disaster if they should advance by the lower road through the open plains.[1] Influenced by this information, the barbarians led their army by a circuitous route over the lower road, and spent above forty days in preparing materials and in constructing a bridge[2] across the river at a point where the banks were quite widely separated and where the building of the bridge, therefore, involved more work. Ventidius utilized this interval for reuniting his forces, and having assembled these, three days before the Parthians arrived, he opened battle, conquered Pacorus, and killed him.[3]

Mithridates, when he was blockaded by Pompey and planned to retreat the next day, wishing to conceal his purpose, made foraging expeditions over a wide territory, and even to the valleys adjacent to the enemy. For the purpose of further averting suspicion, he also arranged conferences for a subsequent date with several of his foes; and ordered numerous fires to be lighted throughout the camp. Then, in the second watch, he led out his forces directly past the camp of the enemy.[4]

When the Emperor Caesar Domitianus Augustus Germanicus wished to crush the Germans, who were in arms, realizing that they would make greater preparations for war if they foresaw the arrival of so eminent a commander as himself, he concealed the reason for his departure from Rome under the pretext of taking a census of the Gallic provinces. Under cover of this he plunged into sudden warfare, crushed the ferocity of these savage tribes, and thus acted for the good of the provinces.[5]

[4] 66 B.C. *Cf.* Appian *Mithr.* 99.   [5] 83 A.D.

SEXTUS JULIUS FRONTINUS

9. Claudius Nero, cum e re publica esset Hasdrubalem copiasque eius, antequam Hannibali fratri iungerentur, excidi idcircoque festinaret se Livio Salinatori collegae suo, cui bellum mandatum fuerat, parum fidens viribus quae sub ipso erant, adiungere neque tamen discessum suum ab Hannibale, cui oppositus erat, sentiri vellet, decem milia fortissimorum militum elegit praecepitque legatis, quos relinquebat, ut eaedem stationes vigiliaeque agerentur, totidem ignes arderent, eadem facies castrorum servaretur, ne quid Hannibal suspicatus auderet adversus paucitatem relictorum. Cum deinde in Umbria occultatis itineribus collegae se iunxisset, vetuit castra ampliari, ne quod signum adventus sui Poeno daret, detractaturo pugnam, si consulum iunctas vires intellexisset. Igitur insciwm duplicatis adgressus copiis superavit et velocius omni nuntio rediit ad Hannibalem. Ita ex duobus callidissimis ducibus Poenorum eodem consilio alterum celavit, alterum oppressit.

10. Themistocles exhortans suos ad suscitandos festinanter muros, quos iussu Lacedaemoniorum deiecerant, legatis Lacedaemone missis, qui interpellarent, respondit, venturum se ad diluendam hanc existimationem; et pervenit Lacedaemonem. Ibi

---

[1] 207 B.C. *Cf.* Val. Max. VII. iv. 4; Livy xxvii. 43 ff.

When it was essential that Hasdrubal and his troops should be destroyed before they joined Hannibal, the brother of Hasdrubal, Claudius Nero, lacking confidence in the troops under his own command, was therefore eager to unite his forces with those of his colleague, Livius Salinator, to whom the direction of the campaign had been committed. Desiring, however, that his departure should be unobserved by Hannibal, whose camp was opposite his, he chose ten thousand of his bravest soldiers, and gave orders to the lieutenants whom he left that the usual number of patrols and sentries should be posted, the same number of fires lighted, and the usual appearance of the camp be maintained, in order that Hannibal might not become suspicious and venture to attack the few troops left behind. Then, when he joined his colleague in Umbria after secret marches, he forbade the enlargement of the camp, lest he give some sign of his arrival to the Carthaginian commander, who would be likely to refuse battle if he knew the forces of the consuls had been united. Accordingly, attacking the enemy unawares with his reinforced troops, he won the day and returned to Hannibal in advance of any news of his exploit. Thus by the same plan he stole a march on one of the two shrewdest Carthaginian generals and crushed the other.[1]

Themistocles, urging upon his fellow-citizens the speedy construction of the walls which, at the command of the Lacedaemonians, they had demolished, informed the envoys sent from Sparta to remonstrate about this matter, that he himself would come, to put an end to this suspicion. Accordingly he came to Sparta. There, by feigning illness, he secured

SEXTUS JULIUS FRONTINUS

simulato morbo aliquantum temporis extraxit; et postquam intellexit suspectam esse tergiversationem suam, contendit falsum allatum ad eos rumorem et rogavit, mitterent aliquos ex principibus, quibus crederent de munitione Athenarum. Suis deinde clam scripsit, ut eos qui venissent retinerent, donec refectis operibus confiteretur Lacedaemoniis, munitas esse Athenas neque aliter principes eorum redire posse, quam ipse remissus foret. Quod facile praestiterunt Lacedaemonii, ne unius interitum multorum morte pensarent.

11. L. Furius, exercitu perducto in locum iniquum, cum constituisset occultare sollicitudinem suam, ne reliqui trepidarent, paulatim se inflectens, tamquam circuitu maiore hostem adgressurus, converso agmine ignarum rei quae agebatur exercitum incolumem reduxit.[1]

12. Metellus Pius in Hispania interrogatus, quid postera die facturus esset, "tunicam meam, si eloqui posset," inquit, "comburerem."

13. M. Licinius Crassus percunctanti, quo tempore castra moturus esset, respondit: "vereris, ne tubam non exaudias?"

II. DE EXPLORANDIS CONSILIIS HOSTIUM

1. SCIPIO AFRICANUS, capta occasione mittendae ad Syphacem legationis, cum Laelio servorum habitu

---

[1] *Identical with* I. v. 13, *and regarded as an interpolation in this place.*

[1] 478 B.C. *Cf.* Thuc. i. 90 ff.
[2] 79–72 B.C. *Cf.* Val. Max. VII. iv. 5.
[3] Plut. *Demetr.* 28 tells the same story of Antigonus and Demetrius.

a considerable delay. But after he realized that his subterfuge was suspected, he declared that the rumour which had come to the Spartans was false, and asked them to send some of their leading men, whose word they would take about the building operations of the Athenians. Then he wrote secretly to the Athenians, telling them to detain those who had come to them, until, upon the restoration of the walls, he could admit to the Spartans that Athens was fortified, and could inform them that their leaders could not return until he himself had been sent back. These terms the Spartans readily fulfilled, that they might not atone for the death of one by that of many.[1]

Lucius Furius, having led his army into an unfavourable position, determined to conceal his anxiety, lest the others take alarm. By gradually changing his course, as though planning to attack the enemy after a wider circuit, he finally reversed his line of march, and led his army safely back, without its knowing what was going on.

When Metellus Pius was in Spain and was asked what he was going to do the next day, he replied: "If my tunic could tell, I would burn it."[2]

When Marcus Licinius Crassus was asked at what time he was going to break camp, he replied: "Are you afraid you'll not hear the trumpet?"[3]

II.—On Finding Out the Enemy's Plans

Scipio Africanus, seizing the opportunity of sending an embassy to Syphax, commanded specially chosen tribunes and centurions to go with Laelius,

SEXTUS JULIUS FRONTINUS

tribunos et centuriones electissimos ire iussit, quibus
curae esset perspicere regias vires. Hi, quo liberius
castrorum positionem scrutarentur, equum de industria dimissum tamquam fugientem persectati
maximam partem munimentorum circumierunt. Quae
cum nuntiassent, incendio confectum bellum est.

2. Q. Fabius Maximus bello Etrusco, cum adhuc
incognitae forent Romanis ducibus sagaciores explorandi viae, fratrem Fabium Caesonem, peritum
linguae Etruscae, iussit Tusco habitu penetrare
Ciminiam silvam, intemptatam ante militi nostro.
Quod is adeo prudenter atque industrie fecit, ut
transgressus silvam Umbros Camertes, cum animadvertisset non alienos nomini Romano, ad societatem
compulerit.

3. Carthaginienses, cum animadvertissent Alexandri ita magnas opes, ut Africae quoque immineret,
unum ex civibus, virum acrem nomine Hamilcarem
Rhodinum, iusserunt simulato exsilio ire ad regem
omnique studio in amicitiam eius pervenire. Qua is
potitus consilia eius nota civibus suis faciebat.

4. Idem Carthaginienses miserunt, qui per speciem
legatorum longo tempore Romae morarentur excipe-
rentque consilia nostrorum.

5. M. Cato in Hispania, quia ad hostium consilia

---

[1] *i.e.* the information furnished by the spies enabled
Scipio to set fire to the camp of Syphax.
[2] 203 B.C.   *Cf.* Livy xxx. 4 ff.
[3] 310 B.C.   *Cf.* Livy ix. 36.
[4] 331 B.C.   *Cf.* Justin. xxi. vi. 1.

disguised as slaves and entrusted with the task of spying out the strength of the king. These men, in order to examine more freely the situation of the camp, purposely let loose a horse and chased it around the greatest part of the fortifications, pretending it was running away. After they had reported the results of their observations, the destruction of the camp by fire[1] brought the war to a close.[2]

During the war with Etruria, when shrewd methods of reconnoitering were still unknown to Roman leaders, Quintus Fabius Maximus commanded his brother, Fabius Caeso, who spoke the Etruscan language fluently, to put on Etruscan dress and to penetrate into the Ciminian Forest, where our soldiers had never before ventured. He showed such discretion and energy in executing these commands, that after traversing the forest and observing that the Umbrians of Camerium were not hostile to the Romans, he brought them into an alliance.[3]

When the Carthaginians saw that the power of Alexander was so great that it menaced even Africa, they ordered one of their citizens, a resolute man named Hamilcar Rhodinus, to go to the king, pretending to be an exile, and to make every effort to gain his friendship. When Rhodinus had succeeded in this, he disclosed to his fellow-citizens the king's plans.[4]

The same Carthaginians sent men to tarry a long time at Rome, in the rôle of ambassadors, and thus to secure information of our plans.

When Marcus Cato was in Spain, being unable otherwise to arrive at a knowledge of the enemy's

SEXTUS JULIUS FRONTINUS

alia via pervenire non potuerat, iussit trecentos milites simul impetum facere in stationem hostium raptumque unum ex his in castra perferre incolumem. Tortus ille omnia suorum arcana confessus est.

6. C. Marius consul bello Cimbrico et Teutonico ad excutiendam Gallorum et Ligurum fidem litteras eis misit, quarum pars prior praecipiebat, ne interiores, quae praesignatae erant, ante certum tempus aperirentur. Easdem postea ante praestitutum diem repetiit et, quia resignatas reppererat, intellexit hostilia agitari.

Est et aliud explorandi genus, quo ipsi duces nullo extrinsecus adiutorio per se provident, sicut:[1]

7. Aemilius Paulus consul, bello Etrusco apud oppidum Vetuloniam demissurus exercitum in planitiem, contemplatus procul avium multitudinem citatiore volatu ex silva consurrexisse, intellexit aliquid illic insidiarum latere, quod et turbatae aves et plures simul evolaverant. Praemissis igitur exploratoribus comperit, decem milia Boiorum excipiendo ibi Romanorum agmini imminere, eaque alio quam exspectabatur latere missis legionibus circumfudit.

---

[1] *The last two lines of No. 6 are probably an interpolation.*

---

[1] 195 B.C. Plut. *Cat. Maj.* 13 attributes this stratagem to Cato at Thermopylae four years later.

[2] The letter was presumably in codex form, with the second and third leaves fastened together by a special seal.

[3] 104 B.C.

plans, he ordered three hundred soldiers to make a simultaneous attack on an enemy post, to seize one of their men, and to bring him unharmed to camp. The prisoner, under torture, revealed all the secrets of his side.[1]

During the war with the Cimbrians and Teutons, the consul Gaius Marius, wishing to test the loyalty of the Gauls and Ligurians, sent them a letter, commanding them in the first part of the letter not to open the inner part,[2] which was specially sealed, before a certain date. Afterwards, before the appointed time had arrived, he demanded the same letter back, and finding all seals broken, he knew that acts of hostility were afoot.[3]

There is also another method of securing intelligence, by which the generals themselves, without calling in any outside help, by their own unaided efforts take precautions, as, for instance:

In the Etruscan war, the consul Aemilius Paulus was on the point of sending his army down into the plain near the town of Vetulonia, when he saw afar off a flock of birds rise in somewhat startled flight from a forest, and realized that some treachery was lurking there, both because the birds had risen in alarm and at the same time in great numbers. He therefore sent some scouts ahead and discovered that ten thousand Boii were lying in wait at that point to meet the Roman army. These he overwhelmed by sending his legions against them at a different point from that at which they were expected.[4]

[4] Q. Aemilius Papus, consul in 282 and 278 B.C., waged war on the Etruscans. Pliny *N.H.* iii. 138 shows a like confusion of these names.

SEXTUS JULIUS FRONTINUS

8. Similiter Tisamenus Orestis filius, cum audisset iugum ab hostibus natura munitum teneri, praemisit sciscitaturos, quid rei foret; ac referentibus eis non esse verum, quod opinaretur, ingressus iter, ubi vidit ex suspecto iugo magnam vim avium simul evolasse neque omnino residere, arbitratus est latere illic agmen hostium. Itaque circumducto exercitu elusit insidiatores.

9. Hasdrubal frater Hannibalis iunctum Livii et Neronis exercitum, quamquam hoc illi non duplicatis castris dissimularent, intellexit, quod ab itinere strigosiores notabat equos et coloratiora hominum, ut ex via, corpora.

III. DE CONSTITUENDO STATU BELLI

1. ALEXANDER MACEDO, cum haberet vehementem exercitum, semper eum statum belli elegit, ut acie confligeret.

2. C. Caesar bello civili cum veteranum exercitum haberet, hostium autem tironem esse sciret, acie semper decertare studuit.

3. Fabius Maximus adversus Hannibalem, successibus proeliorum insolentem, recedere ab ancipiti discrimine et tueri tantummodo Italiam constituit, Cunctatorisque nomen et per hoc summi ducis meruit.

4. Byzantii adversus Philippum omne proeliandi

---

[1] *Cf.* Polyaen. II. xxxvii.
[2] 207 B.C. *Cf.* I. i. 9 and Livy xxvii. 47.
[3] 217 B.C. *Cf.* Livy XXII. xii. 6–12.

In like manner, Tisamenus, the son of Orestes, hearing that a ridge, a natural stronghold, was held by the enemy, sent men ahead to ascertain the facts; and upon their reporting that his impression was without foundation, he began his march. But when he saw a large number of birds all at once fly from the suspected ridge and not settle down at all, he came to the conclusion that the enemy's troops were hiding there; and so, leading his army by a detour, he escaped those lying in wait for him.[1]

Hasdrubal, brother of Hannibal, knew that the armies of Livius and Nero had united (although by avoiding two separate camps they strove to conceal this fact), because he observed horses rather lean from travel and men somewhat sunburned, as naturally results from marching.[2]

III. On Determining the Character of the War

Whenever Alexander of Macedon had a strong army, he chose the sort of warfare in which he could fight in open battle.

Gaius Caesar, in the Civil War, having an army of veterans and knowing that the enemy had only raw recruits, always strove to fight in open battle.

Fabius Maximus, when engaged in war with Hannibal, who was inflated by his success in battle, decided to avoid any dangerous hazards and to devote himself solely to the protection of Italy. By this policy he earned the name of Cunctator ("The Delayer") and the reputation of a consummate general.[3]

The Byzantines in their war with Philip, avoid-

SEXTUS JULIUS FRONTINUS

discrimen evitantes, omissa etiam finium tutela, intra munitiones oppidi se receperunt adsecutique sunt, ut Philippus obsidionalis morae impatiens recederet.

5. Hasdrubal Gisgonis filius secundo Punico bello in Hispania victum exercitum, cum P. Scipio instaret, per urbes divisit. Ita factum est, ut Scipio, ne oppugnatione plurium oppidorum distringeretur, in hiberna suos reduceret.

6. Themistocles adventante Xerxe, quia neque proelio pedestri neque tutelae finium neque obsidioni credebat sufficere Athenienses, auctor fuit eis liberos et coniuges in Troezena et in alias urbes amandandi relictoque oppido statum belli ad navale proelium transferendi.

7. Idem fecit in eadem civitate Pericles adversum Lacedaemonios.[1]

8. Scipio, manente in Italia Hannibale, transmisso in Africam exercitu necessitatem Carthaginiensibus imposuit revocandi Hannibalem. Sic a domesticis finibus in hostiles transtulit bellum.

9. Athenienses, cum Deceliam castellum ipsorum Lacedaemonii communissent et frequentius vexarentur, classem, quae Peloponnesum infestaret, miserunt consecutique sunt, ut exercitus Lacedaemoniorum, qui Deceliae erat, revocaretur.

10. Imperator Caesar Domitianus Augustus, cum Germani more suo e saltibus et obscuris latebris subinde impugnarent nostros tutumque regressum

---

[1] *No. 7 is thought to be an interpolation.*

---

[1] 339 B.C. *Cf.* Justin. ix. 1.
[2] 207 B.C. *Cf.* Livy xxviii. 2–3.
[3] 480 B.C. *Cf.* Herod. viii. 41.
[4] 431 B.C. *Cf.* Thuc. i. 143.

STRATAGEMS, I. III. 4–10

ing all risks of battle, and abandoning even the defence of their territory, retired within the walls of their city and succeeded in causing Philip to withdraw, since he could not endure the delay of a siege.[1]

Hasdrubal, the son of Gisco, in the Second Punic War, distributed his vanquished army among the cities of Spain when Publius Scipio pressed hard upon him. As a result, Scipio, in order not to scatter his forces by laying siege to several towns, withdrew his army into winter quarters.[2]

Themistocles, when Xerxes was approaching, thinking the strength of the Athenians unequal to a land battle, to the defence of their territory, or to the support of a siege, advised them to remove their wives and children to Troezen and other towns, to abandon the city, and to transfer the scene of the war to the water.[3]

Pericles did the same thing in the same state, in the war with the Spartans.[4]

While Hannibal was lingering in Italy, Scipio sent an army into Africa, and so forced the Carthaginians to recall Hannibal. In this way he transferred the war from his own country to that of the enemy.[5]

When the Spartans had fortified Decelea, a stronghold of the Athenians, and were making frequent raids from there, the Athenians sent a fleet to harass the Peloponnesus, and thus secured the recall of the army of Spartans stationed at Decelea.[6]

When the Germans, in accordance with their usual custom, kept emerging from woodland-pastures and unsuspected hiding-places to attack our men, and then finding a safe refuge in the depths of the

---

[5] 204 B.C. *Cf.* Appian *Hann.* 55; Livy xxviii. 40 ff.
[6] 413 B.C. *Cf.* Thuc. vii. 18.

SEXTUS JULIUS FRONTINUS

in profunda silvarum haberent, limitibus per centum viginti milia passuum actis non mutavit tantum statum belli, sed et subiecit dicioni suae hostes, quorum refugia nudaverat.

IV. DE TRANSDUCENDO EXERCITU PER LOCA HOSTI INFESTA

1. AEMILIUS PAULUS consul cum in Lucanis iuxta litus angusto itinere exercitum duceret et Tarentini ei classe insidiati agmen eius scorpionibus adgressi essent, captivos lateri euntium praetexuit, quorum respectu hostes inhibuere tela.

2. Agesilaus Lacedaemonius, cum praeda onustus ex Phrygia rediret insequerenturque hostes et ad locorum opportunitatem lacesserent agmen eius, ordinem captivorum ab utroque latere exercitus sui explicuit; quibus dum parcitur ab hoste, spatium transeundi habuerunt Lacedaemonii.

3. Idem, tenentibus angustias Thebanis, per quas transeundum habebat, flexit iter, quasi Thebas contenderet. Exterritis Thebanis digressisque ad tutanda moenia repetitum iter, quo destinaverat, emensus est nullo obsistente.

4. Nicostratus dux Aetolorum adversus Epirotas, cum ei aditus in fines eorum angusti forent, per

---

[1] 83 A.D.
[2] A military engine for throwing darts, stones, and other missiles.
[3] 282 B.C. *Cf.* Zonar. viii. 2. Since no Aemilius Paulus waged war with the Tarentines, this is probably the Papus referred to in the note to I. ii. 7.
[4] 396 B.C. *Cf.* Polyaen. II. i. 30.

forest, the Emperor Caesar Domitianus Augustus, by advancing the frontier of the empire along a stretch of one hundred and twenty miles, not only changed the nature of the war, but brought his enemies beneath his sway, by uncovering their hiding-places.[1]

IV. ON LEADING AN ARMY THROUGH PLACES INFESTED BY THE ENEMY

WHEN the consul Aemilius Paulus was leading his army along a narrow road near the coast in Lucania, and the fleet of the Tarentines, lying in wait for him, had attacked his troops by means of scorpions,[2] he placed prisoners as a screen to his line of march. Not wishing to harm these, the enemy ceased their attacks.[3]

Agesilaus, the Spartan, when returning from Phrygia laden with booty, was hard pressed by the enemy, who took advantage of their position to harass his line of march. He therefore placed a file of captives on each flank of his army. Since these were spared by the enemy, the Spartans found time to pass.[4]

The same Agesilaus, when the Thebans held a pass through which he had to march, turned his course, as if he were hastening to Thebes. Then, when the Thebans withdrew in alarm to protect their walls, Agesilaus resumed his march and arrived at his goal without opposition.[5]

When Nicostratus, king of the Aetolians, was at war with the Epirotes, and could enter their territory only by narrow defiles, he appeared at one

[5] 394 or 377 B.C. *Cf.* Xen. *Hell.* v. iv. 49 ff.; Polyaen. II. i. 24.

SEXTUS JULIUS FRONTINUS

alterum locum inrupturum se ostendens, omni illa
ad prohibendum occurrente Epirotarum multitudine,
reliquit suos paucos, qui speciem remanentis exercitus
praeberent; ipse cum cetera manu quo non exspectabatur aditu intravit.

5. Autophradates Perses, cum in Pisidiam exercitum duceret et angustias quasdam Pisidae occuparent,
simulata vexatione traiciendi instituit reducere.
Quod cum Pisidae credidissent, ille nocte validissimam manum ad eundem locum occupandum praemisit ac postero die totum traiecit exercitum.

6. Philippus Macedonum rex Graeciam petens,
cum Thermopylas occupatas audiret et ad eum legati
Aetolorum venissent acturi de pace, retentis eis ipse
magnis itineribus ad angustias pertendit securisque
custodibus et legatorum reditus exspectantibus inopinatus Thermopylas traiecit.

7. Iphicrates dux Atheniensium adversus Anaxibium Lacedaemonium in Hellesponto circa Abydon,
cum transducendum exercitum haberet per loca,
quae stationibus hostium tenebantur, alterum autem
latus eius transitus abscisi montes premerent,
alterum mare allueret, aliquamdiu moratus, cum
incidisset frigidior solito dies et ob hoc nemini
suspectus, delegit firmissimos quosque, quibus oleo
ac mero calefactis praecepit, ipsam oram maris

---

[1] 359–330 B.C. *Cf.* Polyaen. VII. xxvii. 1.
[2] 210 B.C.

point, as if intending to break through at that place. Then, when the whole body of Epirotes rushed thither to prevent this, he left a few of his men to produce the impression that his army was still there, while he himself, with the rest of his troops, entered at another place, where he was not expected.

Autophradates, the Persian, upon leading his army into Pisidia, and finding certain passes occupied by the Pisidians, pretended to be thwarted in his plan for crossing, and began to retreat. When the Pisidians were convinced of this, under cover of night he sent a very strong force ahead to seize the same place, and on the following day sent his whole army across.[1]

When Philip of Macedon was aiming at the conquest of Greece, he heard that the Pass of Thermopylae was occupied by Greek troops. Accordingly, when envoys of the Aetolians came to sue for peace, he detained them, while he himself hastened by forced marches to the Pass, and since the guards had relaxed their vigilance while awaiting the return of the envoys, by his unexpected coming he succeeded in marching through the Pass.[2]

When the Athenian general Iphicrates was engaged in a campaign against the Spartan Anaxibius on the Hellespont near Abydus, he had to lead his army on one occasion through places occupied by enemy patrols, hemmed in on the one side by precipitous mountains, and on the other washed by the sea. For some time he delayed, and then on an unusually cold day, when no one suspected such a move, he selected his most rugged men, rubbed them down with oil and warmed them up with wine, and then ordered them to skirt the very edge of the

legerent, abruptiora tranarent, atque ita custodes angustiarum inopinatos oppressit a tergo.

8. Cn. Pompeius, cum flumen transire propter oppositum hostium exercitum non posset, adsidue producere et reducere in castra instituit; deinde, in eam demum persuasionem hoste perducto, ne ullam viam ad progressum Romanorum teneret, repente impetu facto transitum rapuit.

9. Alexander Macedo, prohibente rege Indorum Poro traici exercitum per flumen Hydaspen, adversus aquam adsidue procurrere iussit suos; et ubi eo more exercitationis adsecutus est, ut[1] a Poro adversa ripa caveretur, per superiorem partem subitum transmisit exercitum.

9a. Idem, quia Indi fluminis traiectu prohibebatur ab hoste, diversis locis in flumen equites instituit immittere et transitum minari; cumque exspectatione barbaros intentos teneret, insulam paulo remotiorem primum exiguo, deinde maiore praesidio occupavit atque inde in ulteriorem ripam transmisit. Ad quam manum opprimendam cum universi se hostes effudissent, ipse libero vado transgressus omnes copias coniunxit.

10. Xenophon, ulteriorem ripam Armeniis tenentibus, duos iussit quaeri aditus; et cum a vado

---

[1] *The text is uncertain. The reading gives the obvious sense of the passage.*

---

[1] 389–388 B.C. *Cf.* Polyaen. III. ix. 33.
[2] *Cf.* the Spartan trick at Aegospotami, Xen. *Hell.* II. i. 21 ff.
[3] 326 B.C. *Cf.* Polyaen. IV. iii. 9; Plut. *Alex.* 60; Curt. xiii. 13. Frontinus, misled by different names given to the

sea, swimming across the places that were too precipitous to pass. Thus by an unexpected attack from the rear he overwhelmed the guards of the defile.[1]

When Gnaeus Pompey on one occasion was prevented from crossing a river because the enemy's troops were stationed on the opposite bank, he adopted the device of repeatedly leading his troops out of camp and back again. Then, when the enemy were at last tricked into relaxing their watch on the roads in front of the Roman advance, he made a sudden dash and effected a crossing.[2]

When Porus, a king of the Indians, was keeping Alexander of Macedon from leading his troops across the river Hydaspes, the latter commanded his men to make a practice of running toward the water. When by that sort of manœuvre he had led Porus to guard the opposite bank, he suddenly led his army across at a higher point of the stream.[3]

The same Alexander, prevented by the enemy from crossing the river Indus, began to send horsemen into the water at different points and to threaten to effect a crossing. Then, when he had the barbarians keyed up with expectation, he seized an island a little further off, at first with a small force, then with a larger one, and from there sent troops to the further bank. When the entire force of the enemy rushed away to overwhelm this band, he himself crossed safely by fords left unguarded and reunited all his troops.[3]

Xenophon once ordered his men to attempt a crossing in two places, in the face of Armenians who had possession of the opposite bank. Being repulsed

river, probably took the same story from two different sources. Cf. Introd. p. xxv.

SEXTUS JULIUS FRONTINUS

inferiore repulsus esset, transiit ad superius, inde quoque prohibitus hostium occursu repetit vadum inferius, iussa quidem militum parte subsistere, quae, cum Armenii ad inferioris vadi tutelam redissent, per superius transgrederetur. Armenii, credentes decursuros omnes, decepti sunt a remanentibus; hi cum resistente nullo vadum superassent, transeuntium suorum fuere propugnatores.

11. Appius Claudius consul primo bello Punico, cum a Regio Messanam traicere militem nequiret, custodientibus fretum Poenis, sparsit rumorem, quasi bellum iniussu populi inceptum gerere non posset, classemque in Italiam versus se agere simulavit. Digressis deinde Poenis, qui profectioni eius habuerant fidem, circumactas naves appulit Siciliae.

12. Lacedaemoniorum duces, cum Syracusas navigare destinassent et Poenorum dispositam per litus classem timerent, decem Punicas naves, quas captivas habebant, veluti victrices primas iusserunt agi, aut a latere iunctis aut puppe religatis suis. Qua specie deceptis Poenis transierunt.

13. Philippus, cum angustias maris, quae Στενά appellantur, transnavigare propter Atheniensium classem, quae opportunitatem loci custodiebat, non posset, scripsit Antipatro Thraciam rebellare, praesidiis quae ibi reliquerat interceptis; sequeretur

---

[1] 401 B.C. *Cf.* Xen. *Anab.* IV. iii. 20 ; Polyaen. I. xlix. 4.
[2] 264 B.C. *Cf.* Polyb. I. xi. 9. Zonar. VIII. viii. 6 gives a somewhat different account of this crossing.
[3] 397 B.C. *Cf.* Polyaen. II. xi. [4] *i.e.*, the Hellespont.

at the lower point, he passed to the upper; and when driven back from there also by the enemy's attack, he returned to the lower crossing, but only after ordering a part of his soldiers to remain behind and to cross by the upper passage, so soon as the Armenians should return to protect the lower. The Armenians, supposing that all were proceeding to the lower point, overlooked those remaining above, who, crossing the upper ford without molestation, defended their comrades as they also passed over.[1]

When Appius Claudius, consul in the first Punic War, was unable to transport his soldiers from the neighbourhood of Regium to Messina, because the Carthaginians were guarding the Straits, he caused the rumour to be spread that he could not continue a war which had been undertaken without the endorsement of the people, and turning about he pretended to set sail for Italy. Then, when the Carthaginians dispersed, believing he had gone, Appius turned back and landed in Sicily.[2]

When certain Spartan generals had planned to sail to Syracuse, but were afraid of the Carthaginian fleet anchored along the shore, they commanded that ten Carthaginian ships which they had captured should go ahead as though victors, with their own vessels either lashed to their side or towed behind. Having deceived the Carthaginians by these appearances, the Spartans succeeded in passing by.[3]

When Philip was unable to sail through the straits called Stena,[4] because the Athenian fleet kept guard at a strategic point, he wrote to Antipater that Thrace was in revolt, and that the garrisons which he had left there had been cut off, directing Antipater to leave all other matters and follow him.

omnibus omissis. Quae ut epistulae interciperentur ab hoste, curavit. Athenienses, arcana Macedonum excepisse visi, classem abduxerunt ; Philippus nullo prohibente angustias freti liberavit.

13a. Idem, quia Cherronessum, quae iuris Atheniensium erat, occupare prohibebatur, tenentibus transitum non Byzantiorum tantum, sed Rhodiorum quoque et Chiorum navibus, conciliavit animos eorum reddendo naves, quas ceperat, quasi sequestres futuras [1] ordinandae pacis inter se ac Byzantios, qui causa belli erant. Tractaque per magnum tempus postulatione, cum de industria subinde aliquid in condicionibus retexeret, classem per id tempus praeparavit eaque in angustias freti imparato hoste subitus evasit.

14. Chabrias Atheniensis, cum adire portum Samiorum obstante navali hostium praesidio non posset, paucas e suis navibus praeter portum missas iussit transire, arbitratus, qui in statione erant, persecuturos ; hisque per hoc consilium avocatis, nullo obstante portum cum reliqua adeptus est classe.

V. DE EVADENDO EX LOCIS DIFFICILLIMIS

1. Q. SERTORIUS in Hispania, cum a tergo instante hoste flumen traicere averet,[2] vallum in ripa eius in

[1] futuras *d* ; futuros *Gund.*
[2] haberet *HP* ; haveret, *Hartel, Gund.*

---

[1] 340–339 B.C. Polyaen. IV. ii. 8 attributes this stratagem to Philip on the occasion of his march against Amphissa.
[2] 339 B.C.
[3] 388 B.C. *Cf.* Polyaen. III. xi. 10, 12.

This letter Philip arranged to have fall into the hands of the enemy. The Athenians, imagining they had secured secret intelligence of the Macedonians, withdrew their fleet, while Philip now passed through the straits with no one to hinder him.[1]

The Chersonese happened at one time to be controlled by the Athenians, and Philip was prevented from capturing it, owing to the fact that the strait was commanded by vessels not only of the Byzantines but also of the Rhodians and Chians; but Philip won the confidence of these peoples by returning their captured ships, as pledges of the peace to be arranged between himself and the Byzantines, who were the cause of the war. While the negotiations dragged on for some time and Philip purposely kept changing the details of the terms, in the interval he got ready a fleet, and eluding the enemy while they were off their guard, he suddenly sailed into the straits.[2]

When Chabrias, the Athenian, was unable to secure access to the harbour of the Samians on account of the enemy blockade, he sent a few of his own ships with orders to cross the mouth of the harbour, thinking that the enemy on guard would give chase. When the enemy were drawn away by this ruse, and no one now hindered, he secured possession of the harbour with the remainder of his fleet.[3]

V. On Escaping from Difficult Situations

When Quintus Sertorius, in the Spanish campaign, desired to cross a river while the enemy were harassing him from the rear, he had his men con-

SEXTUS JULIUS FRONTINUS

modum cavae lunae duxit et oneratum materiis incendit; atque ita exclusis hostibus flumen libere transgressus est.

2. Similiter Pelopidas Thebanus bello Thessalico transitum quaesivit. Namque castris ampliorem locum supra ripam complexus, vallum cervolis et alio materiae genere constructum incendit, dumque ignibus submoventur hostes, ipse fluvium superavit.

3. Q. Lutatius Catulus, cum a Cimbris pulsus unam spem salutis haberet, si flumen liberasset, cuius ripam hostes tenebant, in proximo monte copias ostendit, tamquam ibi castra positurus. Ac praecepit suis, ne sarcinas solverent aut onera deponerent neu quis ab ordinibus signisque discederet; et quo magis persuasionem hostium confirmaret, pauca tabernacula in conspectu erigi iussit ignesque fieri et quosdam vallum struere, quosdam in lignationem, ut conspicerentur, exire. Quod Cimbri vere agi existimantes et ipsi castris delegerunt locum dispersique in proximos agros ad comparanda ea, quae mansuris necessaria sunt, occasionem dederunt Catulo non solum flumen traiciendi, sed etiam castra eorum infestandi.

4. Croesus, cum Halyn vado transire non posset neque navium aut pontis faciendi copiam haberet, fossa superiore parte post castra deducta alveum flumini a tergo exercitus sui reddidit.

---

[1] 80–72 B.C   [2] 369–364 B.C. *Cf.* Polyaen. II. iv. 2.
[3] 102 B.C.   [4] 546 B.C. *Cf.* Herod. i. 75.

struct a crescent-shaped rampart on the bank, pile it high with timber, and set fire to it. When the enemy were thus cut off, he crossed the stream without hindrance.[1]

In like manner Pelopidas, the Theban, in the Thessalian war, sought to cross a certain stream. Choosing a site above the bank larger than was necessary for his camp, he constructed a rampart of *chevaux-de-frise* and other materials, and set fire to it. Then, while the enemy were kept off by the fire, he crossed the stream.[2]

When Quintus Lutatius Catulus had been repulsed by the Cimbrians, and his only hope of safety lay in passing a stream the banks of which were held by the enemy, he displayed his troops on the nearest mountain, as though intending to camp there. Then he commanded his men not to loose their packs, or put down their loads, and not to quit the ranks or standards. In order the more effectively to strengthen the impression made upon the enemy, he ordered a few tents to be erected in open view, and fires to be built, while some built a rampart and others went forth in plain sight to collect wood. The Cimbrians, deeming these performances genuine, themselves also chose a place for a camp, scattering through the nearest fields to gather the supplies necessary for their stay. In this way they afforded Catulus opportunity not merely to cross the stream, but also to attack their camp.[3]

When Croesus could not ford the Halys, and had neither boats nor the means of building a bridge, he began up stream and constructed a ditch behind his camp, thus bringing the channel of the river in the rear of his army.[4]

SEXTUS JULIUS FRONTINUS

5. Cn. Pompeius Brundisii, cum excedere Italia et transferre bellum proposuisset, instante a tergo Caesare conscensurus classem quasdam obstruxit vias, alias parietibus intersaepsit, alias intercidit fossis easque sudibus erectis praeacutis [1] operuit cratibus, humo adgesta; quosdam aditus, qui ad portum ferebant, trabibus transmissis et in densum ordinem structis, ingenti mole tutatus est. Quibus perpetratis ad speciem retinendae urbis raros pro moenibus sagittarios reliquit, ceteras copias sine tumultu ad naves deduxit. Navigantem eum mox sagittarii quoque per itinera nota degressi parvis navigiis consecuti sunt.

6. C. Duellius consul in portu Syracusano, quem temere intraverat, obiecta ad ingressum catena clausus universos in puppem rettulit milites atque ita resupina navigia magna remigantium vi concitavit; levatae prorae super catenam processerunt. Qua parte superata transgressi rursus milites proras presserunt, in quas versum pondus decursum super catenam dedit navibus.

7. Lysander Lacedaemonius, cum in portu Atheniensium cum tota classe obsideretur, obrutis [2] hostium navibus ab ea parte, qua faucibus angustissimis

---

[1] praeclusas, *the MSS. and Gund.*; *but Gund. suggests* praeacutis, *comparing Caesar, B.C.* i. 27.
[2] *With this reading of the MSS., Oud. suggests this explanation*: per mersas ab hostibus de industria naves, ab ea scilicet parte qua deberet cum sua classe exire, unde obstructus ei erat exitus.

---

[1] 49 B.C. *Cf.* Caes. *B.C.* i. 27–28.
[2] 260 B.C. Zonar. viii. 16 makes Hippo, rather than Syracuse, the scene of this stratagem.   [3] Piraeus.

STRATAGEMS, I. v. 5-7

When Gnaeus Pompey at Brundisium had planned to leave Italy and to transfer the war to another field, since Caesar was heavy on his heels, just as he was on the point of embarking, he placed obstacles in some roads; others he blocked by constructing walls across them; others he intersected with trenches, setting sharp stakes in the latter, and laying hurdles covered with earth across the openings. Some of the roads leading to the harbour he guarded by throwing beams across and piling them one upon another in a huge heap. After consummating these arrangements, wishing to produce the appearance of intending to retain possession of the city, he left a few archers as a guard on the walls; the remainder of his troops he led out in good order to the ships. Then, when he was under way, the archers also withdrew by familiar roads, and overtook him in small boats.[1]

When the consul Gaius Duellius was caught by a chain stretched across the entrance to the harbour of Syracuse, which he had rashly entered, he assembled all his soldiers in the sterns of the boats, and when the boats were thus tilted up, he propelled them forward with the full force of his oarsmen. Thus lifted up over the chain, the prows moved forward. When this part of the boats had been carried over, the soldiers, returning to the prows, depressed these, and the weight thus transferred to them permitted the boats to pass over the chain.[2]

When Lysander, the Spartan, was blockaded in the harbour[3] of the Athenians with his entire fleet, since the ships of the enemy were sunk at the point where the sea flows in through a very narrow

SEXTUS JULIUS FRONTINUS

influit mare, milites suos clam in litus egredi iussit et subiectis rotis naves ad proximum portum Munychiam traiecit.

8. Hirtuleius legatus Q. Sertorii, cum in Hispania inter duos montes abruptos longum et angustum iter ingressus paucas duceret cohortes comperissetque ingentem manum hostium adventare, fossam transversam inter montes pressit vallumque materia exstructum incendit atque ita intercluso hoste evasit.

9. C. Caesar bello civili, cum adversus Afranium copias educeret et recipiendi se sine periculo facultatem non haberet, sicut constiterat, prima et secunda acie in armis permanente, tertia autem acie furtim a tergo ad opus applicata, quindecim pedum fossam fecit, intra quam sub occasum solis armati se milites eius receperunt.

10. Pericles Atheniensis, a Peloponnesiis in eum locum compulsus, qui undique abruptis cinctus duos tantum exitus habebat, ab altera parte fossam ingentis latitudinis duxit velut hostis excludendi causa, ab altera limitem agere coepit, tamquam per eum erupturus. Hi qui obsidebant, cum per fossam, quam ipse fecerat, exercitum Periclis non crederent evasurum, universi a limite obstiterunt. Pericles, pontibus quos praeparaverat fossae iniectis, suos quis non resistebatur emisit.

11. Lysimachus, ex his unus in quos opes Alexandri transierunt, cum editum collem castris destinasset,

---

[1] 404 B.C.   [2] 79–75 B.C.
[3] 49 B.C. *Cf.* Caes. *B.C.* i. 42.
[4] 430 B.C. *Cf.* III. ix. 9. Variations of the same story. Polyaen. v. x. 3 attributes this stratagem to Himilco.

STRATAGEMS, I. v. 7-11

entrance, he commanded his men to disembark secretly. Then, placing his ships on wheels, he transported them to the neighbouring harbour of Munychia.[1]

When Hirtuleius, lieutenant of Quintus Sertorius, was leading a few cohorts up a long narrow road in Spain between two precipitous mountains, and had learned that a large detachment of the enemy was approaching, he had a ditch dug across between the mountains, fenced it with a wooden rampart, set fire to this, and made his escape, while the enemy were thus cut off from attacking him.[2]

When Gaius Caesar led out his forces against Afranius in the Civil War, and had no means of retreating without danger, he had the first and second lines of battle remain in arms, just as they were drawn up, while the third secretly applied itself to work in the rear, and dug a ditch fifteen feet deep, within the line of which the soldiers under arms withdrew at sunset.[3]

Pericles the Athenian, being driven by the Peloponnesians into a place surrounded on all sides by precipitous cliffs and provided with only two outlets, dug a ditch of great breadth on one side, as if to shut out the enemy; on the other side he began to build a road, as if intending to make a sally by this. The besiegers, not supposing that Pericles' army would make its escape by the ditch which he had constructed, massed to oppose him on the side where the road was. But Pericles, spanning the ditch by bridges which he had made ready, extricated his men without interference.[4]

Lysimachus, one of the heirs to Alexander's power, having determined on one occasion to pitch

41

SEXTUS JULIUS FRONTINUS

imprudentia autem suorum in inferiorem deductus vereretur ex superiore hostium incursum, triplices fossas intra vallum obiecit; deinde simplicibus fossis circa omnia tentoria ductis tota castra confodit et intersaepto hostium aditu, simul humo quoque et frondibus quas fossis superiecerat patefacto,[1] in superiora evasit.

12. C. Fonteius Crassus in Hispania cum tribus milibus hominum praedatum profectus locoque iniquo circumventus ab Hasdrubale, ad primos tantum ordines relato consilio, incipiente nocte, quo tempore minime exspectabatur, per stationes hostium perrupit.[2]

13. L. Furius exercitu perducto in locum iniquum, cum constituisset occultare sollicitudinem suam, ne reliqui trepidarent, paulatim inflexit iter, tamquam circuitu maiore hostem adgressurus; conversoque agmine ignarum rei quae agebatur exercitum incolumem reduxit.[3]

14. P. Decius tribunus bello Samnitico Cornelio Cosso consuli iniquis locis deprehenso ab hostibus suasit, ut ad occupandum collem, qui erat in propinquo, modicam manum mitteret, seque ducem his qui mittebantur obtulit. Avocatus in diversum hostis dimisit consulem, Decium autem cinxit obseditque. Illas quoque angustias noctu eruptione facta cum frustratus esset Decius, incolumis cum militibus consuli accessit.[4]

---

[1] *Text and interpretation of this clause are hopelessly uncertain.*
[2] *Identical with* IV. v. 8.
[3] *Identical with* I. i. 11.
[4] *Practically identical with* IV. v. 9.

---

[1] 323–281 B.C.   [2] 343 B.C.   *Cf.* Livy vii. 34 ff.

his camp on a high hill, was conducted by the inadvertence of his men to a lower one. Fearing that the enemy would attack from above, he dug a triple line of trenches and encircled these with a rampart. Then, running a single trench around all the tents, he thus fortified the entire camp. Having thus shut off the advance of the enemy, he filled in the ditches with earth and leaves, and made his way across them to higher ground.[1]

Gaius Fonteius Crassus, when in Spain, having set out with three thousand men on a foraging expedition, was caught in an awkward position by Hasdrubal. At nightfall, when such a movement was least expected, communicating his plan only to the centurions of the first rank, he burst through the enemy's patrols.

Lucius Furius, having led his army into an unfavourable position, determined to conceal his anxiety, lest the others take alarm. By gradually changing his course, as though planning to attack the enemy after a wider circuit, he finally reversed his line of march, and led his army safely back, without their knowing what was going on.

When the consul Cornelius Cossus had been caught in a disadvantageous position by the enemy in the Samnite War, Publius Decius, tribune of the soldiers, urged him to send a small force to occupy a hill near by, and volunteered as leader of those who should be sent. The enemy, thus diverted to a different quarter, allowed the consul to escape, but surrounded Decius and besieged him. But Decius, extricating himself from this predicament by making a sortie at night, escaped with his men unharmed, and rejoined the consul.[2]

SEXTUS JULIUS FRONTINUS

15. Idem fecit sub Atilio Calatino consule is, cuius varie traditur nomen: alii Laberium, nonnulli Q. Caedicium, plurimi Calpurnium Flammam vocitatum scripserunt. Is cum demissum in eam vallem videret exercitum, cuius latera omniaque superiora hostis insederat, depoposcit et accepit trecentos milites, quos adhortatus, ut virtute sua exercitum servarent, in mediam vallem decucurrit; et ad opprimendos eos undique descendit hostis longoque et aspero proelio retentus occasionem consuli ad extrahendum exercitum dedit.[1]

16. Q. Minucius consul in Liguria, demisso in angustias exercitu, cum iam omnibus obversaretur Caudinae cladis exemplum, Numidas auxiliares, tam propter ipsorum quam propter equorum deformitatem despiciendos, iussit adequitare faucibus quae tenebantur. Primo intenti hostes, ne lacesserentur, stationem obiecerunt. De industria Numidae ad augendum sui contemptum labi equis et per ludibrium spectaculo esse adfectaverunt. Ad novitatem rei laxatis ordinibus barbari in spectaculum usque resoluti sunt. Quod ubi animadverterunt Numidae, paulatim succedentes additis calcaribus per intermissas hostium stationes eruperunt; quorum deinde cum proximos incenderent agros, necesse Liguribus fuit avocari ad defendenda sua inclusosque Romanos emittere.

[1] *Practically identical with* IV. v. 10.

---

[1] 258 B.C. *Cf.* Livy XXII. lx. 11; Zonar. viii. 12. Gell. iii. 7 gives a different account of this incident.
[2] 193 B.C. *Cf.* Livy xxxv. 11.

Under the consul Atilius Calatinus the same thing was done by a man whose name is variously reported. Some say he was called Laberius, and some Quintus Caedicius, but most give it as Calpurnius Flamma. This man, seeing that the army had entered a valley, the sides and all commanding parts of which the enemy had occupied, asked and received from the consul three hundred soldiers. After exhorting these to save the army by their valour, he hastened to the centre of the valley. To crush him and his followers, the enemy descended from all quarters, but, being held in check in a long and fierce battle, they thus afforded the consul an opportunity to extricate his army.[1]

When the army of the consul Quintus Minucius had marched down into a defile of Liguria, and the memory of the disaster of the Caudine Forks occurred to the minds of all, Minucius ordered the Numidian auxiliaries, who seemed of small account because of their own wild appearance and the ungainliness of their steeds, to ride up to the mouth of the defile which the enemy held. The enemy were at first on the alert against attack, and threw out patrols. But when the Numidians, in order to inspire still more contempt for themselves, purposely affected to fall from their horses and to engage in ridiculous antics, the barbarians, breaking ranks at the novel sight, gave themselves up completely to the enjoyment of the show. When the Numidians noticed this, they gradually drew nearer, and putting spurs to their horses, dashed through the lightly held line of the enemy. Then they set fire to the fields near by, so that it became necessary for the Ligurians to withdraw to defend their own territory, thereby releasing the Romans shut up at the pass.[2]

17. L. Sulla, bello sociali apud Aeserniam inter angustias deprehensus ab exercitu hostium, cui Duillius praeerat, conloquio petito de condicionibus pacis agitabat sine effectu. Hostem tamen propter indutias neglegentia resolutum animadvertens, nocte profectus relicto bucinatore, qui vigilias ad fidem remanentium divideret et quarta vigilia commissa consequeretur, incolumes suos cum omnibus impedimentis tormentisque in tuta perduxit.

18. Idem adversus Archelaum praefectum Mithridatis in Cappadocia, iniquitate locorum et multitudine hostium pressus, fecit pacis mentionem interpositoque tempore etiam indutiarum et per haec avocata intentione adversarium evasit.

19. Hasdrubal frater Hannibalis, cum saltum non posset evadere, faucibus eius obsessis, egit cum Claudio Nerone recepitque dimissum Hispania excessurum. Cavillatus deinde condicionibus dies aliquot extraxit, quibus omnibus non omisit per angustos tramites et ob id neglectos dimittere per partes exercitum; ipse deinde cum reliquis expeditis facile effugit.

20. Spartacus fossam, qua erat a M. Crasso circumdatus, caesis captivorum pecorumque corporibus noctu replevit et supergressus est.

---

[1] 90 B.C.   [2] 92 B.C.
[3] 211 B.C.   *Cf.* Livy xxvi. 17; Zonar. ix. 7.
[4] 71 B.C.

In the Social War, Lucius Sulla, surprised in a defile near Aesernia by the army of the enemy under the command of Duillius, asked for a conference, but was unsuccessful in negotiating terms of peace. Noting, however, that the enemy were careless and off their guard as a result of the truce, he marched forth at night, leaving only a trumpeter, with instructions to create the impression of the army's presence by sounding the watches, and to rejoin him when the fourth watch began. In this way he conducted his troops unharmed to a place of safety, with all their baggage and engines of war.[1]

The same Sulla, when fighting in Cappadocia against Archelaus, general of Mithridates, embarrassed by the difficulties of the terrain and the large numbers of the enemy, proposed peace. Then, taking advantage of the opportunity afforded by the truce, which served to divert the watchfulness of his adversary, he slipped out of his hands.[2]

Hasdrubal, brother of Hannibal, when unable to make his way out of a defile the entrance of which was held by the enemy, entered into negotiations with Claudius Nero and promised to withdraw from Spain if allowed to depart. Then, by quibbling over the terms, he dragged out negotiations for several days, during all of which time he was busy sending out his troops in detachments by way of paths so narrow that they were overlooked by the Romans. Finally he himself easily made his escape with the remainder, who were light-armed.[3]

When Marcus Crassus had constructed a ditch around the forces of Spartacus, the latter at night filled it with the bodies of prisoners and cattle that he had slain, and thus marched across it.[4]

21. Idem, in Vesuvio obsessus ea parte, qua mons asperrimus erat ideoque incustoditus, ex vimine silvestri catenas conseruit; quibus demissus non solum evasit, verum etiam ex alio latere Clodium ita terruit, ut aliquot cohortes gladiatoribus quattuor et septuaginta cesserint.

22. Idem, cum a P. Varinio proconsule praeclusus esset, palis per modica intervalla fixis ante portam erecta cadavera, adornata veste atque armis, alligavit, ut procul intuentibus stationis species esset, ignibus per tota castra factis. Imagine vana deluso hoste copias silentio noctis eduxit.

23. Brasidas dux Lacedaemoniorum, circa Amphipolim ab Atheniensium multitudine numero impar deprehensus, claudendum se praestitit, ut per longum coronae ambitum extenuaret hostilem frequentiam, quaque rarissimi obstabant, erupit.

24. Iphicrates in Thracia, cum depresso loco castra posuisset, explorasset autem ab hoste proximum teneri collem, ex quo unus ad opprimendos ipsos descensus erat, nocte paucis intra castra relictis imperavit, multos ignes facerent, eductoque exercitu

---

[1] 73 B.C.   *Cf.* Plut. *Crassus* 9; Flor. iii. 20.
[2] 73 B.C.
[3] 424 or 422 B.C.   *Cf.* Thuc. iv. 102, 106 ff.; v. 6–11.

The same Spartacus, when besieged on the slopes of Vesuvius at the point where the mountain was steepest and on that account unguarded, plaited ropes of osiers from the woods. Letting himself down by these, he not only made his escape, but by appearing in another quarter struck such terror into Clodius that several cohorts gave way before a force of only seventy-four gladiators.[1]

The same Spartacus, when enveloped by the troops of the proconsul Publius Varinius, placed stakes at short intervals before the gate of the camp; then setting up corpses, dressed in clothes and furnished with weapons, he tied these to the stakes, to give the appearance of sentries when viewed from a distance. He also lighted fires throughout the whole camp. Deceiving the enemy by this empty show, Spartacus by night silently led out his troops.[2]

When Brasidas, a general of the Spartans, was surprised near Amphipolis by a host of Athenians who outnumbered him, he allowed himself to be enveloped, in order to diminish the density of the enemy's ranks by lengthening the line of besiegers. Then he broke through at the point where the line was most lightly held.[3]

Iphicrates, when campaigning in Thrace, having on one occasion pitched his camp on low ground, discovered through scouts that the neighbouring hill was held by the enemy, and that from it came down a single road which might be utilized to overwhelm him and his men. Accordingly he left a few men in camp at night, and commanded them to light a number of fires. Then leading forth his troops and ranging them along the sides of the

et disposito circa latera praedictae viae passus est transire barbaros; locorumque iniquitate, in qua ipse fuerat, in illos conversa, parte exercitus tergo eorum cecidit, parte castra cepit.

25. Darius, ut falleret Scythas, discessu suo[1] canes atque asinos in castris reliquit. Quos cum latrantes rudentesque hostis audiret, remanere Darium credidit.

26. Eundem errorem obiecturi nostris Ligures per diversa loca buculos laqueis ad arbores alligaverunt, qui diducti frequentiore mugitu speciem remanentium praebebant hostium.

27. Hanno, ab hostibus clausus, locum eruptioni maxime aptum adgestis levibus materiis incendit; tum hoste ad ceteros exitus custodiendos avocato milites per ipsam flammam eduxit, admonitos ora scutis, crura veste contegere.

28. Hannibal, ut iniquitatem locorum et inopiam instante Fabio Maximo effugeret, noctu boves, quibus ad cornua fasciculos alligaverat sarmentorum, subiecto igne dimisit; cumque ipso motu adolescente flamma turbaretur pecus, magna discursatione montes, in quos actum erat, conlustravit. Romani, qui ad speculandum concurrerant, primo prodigium opinati sunt; dein cum certa Fabio renuntiassent,

---

[1] ut falleret Scythas, discessu suo *Bennett*; ut falleret Scythas discessu, canes *Gund*. *The MSS. lack* suo.

---

[1] 389 B.C. This same story is told in II. xii. 4. *Cf.* also Polyaen. III. ix. 41, 46, 50.
[2] 513 B.C. *Cf.* Herod. iv. 135; Polyaen. VII. xi. 4.

road just mentioned, he suffered the barbarians to pass by. When in this way the disadvantage of terrain from which he himself suffered had been turned against them, with part of his army he overwhelmed their rear, while with another part he captured their camp.[1]

Darius, in order to deceive the Scythians, left dogs and asses in camp at his departure. When the enemy heard these barking and braying, they imagined that Darius was still there.[2]

To produce a like misconception in the minds of our men, the Ligurians, in various places, tied bullocks to trees with halters. The animals, being thus separated, bellowed incessantly and produced the impression that the Ligurians were still there.

Hanno, when enveloped by the enemy, selected the point in the line best suited for a sortie, and, piling up light stuff, set fire to it. Then, when the enemy withdrew to guard the other exits, he marched his men straight through the fire, directing them to protect their faces with their shields, and their legs with their clothing.

Hannibal on one occasion was embarrassed by difficulties of terrain, by lack of supplies, and by the circumstance that Fabius Maximus was heavy on his heels. Accordingly he tied bundles of lighted fagots to the horns of oxen, and turned the animals loose at night. When the flames spread, fanned by the motion, the panic-stricken oxen ran wildly hither and thither over the mountains to which they had been driven, illuminating the whole scene. The Romans, who had gathered to witness the sight, at first thought a prodigy had occurred. Then, when scouts reported the facts, Fabius, fearing an ambush,

ille insidiarum metu suos castris continuit. Barbari obsistente nullo profecti sunt.

VI. De Insidiis in Itinere Factis

1. Fulvius Nobilior, cum ex Samnio in Lucanos exercitum duceret et cognovisset a perfugis hostes novissimum agmen eius adgressuros, fortissimam legionem primo ire, ultima sequi iussit impedimenta. Ita factum pro occasione amplexi hostes diripere sarcinas coeperunt. Fulvius legionem, de qua supra dictum est, quinque cohortes in dextram viae partem direxit, quinque ad sinistram, atque ita praedationi intentos hostes explicato per utraque latera milite clausit ceciditque.

2. Idem, hostibus tergum eius in itinere prementibus, flumine interveniente non ita magno, ut transitum prohiberet, moraretur tamen rapiditate, alteram legionem in occulto citra flumen conlocavit, ut hostes paucitate contempta audacius sequerentur. Quod ubi factum est, legio, quae ob hoc disposita erat, ex insidiis hostem adgressa vastavit.

3. Iphicrates in Thracia, cum propter condicionem locorum longum agmen deduceret et nuntiatum esset ei hostes summum id adgressuros, cohortes in utra-

---

[1] 217 B.C. *Cf.* Livy xxii. 16–17; Appian *Hann.* 14–15.
[2] Thus giving him time to set this ambush.

kept his men in camp. Meanwhile the barbarians marched away, as no one prevented them.[1]

VI.—On Laying and Meeting Ambushes while on the March

When Fulvius Nobilior was leading his army from Samnium against the Lucanians, and had learned from deserters that the enemy intended to attack his rearguard, he ordered his bravest legion to go in advance, and the baggage train to follow in the rear. The enemy, regarding this circumstance as a favourable opportunity, began to plunder the baggage. Fulvius then marshalled five cohorts of the legion I have mentioned above on the right side of the road, and five on the left. Then, when the enemy were intent on plundering, Fulvius, deploying his troops on both flanks, enveloped the foe and cut them to pieces.

The same Nobilior on one occasion was hard pressed from the rear by the enemy, as he was on the march. Across his route ran a stream, not so large as to prevent passage, but large enough to cause delay by the swiftness of the current.[2] On the nearer side of this, Nobilior placed one legion in hiding, in order that the enemy, despising his small numbers, might follow more boldly. When this expectation was realized, the legion which had been posted for the purpose attacked the enemy from ambush and destroyed them.

When Iphicrates was leading his army in Thrace in a long file on account of the nature of the terrain, and the report was brought to him that the enemy planned to attack his rearguard, he ordered some

SEXTUS JULIUS FRONTINUS

que latera secedere et consistere iussit, ceteros suffugere et iter maturare ; transeunte autem toto agmine lectissimos quosque retinuit et ita passim circa praedam occupatos hostes, iam etiam fatigatos, ipse requietis et ordinatis suis adgressus fudit exuitque praeda.

4. Boii in silva Litana, quam transiturus erat noster exercitus, succiderunt arbores ita, ut parte exigua sustentatae starent, donec impellerentur ; delituerunt deinde ad extremas ipsi ingressoque silvam hoste per proximas ulteriores impulerunt. Eo modo propagata pariter supra Romanos ruina magnam manum eliserunt.

VII. Quemadmodum ea, quibus Deficiemur, Videantur non Deesse aut Usus eorum Expleatur

1. L. Caecilius Metellus, quia usu navium, quibus elephantos transportaret, deficiebatur, iunxit dolia constravitque tabulatis ac super ea positos per Siculum fretum transmisit.

2. Hannibal, cum in praealti fluminis transitum elephantos non posset compellere nec navium aut materiarum, quibus rates construerentur, copiam haberet, iussit ferocissimum elephantum sub aure vulnerari et eum qui vulnerasset tranato statim flumine procurrere. Elephantus exasperatus ad per-

---

[1] *i.e.* to give the appearance of flight.
[2] 389 B.C. *Cf.* Polyaen. III. ix. 49, 54.
[3] 216 B.C. Twenty-five thousand, according to Livy xxiii. 24.   [4] 250 B.C. *Cf.* Zonar. viii. 14.

cohorts to withdraw to both flanks and halt, while the rest were to quicken their pace and flee.[1] But from the complete line as it passed by, he kept back all the choicest soldiers. Thus, when the enemy were busy with promiscuous pillaging, and in fact were already exhausted, while his own men were refreshed and drawn up in order, he attacked and routed the foe and stripped them of their booty.[2]

When our army was about to pass through the Litana Forest, the Boii cut into the trees at the base, leaving them only a slender support by which to stand, until they should be pushed over. Then the Boii hid at the further edge of the woods and by toppling over the nearest trees caused the fall of those more distant, as soon as our men entered the forest. In that way they spread general disaster among the Romans, and destroyed a large force.[3]

VII. How to conceal the Absence of the Things we lack, or to supply Substitutes for Them

Lucius Caecilius Metellus, lacking ships with which to transport his elephants, fastened together large earthen jars, covered them with planking, and then, loading the elephants on these, ferried them across the Sicilian Straits.[4]

When Hannibal on one occasion could not force his elephants to ford an especially deep stream, having neither boats nor material of which to construct them, he ordered one of his men to wound the most savage elephant under the ear, and then straightway to swim across the stream and take to his heels. The infuriated elephant, eager to pursue

SEXTUS JULIUS FRONTINUS

sequendum doloris sui auctorem tranavit amnem et reliquis idem audendi fecit exemplum.

3. Carthaginiensium duces instructuri classem, quia sparto deficiebantur, crinibus tonsarum mulierum ad funes efficiendos usi sunt.

4. Idem Massilienses et Rhodii fecerunt.[1]

5. M. Antonius a Mutina profugus cortices pro scutis militibus suis dedit.

6. Spartaco copiisque eius scuta ex vimine fuerunt, quae coriis tegebantur.

7. Non alienus, ut arbitror, hic locus est referendi factum Alexandri Macedonis illud nobile, qui per deserta Africae itinera gravissima siti cum exercitu adfectus oblatam sibi a milite in galea aquam spectantibus universis effudit, utilior exemplo temperantiae, quam si communicare potuisset.[2]

VIII. DE DISTRINGENDIS HOSTIBUS

1. CORIOLANUS, cum ignominiam damnationis suae bello ulcisceretur, populationem patriciorum agrorum inhibuit, deustis vastatisque plebeiorum, ut discordiam moveret, qua consensus Romanorum distringeret.

---

[1] *No. 4 is regarded as an interpolation.*
[2] *No. 7 is thought to be an interpolation.*

---

[1] 218 B.C. *Cf.* Livy XXI. xxviii. 5–12.
[2] The Spanish broom, used for making rope.
[3] 146 B.C. *Cf.* Flor. II. xv. 10.   [4] 43 B.C.
[5] 73 B.C. *Cf.* Flor. III. xx. 6.

the author of his suffering, swam the stream, and thus set an example for the rest to make the same venture.[1]

When the Carthaginian admirals were about to equip their fleet, but lacked broom,[2] they cut off the hair of their women and employed it for making cordage.[3]

The Massilians and Rhodians did the same.

Marcus Antonius, when a refugee from Mutina, gave his soldiers bark to use as shields.[4]

Spartacus and his troops had shields made of osiers and covered with hides.[5]

This place, I think, is not inappropriate for recounting that famous deed of Alexander of Macedon. Marching along the desert roads of Africa, and suffering in common with his men from most distressing thirst, when some water was brought him in a helmet by a soldier, he poured it out upon the ground in the sight of all, in this way serving his soldiers better by his example of restraint than if he had been able to share the water with the rest.[6]

VIII. ON DISTRACTING THE ATTENTION OF THE ENEMY

WHEN Coriolanus was seeking to avenge by war the shame of his own condemnation, he prevented the ravaging of the lands of the patricians, while burning and harrying those of the plebeians, in order to arouse discord whereby to destroy the harmony of the Romans.[7]

---

[6] 332–331 B.C. *Cf.* Polyaen. IV. iii. 25. Curt. VII. v. 9–12 and Plut. *Alex.* 42 have a slightly different version.
[7] 489 B.C. *Cf.* Livy II. xxxix. 5–8; Plut. *Coriol.* 27.

SEXTUS JULIUS FRONTINUS

2. Hannibal Fabium, cui neque virtute neque artibus bellandi par erat, ut infamia distringeret, agris eius abstinuit, ceteros populatus. Contra ille, ne suspecta civibus fides esset, magnitudine animi effecit, publicatis possessionibus suis.

3. Fabius Maximus quinto consul, cum Gallorum et Umbrorum, Etruscorum, Samnitium adversus populum Romanum exercitus coissent, contra quos et ipse trans Appenninum in Sentinate castra communiebat, scripsit Fulvio et Postumio, qui in praesidio urbi erant, copias ad Clusium moverent. Quibus obsecutis[1] ad sua defendenda Etrusci Umbrique deverterunt; relictos Samnites Gallosque Fabius et collega Decius adgressi vicerunt.

4. M'. Curius adversus Sabinos, qui ingenti exercitu conscripto relictis finibus suis nostros occupaverant, occultis itineribus manum misit, quae desolatos agros eorum vicosque per diversa incenderunt. Sabini ad arcendam domesticam vastitatem recesserunt; Curio contigit et vacuos infestare hostium fines et exercitum sine proelio avertere sparsumque caedere.

5. T. Didius, paucitate suorum diffidens, cum in adventum earum legionum, quas exspectabat, traheret bellum et occurrere eis hostem comperisset,

---

[1] obsecutis *Oudendorp*; adsecutis *Gund. with MSS.*

[1] 217 B.C. *Cf.* Livy XXII. xxiii. 1-8; Plut. *Fab.* 7. Polyaen. I. xxxvi. 2 attributes a like act to Pericles.
[2] 295 B.C. *Cf.* Livy x. 27.   [3] 290 B.C.

When Hannibal had proved no match for Fabius either in character or in generalship, in order to smirch him with dishonour, he spared his lands, when he ravaged all others. To meet this assault, Fabius transferred the title to his property to the State, thus, by his loftiness of character, preventing his honour from falling under the suspicion of his fellow-citizens.[1]

In the fifth consulship of Fabius Maximus, the Gauls, Umbrians, Etruscans, and Samnites had formed an alliance against the Roman people. Against these tribes Fabius first constructed a fortified camp beyond the Apennines in the region of Sentinum. Then he wrote to Fulvius and Postumius, who were guarding the City, directing them to move on Clusium with their forces. When these commanders complied, the Etruscans and Umbrians withdrew to defend their own possessions, while Fabius and his colleague Decius attacked and defeated the remaining forces of Samnites and Gauls.[2]

When the Sabines levied a large army, left their own territory, and invaded ours, Manius Curius by secret routes sent against them a force which ravaged their lands and villages and set fire to them in divers places. In order to avert this destruction of their country, the Sabines thereupon withdrew. But Curius succeeded in devastating their country while it was unguarded, in repelling their army without an engagement, and then in slaughtering it piecemeal.[3]

Titus Didius at one time lacked confidence because of the small number of his troops, but continued the war in hope of the arrival of certain legions which he was awaiting. On hearing that the

SEXTUS JULIUS FRONTINUS

contione advocata aptari iussit milites ad pugnam ac de industria neglegentius custodiri captivos. Ex quibus pauci, qui profugerant, nuntiaverunt suis pugnam imminere; et illi, ne sub exspectatione proelii diducerent viris, omiserunt occurrere eis, quibus insidiabantur; legiones tutissime nullo excipiente ad Didium pervenerunt.

6. Bello Punico quaedam civitates, quae a Romanis deficere ad Poenos destinaverant, cum obsides dedissent, quos recipere antequam descisferent studebant, simulaverunt seditionem inter finitimos ortam, quam Romanorum legati dirimere deberent, missosque eos velut contraria pignora retinuerunt nec ante reddiderunt, quam ipsi reciperarent suos.

7. Legati Romanorum, cum missi essent ad Antiochum regem, qui secum Hannibalem victis iam Carthaginiensibus habebat consiliumque eius adversus Romanos instruebat, crebris cum Hannibale conloquiis effecerunt, ut is regi fieret suspectus, cui gratissimus alioquin et utilis erat propter calliditatem et peritiam bellandi.

8. Q. Metellus adversus Iugurtham bellum gerens missos ad se legatos eius corrupit, ut sibi proderent regem; cum et alii venissent, idem fecit; eodem consilio usus est et adversus tertios. Sed de capti-

---

[1] 98–93 B.C. In Spain.
[2] 192 B.C. *Cf.* Livy xxxv. 14; Nep. *Hann.* 2.

enemy planned to attack these legions, he called an assembly of the soldiers and ordered them to get ready for battle, and purposely to exercise a careless supervision over their prisoners. As a result, a few of the latter escaped and reported to their people that battle was imminent. The enemy, to avoid dividing their strength when expecting battle, abandoned their plan of attacking those for whom they were lying in wait, so that the legions arrived without hindrance and in perfect safety at the camp of Didius.[1]

In the Punic War certain cities had resolved to revolt from the Romans to the Carthaginians, but wishing, before they revolted, to recover the hostages they had given, they pretended that an uprising had broken out among their neighbours which Roman commissioners ought to come and suppress. When the Romans sent these envoys, the cities detained them as counter-pledges, and refused to restore them until they themselves recovered their own hostages.

After the defeat of the Carthaginians, King Antiochus sheltered Hannibal and utilized his counsel against the Romans. When Roman envoys were sent to Antiochus, they held frequent conferences with Hannibal, and thus caused him to become an object of suspicion to the king, to whom he was otherwise most agreeable and useful, in consequence of his cleverness and experience in war.[2]

When Quintus Metellus was waging war against Jugurtha, he bribed the envoys sent him to betray the king into his hands. When other envoys came, he did the same; and with a third embassy he adopted the same policy. But his efforts to take

SEXTUS JULIUS FRONTINUS

vitate Iugurthae res parum processit; vivum enim tradi sibi volebat. Plurimum tamen consecutus est, nam cum interceptae fuissent epistulae eius ad regios amicos scriptae, in omnis eos rex animadvertit spoliatusque consiliis amicos postea parare non potuit.

9. C. Caesar, per exceptum quendam aquatorem cum comperisset Afranium Petreiumque castra noctu moturos, ut citra vexationem suorum hostilia impediret consilia, initio statim noctis vasa conclamare milites et praeter adversariorum castra agi mulos cum fremitu et sono iussit. Continuere se, quos retentos volebat, arbitrati castra Caesarem movere.

10. Scipio Africanus ad excipienda auxilia cum commeatibus Hannibali venientia Minucium Thermum dimisit, ipse subventurus.

11. Dionysius Syracusanorum tyrannus, cum Afri ingenti multitudine traiecturi essent in Siciliam ad eum oppugnandum, castella pluribus locis communiit custodibusque praecepit, ut ea advenienti hosti dederent dimissique Syracusas occulte redirent. Afris necesse fuit capta castella praesidio obtinere; quos Dionysius, redactos ad quam voluerat paucitatem,

---

[1] 108 B.C. *Cf.* Sall. *Jug.* 61, 62, 70, 72.
[2] 49 B.C. *Cf.* Caes. *B.C.* i. 66.
[3] 202 B.C. *Cf.* Appian *Pun.* 36.

STRATAGEMS, I. viii. 8–11

Jugurtha prisoner met with small success, for Metellus wished the king to be delivered into his hands alive. And yet he accomplished a great deal, for when his letters addressed to the friends of the king were intercepted, the king punished all these men, and, being thus deprived of advisers, was unable to secure any friends for the future.[1]

Gaius Caesar on one occasion caught a soldier who had gone to procure water, and learned from him that Afranius and Petreius planned to break camp that night. In order to hamper the plans of the enemy, and yet not cause alarm to his own troops, Caesar early in the evening gave orders to sound the signal for breaking camp, and commanded mules to be driven past the camp of the enemy with noise and shouting. Thinking that Caesar was breaking camp, his adversaries stayed where they were, precisely as Caesar desired.[2]

When, on one occasion, reinforcements and provisions were on the way to Hannibal, Scipio, wishing to intercept these, sent ahead Minucius Thermus, and arranged to come himself to lend his support.[3]

When the Africans were planning to cross over to Sicily in vast numbers in order to attack Dionysius, tyrant of Syracuse, the latter constructed strongholds in many places and commanded their defenders to surrender them at the coming of the enemy, and then, when they retired, to return secretly to Syracuse. The Africans were forced to occupy the captured strongholds with garrisons, whereupon Dionysius, having reduced the army of his opponents to the scanty number which he desired, and being now approximately on an equality, attacked and

paene iam par numero adgressus vicit, cum suos contraxisset et adversarios sparsisset.

12. Agesilaus Lacedaemonius, cum inferret bellum Tissaphernae, Cariam se petere simulavit, quasi aptius locis montuosis adversus hostem equitatu praevalentem pugnaturus. Per hanc consilii ostentationem avocato in Cariam Tissapherne, ipse Lydiam, ubi caput hostium regni erat, inrupit oppressisque, qui illic agebant, pecunia regia potitus est.

VIIII. DE SEDITIONE MILITUM COMPESCENDA

1. AULUS MANLIUS consul, cum comperisset coniurasse milites in hibernis Campaniae, ut iugulatis hospitibus ipsi res invaderent eorum, rumorem sparsit, eodem loco hibernaturos; atque ita dilato coniuratorum consilio Campaniam periculo liberavit et ex occasione nocentes puniit.

2. L. Sulla, cum legiones civium Romanorum perniciosa seditione furerent, consilio restituit sanitatem efferatis. Propere enim adnuntiari iussit, hostem adesse, et ad arma vocantium clamorem tolli, signa canere. Discussa seditio est universis adversus hostem consentientibus.

3. Cn. Pompeius, trucidato ab exercitu Mediolani senatu, ne tumultum moveret, si solos evocasset

---

[1] 396 B.C. *Cf.* Polyaen. v. ii. 9.
[2] 395 B.C. *Cf.* Xen. *Hell.* III. iv. 20; Plut. *Ages.* 9–10; Nep. *Ages.* 3.
[3] Livy vii. 38–39 attributes this stratagem to C. Marcius Rutilius, consul in 342 B.C.

defeated them, since he had concentrated his own forces, and had separated those of his adversaries.[1]

When Agesilaus, the Spartan, was waging war against Tissaphernes, he pretended to make for Caria, as though likely to fight more advantageously in mountainous districts against an enemy strong in cavalry. When he had advertised this purpose, and had thus drawn Tissaphernes off to Caria, he himself invaded Lydia, where the capital of the enemy's kingdom was situated, and having crushed those in command at that place, he obtained possession of the king's treasure.[2]

IX. On Quelling a Mutiny of Soldiers

When the consul, Aulus Manlius, had learned that the soldiers had formed a plot in their winter-quarters in Campania to murder their hosts and seize their property, he disseminated the report that they would winter next season in the same place. Having thus postponed the plans of the conspirators, he rescued Campania from peril, and, so soon as occasion offered, inflicted punishment on the guilty.[3]

When on one occasion legions of Roman soldiers had broken out in a dangerous mutiny, Lucius Sulla shrewdly restored sanity to the frenzied troops; for he ordered a sudden announcement to be made that the enemy were at hand, bidding a shout to be raised by those summoning the men to arms, and the trumpets to be sounded. Thus the mutiny was broken up by the union of all forces against the foe.

When the senate of Milan had been massacred by Pompey's troops, Pompey, fearing that he might

SEXTUS JULIUS FRONTINUS

nocentes, mixtos eis qui extra delictum erant venire iussit. Ita et noxii minore cum metu, quia non segregati ideoque non ex causa culpae videbantur arcessiri, paruerunt et illi, quibus integra erat conscientia, custodiendis quoque nocentibus adtenderunt, ne illorum fuga inquinarentur.

4. C. Caesar, cum quaedam legiones eius seditionem movissent, adeo ut in perniciem quoque ducis viderentur consurrecturae, dissimulato metu processit ad milites postulantibusque missionem ultro minaci vultu dedit. Exauctoratos paenitentia coegit satisfacere imperatori obsequentioresque in reliqua opera se dare.

X. Quemadmodum Intempestiva Postulatio Pugnae Inhibeatur

1. Q. Sertorius, quod experimento didicerat imparem se universo Romanorum exercitui, ut barbaros quoque inconsulte pugnam exposcentes doceret, adductis in conspectum duobus equis, praevalido alteri, alteri admodum exili, duos admovit iuvenes similiter adfectos, robustum et gracilem. Ac robustiori imperavit equo exili universam caudam abrumpere, gracili autem valentiorem per singulos pilos vellere. Cumque gracilis fecisset quod imperatum erat, validissi-

---

[1] 47 B.C. *Cf.* Suet. *Caes.* 70; Appian *B.C.* ii. 92.

cause a mutiny if he should call out the guilty alone, ordered certain ones who were innocent to come interspersed among the others. In this way the guilty came with less fear, because they had not been singled out, and so did not seem to be sent for in consequence of any wrong-doing; while those whose conscience was clear kept watch on the guilty, lest by the escape of these the innocent should be disgraced.

When certain legions of Gaius Caesar mutinied, and in such a way as to seem to threaten even the life of their commander, he concealed his fear, and, advancing straight to the soldiers, with grim visage, readily granted discharge to those asking it. But these men were no sooner discharged than penitence forced them to apologize to their commander and to pledge themselves to greater loyalty in future enterprises.[1]

X. How to Check an Unseasonable Demand for Battle

After Quintus Sertorius had learned by experience that he was by no means a match for the whole Roman army, in order to prove this to the barbarians also, who were rashly demanding battle, he brought into their presence two horses, one very strong, the other very feeble. Then he brought up two youths of corresponding physique, one robust, the other slight. The stronger youth was commanded to pull out the entire tail of the feeble horse, while the slight youth was commanded to pull out the hairs of the strong horse one by one. Then, when the slight youth had succeeded in his

mus cum infirmi equi cauda sine effectu luctaretur, "naturam," inquit Sertorius, " Romanarum cohortium per hoc vobis exemplum ostendi, milites; insuperabiles sunt universas adgredienti; easdem lacerabit et carpet, qui per partes adtemptaverit."[1]

2. Idem, cum videret suos pugnae signum inconsulte flagitantes crederetque rupturos imperium, nisi congrederentur, permisit turmae equitum ad lacessendos hostes ire laborantique submisit alias et sic recepit omnes tutiusque et sine noxa ostendit, quis exitus flagitatam pugnam mansisset. Obsequentissimis inde eis usus est.

3. Agesilaus Lacedaemonius, cum adversus Thebanos castra super ripam posuisset multoque maiorem hostium manum esse intellegeret et ideo suos arcere a cupiditate decernendi vellet, dixit responso deum se ex collibus pugnare iussum et ita exiguo praesidio ad ripam posito accessit in colles. Quod Thebani pro metu interpretati transierunt flumen et, cum facile depulissent praesidium, ceteros insecuti avidius iniquitate locorum a paucioribus victi sunt.

4. Scorylo dux Dacorum, cum sciret dissociatum armis civilibus populum Romanum neque tamen sibi temptandum arbitraretur, quia externo bello posset

---

[1] *Practically identical with* IV. vii. 6.

---

[1] 80–72 B.C. *Cf.* Val. Max. VII. iii. 6 ; Plut. *Sert.* 16 ; Hor. *Epist.* II. i. 45 ff ; Plin. *Epist.* III. ix. 11.
[2] 80–72 B.C. *Cf.* Plut. *Sert.* 16.
[3] 369 B.C. *Cf.* Polyaen. II. i. 27.

task, while the strong one was still vainly struggling with the tail of the weak horse, Sertorius observed: "By this illustration I have exhibited to you, my men, the nature of the Roman cohorts. They are invincible to him who attacks them in a body; yet he who assails them by groups will tear and rend them."[1]

When the same Sertorius saw his men rashly demanding the signal for battle and thought them in danger of disobeying orders unless they should engage the enemy, he permitted a squadron of cavalry to advance to harass the foe. When these troops became involved in difficulties, he sent others to their relief, and thus rescued all, showing more safely, and without injury, what would have been the outcome of the battle they had demanded. After that he found his men most amenable.[2]

When Agesilaus, the Spartan, was fighting against the Thebans and had encamped on the bank of a stream, being aware that the forces of the enemy far outnumbered his own, and wishing therefore to keep his men from the desire of fighting, he announced that he had been bidden by a response of the gods to fight on high ground. Accordingly, posting a small guard on the bank, he withdrew to the hills. The Thebans, interpreting this as a mark of fear, crossed the stream, easily dislodged the defending troops, and, following the rest too eagerly, were defeated by a smaller force, owing to the difficulties of the terrain.[3]

Scorylo, a chieftain of the Dacians, though he knew that the Romans were torn with the dissensions of the civil wars, yet did not think he ought to venture on any enterprise against them, inasmuch

concordia inter cives coalescere, duos canes in conspectu popularium commisit iisque acerrime inter ipsos pugnantibus lupum ostendit, quem protinus canes omissa inter se ira adgressi sunt. Quo exemplo prohibuit barbaros ab impetu Romanis profuturo.

XI. Quemadmodum Incitandus Sit ad Proelium Exercitus

1. M. Fabius et Cn. Manlius consules adversus Etruscos propter seditiones detractante proelium exercitu ultro simulaverunt cunctationem, donec milites probris hostium coacti pugnam deposcerent iurarentque se ex ea victores redituros.

2. Fulvius Nobilior, cum adversus Samnitium numerosum exercitum et successibus tumidum parvis copiis necesse haberet decertare, simulavit unam legionem hostium corruptam a se ad proditionem imperavitque ad eius rei fidem tribunis et primis ordinibus et centurionibus, quantum quisque numeratae pecuniae aut auri argentique haberet conferret, ut repraesentari merces proditoribus posset; se autem his qui contulissent pollicitus est consummata victoria ampla insuper praemia daturum. Quae persuasio Romanis alacritatem adtulit et fiduciam, unde etiam praeclara victoria commisso statim bello parata est.

---

[1] 480 b.c. *Cf.* Livy ii. xliii. 11–xlv; Dionys. ix. 7–10.
[2] These were a special class of centurions.

as a foreign war might be the means of uniting the citizens in harmony. Accordingly he pitted two dogs in combat before the populace, and when they became engaged in a desperate encounter, exhibited a wolf to them. The dogs straightway abandoned their fury against each other and attacked the wolf. By this illustration, Scorylo kept the barbarians from a movement which could only have benefited the Romans.

XI. How to arouse an Army's Enthusiasm for Battle

When the consuls Marcus Fabius and Gnaeus Manlius were warring against the Etruscans, and the soldiers mutinied against fighting, the consuls on their side feigned a policy of delay, until the soldiers, wrought upon by the taunts of the enemy, demanded battle and swore to return from it victorious.[1]

Fulvius Nobilior, deeming it necessary to fight with a small force against a large army of the Samnites who were flushed with success, pretended that one legion of the enemy had been bribed by him to turn traitor; and to strengthen belief in this story, he commanded the tribunes, the "first ranks,"[2] and the centurions to contribute all the ready money they had, or any gold and silver, in order that the price might be paid the traitors at once. He promised that, when victory was achieved, he would give generous presents besides to those who contributed for this purpose. This assurance brought such ardour and confidence to the Romans that they straightway opened battle and won a glorious victory.

SEXTUS JULIUS FRONTINUS

3. C. Caesar adversus Germanos et Ariovistum pugnaturus confusis suorum animis pro contione dixit, nullius se eo die opera nisi decimae legionis usurum. Quo consecutus est, ut decimani tamquam praecipuae fortitudinis testimonio concitarentur et ceteri pudore, ne penes alios gloria virtutis esset.[1]

4. Q. Fabius, quia egregie sciebat et Romanos eius esse libertatis, quae contumelia exasperaretur, et a Poenis nihil iustum aut moderatum exspectabat, misit legatos Carthaginem de condicionibus pacis. Quas cum illi iniquitatis et insolentiae plenas rettulissent, exercitus Romanorum ad pugnandum concitatus est.

5. Agesilaus Lacedaemoniorum dux, cum prope ab Orchomeno socia civitate castra haberet comperissetque plerosque ex militibus pretiosissima rerum deponere intra munimenta, praecepit oppidanis, ne quid ad exercitum suum pertinens reciperetur, quo ardentius dimicaret miles, qui sciret sibi pro omnibus suis pugnandum.

6. Epaminondas dux Thebanorum adversus Lacedaemonios dimicaturus, ut non solum viribus milites sui, verum etiam adfectibus adiuvarentur, pronuntiavit in contione destinatum Lacedaemoniis, si victoria potirentur, omnes virilis sexus interficere, uxoribus autem eorum et liberis in servitutem

[1] *Practically identical with* IV. v. 11.

---

[1] 58 B.C. *Cf.* Caes. *B.G.* I. xxxix. 7, XL. 1, 14 ff.
[2] 217–203 B.C.   [3] *Cf.* Polyaen. II. i. 18.

Gaius Caesar, when about to fight the Germans and their king, Ariovistus, at a time when his own men had been thrown into panic, called his soldiers together and declared to the assembly that on that day he proposed to employ the services of the tenth legion alone. In this way he caused the soldiers of this legion to be stirred by his tribute to their unique heroism, while the rest were overwhelmed with mortification to think that reputation for courage should rest with others.[1]

Quintus Fabius Maximus, since he knew full well that the Romans possessed a spirit of independence which was roused by insult, and since he expected nothing just or reasonable from the Carthaginians, sent envoys to Carthage to inquire about terms of peace. When the envoys brought back proposals full of injustice and arrogance, the army of the Romans was stirred to combat.[2]

When Agesilaus, general of the Spartans, had his camp near the allied city of Orchomenos and learned that very many of his soldiers were depositing their valuables within the fortifications, he commanded the townspeople to receive nothing belonging to his troops, in order that his soldiers might fight with more spirit, when they realised that they must fight for all their possessions.[3]

Epaminondas, general of the Thebans, on one occasion, when about to engage in battle with the Spartans, acted as follows. In order that his soldiers might not only exercise their strength, but also be stirred by their feelings, he announced in an assembly of his men that the Spartans had resolved, in case of victory, to massacre all males, to lead the wives and children of those executed into bondage,

SEXTUS JULIUS FRONTINUS

abductis Thebas diruere. Qua denuntiatione concitati primo impetu Thebani Lacedaemonios expugnaverunt.

7. Leotychidas dux Lacedaemoniorum classe pugnaturus eodem die, quo vicerant socii, quamvis ignarus actae rei vulgavit nuntiatam sibi victoriam partium, quo constantiores ad pugnam milites haberet.

8. Aulus Postumius proelio, quo cum Latinis conflixit, oblata specie duorum in equis iuvenum animos suorum erexit, Pollucem et Castorem adesse dicens, ac sic proelium restituit.

9. Archidamus Lacedaemonius adversus Arcadas bellum gerens arma in castris statuit et circa ea duci equos noctu clam imperavit. Quorum vestigia mane, tamquam Castor et Pollux perequitassent, ostendens adfuturos eosdem ipsis proeliantibus persuasit.

10. Pericles dux Atheniensium initurus proelium, cum animadvertisset lucum, ex quo utraque acies conspici poterat, densissimae opacitatis, vastum alioquin et Diti patri sacrum, ingentis illic staturae hominem, altissimis coturnis et veste purpurea ac coma venerabilem, in curru candidorum equorum sublimem constituit, qui dato signo pugnae proveheretur et voce Periclem nomine appellans cohorta-

---

[1] 371 B.C.
[2] 479 B.C. *Cf.* Diodor. xi. 34–35; Polyaen. i. 33. Herod. ix. 100–101 has a different version.
[3] 496 B.C. *Cf.* Val. Max. I. viii. 1; Cic. *De Nat. Deor.* II. ii. 6; Dionys. vi. 13. [4] 467 B.C. *Cf.* Polyaen. I. xli. 1.

and to raze Thebes to the ground. By this announcement the Thebans were so roused that they overwhelmed the Spartans at the first onset.[1]

When Leotychides, the Spartan admiral, was on the point of fighting a naval battle on the very day when the allies had been victorious, although he was ignorant of the fact, he nevertheless announced that he had received news of the victory of their side, in order that in this way he might find his men more resolute for the encounter.[2]

When two youths, mounted on horseback, appeared in the battle which Aulus Postumius fought with the Latins, Postumius roused the drooping spirits of his men by declaring that the strangers were Castor and Pollux. In this way he inspired them to fresh combat.[3]

Archidamus, the Spartan, when waging war against the Arcadians, set up weapons in camp, and ordered horses to be led around them secretly at night. In the morning, pointing to their tracks and claiming that Castor and Pollux had ridden through the camp, he convinced his men that the same gods would also lend them aid in the battle itself.[4]

On one occasion when Pericles, general of the Athenians, was about to engage in battle, noticing a grove from which both armies were visible, very dense and dark, but unoccupied and consecrated to Father Pluto, he took a man of enormous stature, made imposing by high buskins, purple robes, and flowing hair, and placed him in the grove, mounted high on a chariot drawn by gleaming white horses. This man was instructed to drive forth, when the signal for battle should be given, to call Pericles by name, and to encourage him by declaring that the

SEXTUS JULIUS FRONTINUS

retur eum diceretque deos Atheniensibus adesse. Quo paene ante coniectum teli hostes terga verterunt.

11. L. Sulla, quo paratiorem militem ad pugnandum haberet, praedici sibi a diis futura simulavit. Postremo etiam in conspectu exercitus, priusquam in aciem descenderet, signum modicae amplitudinis, quod Delphis sustulerat, orabat petebatque, promissam victoriam maturaret.

12. C. Marius sagam quandam ex Syria habuit, a qua se dimicationum eventus praediscere simulabat.

13. Q. Sertorius, cum barbaro et rationis indocili milite uteretur, cervam candidam insignis formae per Lusitaniam ducebat et ab ea se quae agenda aut vitanda essent, praenoscere adseverabat, ut barbari ad omnia tamquam divinitus imperata oboedirent.

Hoc genere strategematon non tantum ea parte utendum est, qua imperitos existimabimus esse, apud quos his utemur, sed multo magis ea, qua talia erunt, quae excogitabuntur, ut a diis monstrata credantur.[1]

14. Alexander Macedo sacrificaturus inscripsit medicamento haruspicis manum, quam ille extis erat suppositurus. Litterae significabant victoriam Alexandro dari. Quas cum iecur calidum rapuisset et a rege militi esset ostensum, auxit animum tamquam deo spondente victoriam.

---

[1] *The second paragraph of 13 is regarded as an interpolation.*

[1] *Cf.* Val. Max. I. ii. 3; Plut. *Sulla* 29.
[2] *Cf.* Val. Max. I. ii. 3ᵃ; Plut. *Mar.* 17.
[3] *Cf.* Val. Max. I. ii. 4; Plut. *Sertor.* 11; Gell. xv. 22.
[4] *Cf.* Polyaen. IV. iii. 14 and IV. xx. Plut. *Apophth. Lacon. Ages. Magni* 77 attributes this stratagem to Agesilaus.

gods were lending their aid to the Athenians. As a result, the enemy turned and fled almost before a dart was hurled.

Lucius Sulla, in order to make his soldiers readier for combat, pretended that the future was foretold him by the gods. His last act, before engaging in battle, was to pray, in the sight of his army, to a small image which he had taken from Delphi, entreating it to speed the promised victory.[1]

Gaius Marius had a certain wisewoman from Syria, from whom he pretended to learn in advance the outcome of battles.[2]

Quintus Sertorius, employing barbarian troops who were not amenable to reason, used to take with him through Lusitania a beautiful white deer, and claimed that from it he knew in advance what ought to be done, and what avoided. In this way he aimed to induce the barbarians to obey all his commands as though divinely inspired.[3]

This sort of stratagem is to be used not merely in cases when we deem those to whom we apply it simple-minded, but much more when the ruse invented is such as might be thought to have been suggested by the gods.

Alexander of Macedon on one occasion, when about to make sacrifice, used a preparation to inscribe certain letters on the hand which the priest was about to place beneath the vitals. These letters indicated that victory was vouchsafed to Alexander. When the steaming liver had received the impress of these characters and had been displayed by the king to the soldiers, the circumstances raised their spirits, since they thought that the god gave them assurance of victory.[4]

SEXTUS JULIUS FRONTINUS

15. Idem fecit Sudines haruspex proelium Eumene cum Gallis commissuro.[1]

16. Epaminondas Thebanus adversus Lacedaemonios, fiduciam suorum religione adiuvandam ratus, arma quae ornamentis adfixa in templis erant nocte subtraxit persuasitque militibus deos iter suum sequi, ut proeliantibus ipsis adessent.

17. Agesilaus Lacedaemonius, cum quosdam Persarum cepisset, quorum habitus multum terroris praefert, quotiens veste tegitur, nudatos militibus suis ostendit, ut alba corpora et umbratica contemnerentur.

18. Gelo Syracusarum tyrannus bello adversum Poenos suscepto, cum multos cepisset, infirmissimum quemque praecipue ex auxiliaribus, qui nigerrimi erant, nudatum in conspectum suorum produxit, ut persuaderet contemnendos.

19. Cyrus rex Persarum, ut concitaret animos popularium, tota die in excidenda silva quadam eos fatigavit; deinde postridie praestitit eis liberalissimas epulas et interrogavit, utro magis gauderent. Cumque ei praesentia probassent, "atqui per haec," inquit, " ad illa perveniendum est; nam liberi beatique esse, nisi Medos viceritis, non potestis." Atque ita eos ad cupiditatem proelii concitavit.

---

[1] *No. 15 is thought to be an interpolation.*

---

[1] *Cf*. Polyaen. IV. xx.
[2] 371 B.C. *Cf*. Polyaen. II. iii. 8 and 12.
[3] 395 B.C. *Cf*. Polyaen. II. i. 6; Xen. *Hell*. III. iv. 19; Plut. *Ages*. 9.
[4] 480 B.C.
[5] 558 B.C. *Cf.* Herod. i. 126; Polyaen. VII. vi. 7; Justin. I. vi. 4–6.

The soothsayer Sudines did the same thing when Eumenes was about to engage in battle with the Gauls.[1]

Epaminondas, the Theban, in his contest against the Spartans, thinking that the confidence of his troops needed strengthening by an appeal to religious sentiment, removed by night the weapons which were attached to the decorations of the temples, and convinced his soldiers that the gods were attending his march, in order to lend their aid in the battle itself.[2]

Agesilaus, the Spartan, on one occasion captured certain Persians. The appearance of these people, when dressed in uniform, inspired great terror. But Agesilaus stripped his prisoners and exhibited them to his soldiers, in order that their delicate white bodies might excite contempt.[3]

Gelo, tyrant of Syracuse, having undertaken war against the Carthaginians, after taking many prisoners, stripped all the feeblest, especially from among the auxiliaries, who were very swarthy, and exhibited them nude before the eyes of his troops, in order to convince his men that their foes were contemptible.[4]

Cyrus, king of the Persians, wishing to rouse the ambition of his men, employed them an entire day in the fatiguing labour of cutting down a certain forest. Then on the following day he gave them a most generous feast, and asked them which they liked better. When they had expressed their preference for the feast, he said: "And yet it is only through the former that we can arrive at the latter; for unless you conquer the Medes, you cannot be free and happy." In this way he roused them to the desire for combat.[5]

SEXTUS JULIUS FRONTINUS

20. L. Sulla, quia adversus Archelaum praefectum Mithridatis apud Piraeea pigrioribus ad proelium militibus utebatur, opere eos fatigando compulit ad poscendum ultro pugnae signum.

21. Fabius Maximus veritus, ne qua fiducia navium, ad quas refugium erat, minus constanter pugnaret exercitus, incendi eas, priusquam iniret proelium, iussit.

XII. DE DISSOLVENDO METU, QUEM MILITES EX ADVERSIS CONCEPERINT OMINIBUS

1. SCIPIO, ex Italia in Africam transportato exercitu, cum egrediens nave prolapsus esset et ob hoc attonitos milites cerneret, id quod trepidationem adferebat, constantia et magnitudine animi in hortationem convertit et "plaudite," inquit, "milites, Africam oppressi."

2. C. Caesar, cum forte conscendens navem lapsus esset, "teneo te, terra mater," inquit. Qua interpretatione effecit, ut repetiturus illas a quibus proficiscebatur terras videretur.

3. T. Sempronius Gracchus consul, acie adversus Picentes directa, cum subitus terrae motus utrasque partes confudisset, exhortatione confirmavit suos et impulit, consternatum superstitione invaderent hostem, adortusque devicit.

---

[1] 86 B.C. [2] 315 B.C. *Cf.* Livy ix. 23.
[3] 204 B.C. [4] *Cf.* Suet. *Caes.* 59.
[5] P. Sempronius Sophus, consul, defeated the Picentines in 268 B.C. *Cf.* Flor. i. 19.

Lucius Sulla, in the campaign against Archelaus, general of Mithridates, found his troops somewhat disinclined for battle at the Piraeus. But by imposing tiresome tasks upon his men he brought them to the point where they demanded the signal for battle of their own accord.[1]

Fabius Maximus, fearing that his troops would fight less resolutely in consequence of their reliance on their ships, to which it was possible to retreat, ordered the ships to be set on fire before the battle began.[2]

XII. On Dispelling the Fears Inspired in Soldiers by Adverse Omens

Scipio, having transported his army from Italy to Africa, stumbled as he was disembarking. When he saw the soldiers struck aghast at this, by his steadiness and loftiness of spirit he converted their cause of concern into one of encouragement, by saying: "Congratulate me, my men! I have hit Africa hard."[3]

Gaius Caesar, having slipped as he was about to embark on ship, exclaimed: "I hold thee fast, Mother Earth." By this interpretation of the incident he made it seem that he was destined to come back to the lands from which he was setting out.[4]

When the consul Tiberius Sempronius Gracchus was engaged in battle with the Picentines, a sudden earthquake threw both sides into panic. Thereupon Gracchus put new strength and courage into his men by urging them to attack the enemy while the latter were overwhelmed with superstitious awe. Thus he fell upon them and defeated them.[5]

SEXTUS JULIUS FRONTINUS

4. Sertorius, cum equitum scuta extrinsecus equorumque pectora cruenta subito prodigio apparuissent, victoriam portendi interpretatus est, quoniam illae partes solerent hostili cruore respergi.

5. Epaminondas Thebanus contristatis militibus, quod ex hasta eius ornamentum infulae more dependens ventus ablatum in sepulchrum Lacedaemonii cuiusdam depulerat, "nolite," inquit, "milites, trepidare; Lacedaemoniis significatur interitus; sepulchra enim funeribus ornantur."

6. Idem, cum fax de caelo nocte delapsa eos qui animadverterunt terruisset, "lumen," inquit, "hoc numina ostendunt."

7. Idem, instante adversus Lacedaemonios pugna, cum sedile in quo resederat succubuisset et id vulgo pro tristi exciperetur significatione confusi milites interpretarentur, "immo," inquit, "vetamur sedere."

8. C. Sulpicius Gallus defectum lunae imminentem, ne pro ostento exciperent milites, praedixit futurum, additis rationibus causisque defectionis.

9. Agathocles Syracusanus adversus Poenos, simili eiusdem sideris deminutione quia sub diem pugnae ut prodigio milites sui consternati erant, ratione qua id accideret exposita docuit, quidquid illud foret, ad rerum naturam, non ad ipsorum propositum pertinere.

---

[1] 371 B.C. *Cf.* Diodor. xv. lii. 5 ff.
[2] *i.e.*, "we must be up and doing."
[3] 168 B.C. *Cf.* Livy xliv. 37 ; Cic. *De Senect.* xiv. 49 ; Val. Max. VIII. xi. 1.
[4] 310 B.C. *Cf.* Justin. XXII. vi. 1–5 ; Diodor. xx. v. 5.

Sertorius, when by a sudden prodigy the outsides of the shields of his cavalrymen and the breasts of their horses showed marks of blood, interpreted this as a mark of victory, since those were the parts which were wont to be spattered with the blood of the enemy.

Epaminondas, the Theban, when his soldiers were depressed because the decoration hanging from his spear like a fillet had been torn away by the wind and carried to the tomb of a certain Spartan, said: "Do not be concerned, comrades! Destruction is foretold for the Spartans. Tombs are not decorated except for funerals." [1]

The same Epaminondas, when a meteor fell from the sky by night and struck terror to the hearts of those who noticed it, exclaimed: "It is a light sent us from the powers above."

When the same Epaminondas was about to open battle against the Spartans, the chair on which he had sat down gave way beneath him, whereat all the soldiers, greatly troubled, interpreted this as an unlucky omen. But Epaminondas exclaimed: "Not at all; we are simply forbidden to sit." [2]

Gaius Sulpicius Gallus not only announced an approaching eclipse of the moon, in order to prevent the soldiers from taking it as a prodigy, but also gave the reasons and causes of the eclipse.[3]

When Agathocles, the Syracusan, was fighting against the Carthaginians, and his soldiers on the eve of battle were thrown into panic by a similar eclipse of the moon, which they interpreted as a prodigy, he explained the reason why this happened, and showed them that, whatever it was, it had to do with nature, and not with their own purposes.[4]

SEXTUS JULIUS FRONTINUS

10. Pericles, cum in castra eius fulmen decidisset terruissetque milites, advocata contione lapidibus in conspectu omnium conlisis ignem excussit sedavitque conturbationem, cum docuisset similiter nubium adtritu excuti fulmen.

11. Timotheus Atheniensis adversus Corcyraeos navali proelio decertaturus gubernatori suo, qui proficiscenti iam classi signum receptui coeperat dare, quia ex remigibus quendam sternutantem audierat, "miraris," inquit, "ex tot milibus unum perfrixisse?"

12. Chabrias Atheniensis classe dimicaturus, excusso ante navem ipsius fulmine, exterritis per tale prodigium militibus "nunc," inquit, "potissimum ineunda pugna est, cum deorum maximus Iuppiter adesse numen suum classi nostrae ostendit."

---

[1] 375 B.C. *Cf.* Polyaen. III. x. 2.
[2] 391–357 B.C.

Pericles, when a thunderbolt struck his camp and terrified his soldiers, calling an assembly, struck fire by knocking two stones together in the sight of all his men. He thus allayed their panic by explaining that the thunderbolt was similarly produced by the contact of the clouds.

When Timotheus, the Athenian, was about to contend against the Corcyreans in a naval battle, his pilot, hearing one of the rowers sneeze, started to give the signal for retreat, just as the fleet was setting out; whereupon Timotheus exclaimed: "Do you think it strange if one out of so many thousands has had a chill?"[1]

As Chabrias, the Athenian, was about to fight a naval battle, a thunderbolt fell directly across the path of his ship. When the soldiers were filled with dismay at such a portent, he said: "Now is the very time to begin battle, when Jupiter, mightiest of the gods, reveals that his power is present with our fleet."[2]

BOOK II

LIBER SECUNDUS

Dispositis primo libro exemplis instructuris, ut mea fert opinio, ducem in his, quae ante commissum proelium agenda sunt, deinceps reddemus pertinentia ad ea, quae in ipso proelio agi solent, et deinde ea, quae post proelium.

Eorum, quae ad proelium pertinent, species sunt:

    I. De tempore ad pugnam eligendo.
   II. De loco ad pugnam eligendo.
  III. De acie ordinanda.
 IIII. De acie hostium turbanda.
    V. De insidiis.
  VI. De emittendo hoste, ne clausus proelium ex desperatione redintegret.
 VII. De dissimulandis adversis.
VIII. De restituenda per constantiam acie.

Eorum deinde, quae post proelium agenda sunt, has esse species existimaverim:

VIIII. Si res prospere cesserit, de consummandis reliquiis belli.
    X. Si res durius cesserit, de adversis emendandis.
  XI. De dubiorum animis in fide retinendis.
 XII. Quae facienda sint pro castris, si satis fiduciae in praesentibus copiis non habeamus.
XIII. De effugiendo.

BOOK II

Having in Book I given classes of examples which, as I believe, will suffice to instruct a general in those matters which are to be attended to before beginning battle, I will next in order present examples which bear on those things that are usually done in the battle itself, and then those that come subsequent to the engagement.

Of those which concern the battle itself, there are the following classes:

    I. On choosing the time for battle.
   II. On choosing the place for battle.
  III. On the disposition of troops for battle.
  IV. On creating panic in the enemy's ranks.
   V. On ambushes.
  VI. On letting the enemy escape, lest, brought to bay, he renew the battle in desperation.
 VII. On concealing reverses.
VIII. On restoring morale by firmness.

Of the matters which deserve attention after battle, I consider that there are the following classes:

  IX. On bringing the war to a close after a successful engagement.
   X. On repairing one's losses after a reverse.
  XI. On ensuring the loyalty of those whom one mistrusts.
 XII. What to do for the defence of the camp, in case a commander lacks confidence in his present forces.
XIII. On retreating.

SEXTUS JULIUS FRONTINUS

I. De Tempore ad Pugnam Eligendo

1. P. Scipio in Hispania, cum comperisset Hasdrubalem Poenorum ducem ieiuno exercitu mane processisse in aciem, continuit in horam septimam suos, quibus praeceperat, ut quiescerent et cibum caperent; cumque hostes inedia, siti, mora sub armis fatigati repetere castra coepissent, subito copias eduxit et commisso proelio vicit.

2. Metellus Pius in Hispania adversus Hirtuleium, cum ille oriente protinus die instructam aciem vallo eius admovisset, fervidissimo tunc tempore anni intra castra continuit suos in horam diei sextam. Atque ita fatigatos aestu facile integris et recentibus suorum viribus vicit.

3. Idem, iunctis cum Pompeio castris adversus Sertorium in Hispania, cum saepe instruxisset aciem, hoste qui imparem se duobus credebat pugnam detrectante, quodam deinde tempore Sertorianos milites animadvertisset magno impetu instinctos, deposcentes pugnam umerosque exserentes et lanceas vibrantes, existimavit ardori cedendum in tempore recepitque exercitum et Pompeio idem faciendi auctor fuit.

4. Postumius consul in Sicilia, cum castra eius a Punicis trium milium passuum spatio distarent et dictatores Carthaginiensium cotidie ante ipsa muni-

---

[1] 206 b.c. *Cf.* Polyaen. viii. xvi. 1.   [2] 76 b.c.

STRATAGEMS, II. i. 1-4

I. On Choosing the Time for Battle

When Publius Scipio was in Spain and had learned that Hasdrubal, leader of the Carthaginians, had marched out and drawn up his troops in battle array early in the morning before they had had breakfast, he kept back his own men till one o'clock, having ordered them to rest and eat. When the enemy, exhausted with hunger, thirst, and waiting under arms, had begun to return to camp, Scipio suddenly led forth his troops, opened battle, and won the day.[1]

When Metellus Pius was waging war against Hirtuleius in Spain, and the latter had drawn up his troops immediately after daybreak and marched them against Metellus's entrenchments, Metellus held his own forces in camp till noon, as the weather at that time of year was extremely hot. Then, when the enemy were overcome by the heat, he easily defeated them, since his own men were fresh and their strength unimpaired.[2]

When the same Metellus had joined forces with Pompey against Sertorius in Spain, and had repeatedly offered battle, the enemy declined combat, deeming himself unequal to two. Later on, however, Metellus, noticing that the soldiers of the enemy, fired with great enthusiasm, were calling for battle, baring their arms, and brandishing their spears, thought it best to retreat betimes before their ardour. Accordingly he withdrew and caused Pompey to do the same.

When Postumius was in Sicily in his consulate, his camp was three miles distant from the Carthaginians. Every day the Punic chieftains drew up their line of battle directly in front of the fortifica-

SEXTUS JULIUS FRONTINUS

menta Romanorum dirigerent aciem, exigua manu levibus adsidue proeliis pro vallo resistebat. Quam consuetudinem contemnente iam Poeno, reliquis omnibus per quietem intra vallum praeparatis, ex more pristino cum paucis sustentavit incursum adversariorum ac solito diutius detinuit. Quibus fatigatis post sextam horam et iam se recipientibus, cum inedia quoque laborarent, per recentes suos hostem, quem praedicta profligaverant incommoda, fugavit.

5. Iphicrates Atheniensis, quia exploraverat eodem adsidue tempore hostes cibum capere, maturius vesci suos iussit et eduxit in aciem egressumque hostem ita detinuit, ut ei neque confligendi neque abeundi daret facultatem. Inclinato deinde iam die reduxit suos et nihilominus in armis retinuit. Fatigati hostes non statione magis quam inedia statim ad curam corporis et cibum capiendum festinaverunt. Iphicrates rursus eduxit et incompositi hostis adgressus est castra.

6. Idem, cum adversus Lacedaemonios pluribus diebus castra comminus haberet et utraque pars certis temporibus adsidue pabulatum lignatumque procederet, quodam die militum habitu servos lixasque dimisit ad munera, milites retinuit; et cum hostes dispersi essent ad similia munera, expugnavit castra eorum inermesque cum fasciculis passim ad tumultum recurrentes facile aut occidit aut cepit.

---

[1] 262 B.C.   [2] *Cf.* Polyaen. III. ix. 53.
[3] 393-392 B.C.   *Cf.* Polyaen. III. ix. 52.

tions of the Romans, while Postumius offered resistance by way of constant skirmishes, conducted by a small band before his entrenchments. As soon as the Carthaginian commander came to regard this as a matter of course, Postumius quietly made ready all the rest of his troops within the ramparts, meeting the assault of the foe with a few, according to his former practice, but keeping them engaged longer than usual. When, after noon was past, they were retreating, weary and suffering from hunger, Postumius, with fresh troops, put them to rout, exhausted as they were by the aforementioned embarrassments.[1]

Iphicrates, the Athenian, having discovered that the enemy regularly ate at the same hour, commanded his own troops to eat at an earlier hour, and then led them out to battle. When the enemy came forth, he so detained them as to afford them no opportunity either of fighting or of withdrawing. Then, as the day drew to a close, he led his troops back, but nevertheless held them under arms. The enemy, exhausted both by standing in the line and by hunger, straightway hurried off to rest and eat, whereupon Iphicrates again led forth his troops, and finding the enemy disorganized, attacked their camp.[2]

When the same Iphicrates had his camp for several days near the Lacedaemonians, and each side was in the habit of going forth at a regular hour for forage and wood, he one day sent out slaves and camp-followers in the dress of soldiers for this service, holding back his fighting men; and as soon as the enemy had dispersed on similar errands, he captured their camp. Then as they came running back from all quarters to the *mêlée*, unarmed and carrying their bundles, he easily slew or captured them.[3]

SEXTUS JULIUS FRONTINUS

7. Verginius consul in Volscis, cum procurrere hostes effuse ex longinquo vidisset, quiescere suos ac defixa tenere pila iussit. Tum anhelantes integris viribus exercitus sui adgressus avertit.

8. Fabius Maximus non ignarus, Gallos et Samnites primo impetu praevalere, suorum autem infatigabiles spiritus inter moras decertandi etiam incalescere, imperavit militibus, contenti primo congressu sustinere hostem mora fatigarent. Quod ubi successit, admoto etiam subsidio suis in prima acie, universis viribus oppressum fudit hostem.

9. Philippus ad Chaeroneam memor, sibi esse militem longo usu duratum, Atheniensibus acrem quidem, sed inexercitatum et in impetu tantum violentum, ex industria proelium traxit, moxque languentibus iam Atheniensibus concitatius intulit signa et ipsos cecidit.

10. Lacedaemonii certiores ab exploratoribus facti, Messenios in eam exarsisse rabiem, ut in proelium cum coniugibus ac liberis descenderent, pugnam distulerunt.

11. C. Caesar bello civili, cum exercitum Afrani et Petrei circumvallatum siti angeret isque ob hoc exasperatus interfectis omnibus impedimentis ad pugnam descendisset, continuit suos, arbitratus

---

[1] 494 B.C.   *Cf.* Livy II. xxx. 10 ff.
[2] 295 B.C.   *Cf.* Livy x. 28 ff.
[3] 338 B.C.   *Cf.* Polyaen. IV. ii. 7: Justin. IX. iii. 9.
[4] The motive was to be less encumbered on the march.

When the consul Verginius, in the war with the Volscians, saw the enemy run forward at full stretch from a distance, he commanded his own men to keep steady and hold their javelins at rest. Then, when the enemy were out of breath, while his own army was still strong and fresh, he attacked and routed them.[1]

Since Fabius Maximus was well aware that the Gauls and Samnites were strong in the initial attack, while the tireless spirits of his own men actually waxed hotter as the struggle continued, he commanded his soldiers to rest content with holding the foe at the first encounter and to wear them out by delay. When this succeeded, bringing up reinforcements to his men in the van, and attacking with his full strength, he crushed and routed the enemy.[2]

At Chaeronea, Philip purposely prolonged the engagement, mindful that his own soldiers were seasoned by long experience, while the Athenians were ardent but untrained, and impetuous only in the charge. Then, as the Athenians began to grow weary, Philip attacked more furiously and cut them down.[3]

When the Spartans learned from scouts that the Messenians had broken out into such fury that they had come down to battle attended by their wives and children, they postponed the engagement.

In the Civil War, when Gaius Caesar held the army of Afranius and Petreius besieged and suffering from thirst, and when their troops, infuriated because of this, had slain all their beasts of burden [4] and come out for battle, Caesar held back his own soldiers, deeming the occasion ill-suited for an engagement,

SEXTUS JULIUS FRONTINUS

alienum dimicationi tempus, quod adversarios ira et desperatio incenderet.

12. Cn. Pompeius, fugientem Mithridatem cupiens ad proelium compellere, elegit tempus dimicationi nocturnum, ut abeunti se opponeret. Atque ita praeparatus subitam hostibus necessitatem decernendi iniecit. Praeterea sic constituit aciem, ut Ponticorum quidem oculos adversa luna praestringeret, suis autem inlustrem et conspicuum praeberet hostem.

13. Iugurtham constat, memorem virtutis Romanorum, semper inclinato die committere proelia solitum, ut, si fugarentur sui, opportunam noctem haberent ad delitiscendum.

14. Lucullus adversus Mithridatem et Tigranem in Armenia Maiore apud Tigranocertam, cum ipse non amplius quindecim milia armatorum haberet, hostis autem innumerabilem multitudinem eoque ipso inhabilem, usus hoc eius incommodo nondum ordinatam hostium aciem invasit atque ita protinus dissipavit, ut ipsi quoque reges abiectis insignibus fugerent.

15. Ti. Nero adversus Pannonios, cum barbari feroces in aciem oriente statim die processissent, continuit suos passusque est hostem nebula et imbribus, qui forte illo die crebri erant, verberari. Ac deinde, ubi fessum stando et pluvia non solum sed et lassitudine deficere animadvertit, signo dato adortus superavit.

---

[1] 49 B.C. *Cf.* Caes. *B.C.* i. 81 ff.
[2] 66 B.C *Cf.* Flor. III. v. 22–24; Plut. *Pomp.* 32 ff.
[3] 111–106 B.C. *Cf.* Sall. *Jug.* xcviii. 2.
[4] 69 B.C. *Cf.* Plut. *Lucul.* 26–28; Appian *Mithr.* 84–85
[5] 12–10 B.C. or 6–9 A.D. The future Emperor Tiberius.

since his opponents were so inflamed with wrath and desperation.[1]

Gnaeus Pompey, desiring to check the flight of Mithridates and force him to battle, chose night as the time for the encounter, arranging to block his march as he withdrew. Having made his preparations accordingly, he suddenly forced his enemy to fight. In addition to this, he so drew up his force that the moonlight falling in the faces of the Pontic soldiers blinded their eyes, while it gave his own troops a distinct and clear view of the enemy.[2]

It is well known that Jugurtha, aware of the courage of the Romans, was always wont to engage in battle as the day was drawing to a close, so that, in case his men were routed, they might have the advantage of night for getting away.[3]

At Tigranocerta in Greater Armenia, Lucullus, in the campaign against Mithridates and Tigranes, did not have above 15,000 armed men, while the enemy had an innumerable host, which for this very reason was unwieldy. Taking advantage, accordingly, of this handicap of the foe, Lucullus attacked their line before it was in order, and straightway routed it so completely that even the kings themselves discarded their trappings and fled.[4]

In the campaign against the Pannonians, when the barbarians in warlike mood had formed for battle at the very break of day, Tiberius Nero held back his own troops, and allowed the enemy to be hampered by the fog and be drenched with the showers, which happened to be frequent that day. Then, when he noticed that they were weary with standing, and faint not only from exposure but also from exhaustion, he gave the signal, attacked and defeated them.[5]

SEXTUS JULIUS FRONTINUS

16. C. Caesar in Gallia, quia compererat Ariovisto Germanorum regi institutum et quasi legem esse non pugnandi decrescente luna, tum potissimum acie commissa impeditos religione hostes vicit.

17. Divus Augustus Vespasianus Iudaeos Saturni die, quo eis nefas est quicquam seriae rei agere, adortus superavit.

18. Lysander Lacedaemonius adversus Athenienses apud Aegospotamos instituit certo tempore infestare naves Atheniensium, dein revocare classem. Ea re in consuetudinem perducta, cum Athenienses post digressum eius ad contrahendas copias dispergerentur, extendit ex consuetudine classem et recepit. Tum hostium maxima parte ex more dilapsa, reliquos adortus occidit et universas naves cepit.

II. DE LOCO AD PUGNAM ELIGENDO

1. M'. CURIUS, quia phalangi regis Pyrrhi explicitae resisti non posse animadvertebat, dedit operam, ut in angustiis confligeret, ubi conferta sibi ipsa esset impedimento.

2. Cn. Pompeius in Cappadocia elegit castris locum editum. Unde adiuvante proclivi impetum militum facile ipso decursu Mithridatem superavit.

3. C. Caesar adversus Pharnacem Mithridatis

---

[1] 58 B.C. *Cf.* Caes. *B.G.* i. 50; Plut. *Caes.* 19.
[2] 70 A.D.
[3] 405 B C. *Cf.* Xen. *Hell.* II. i. 21 ff; Plut. *Lysand.* 10–11.
[4] 281–275 B.C.   [5] 66 B.C.

Gaius Caesar, when in Gaul, learned that it was a principle and almost a law with Ariovistus, king of the Germans, not to fight when the moon was waning. Caesar therefore chose that time above all others for engaging in battle, when the enemy were embarrassed by their superstition, and so conquered them.[1]

The deified Vespasian Augustus attacked the Jews on their sabbath, a day on which it is sinful for them to do any business, and so defeated them.[2]

When Lysander, the Spartan, was fighting against the Athenians at Aegospotami, he began by attacking the vessels of the Athenians at a regular hour and then calling off his fleet. After this had become an established procedure, as the Athenians on one occasion, after his withdrawal, were dispersing to collect their troops, he deployed his fleet as usual and withdrew it. Then, when most of the enemy had scattered according to their wont, he attacked and slew the rest, and captured all their vessels.[3]

II. ON CHOOSING THE PLACE FOR BATTLE

MANIUS CURIUS, observing that the phalanx of King Pyrrhus could not be resisted when in extended order, took pains to fight in confined quarters, where the phalanx, being massed together, would embarrass itself.[4]

In Cappadocia Gnaeus Pompey chose a lofty site for his camp. As a result the elevation so assisted the onset of his troops that he easily overcame Mithridates by the sheer weight of his assault.[5]

When Gaius Caesar was about to contend with Pharnaces, son of Mithridates, he drew up his line

SEXTUS JULIUS FRONTINUS

filium dimicaturus in colle instruxit aciem; quae res expeditam ei victoriam fecit, nam pila ex edito in subeuntis barbaros emissa protinus eos averterunt.

4. Lucullus adversus Mithridatem et Tigranem in Armenia Maiore apud Tigranocertam dimicaturus, collis proximi planum verticem raptim cum parte copiarum adeptus, in subiectos hostes decucurrit et equitatum eorum a latere invasit; aversumque et eorundem protinus pedites proturbantem insecutus clarissimam victoriam rettulit.

5. Ventidius adversus Parthos non ante militem eduxit, quam illi quingentis non amplius passibus abessent, atque ita procursione subita adeo se admovit, ut sagittas, quibus ex longinquo usus est, comminus applicitus eluderet. Quo consilio, quia quandam etiam fiduciae speciem ostentaverat, celeriter barbaros debellavit.

6. Hannibal apud Numistronem contra Marcellum pugnaturus cavas et praeruptas vias obiecit a latere, ipsaque loci natura pro munimentis usus clarissimum ducem vicit.

7. Idem apud Cannas, cum comperisset Volturnum amnem ultra reliquorum naturam fluminum ingentis auras mane proflare, quae arenarum et pulveris vertices agerent, sic direxit aciem, ut tota vis a tergo suis, Romanis in ora et oculos incederet. Quibus

---

[1] 47 B.C. *Cf. Bell. Alexandr.* 73–76.
[2] 69 B.C. *Cf.* Plut. *Lucul.* 28; Appian *Mithr.* 85.
[3] 38 B.C. *Cf.* Flor. IV. ix. 6.
[4] 210 B.C. According to Livy XXVII. ii. 4 and Plut. *Marc.* 24, the result of the battle was indecisive.

of battle on a hill. This move made victory easy for him, since the darts, hurled from higher ground against the barbarians charging from below, straightway put them to flight.[1]

When Lucullus was planning to fight Mithridates and Tigranes at Tigranocerta in Greater Armenia, he himself swiftly gained the level top of the nearest hill with a part of his troops, and then rushed down upon the enemy posted below, at the same time attacking their cavalry on the flank. When the cavalry broke and straightway threw the infantry into confusion, Lucullus followed after them and gained a most notable victory.[2]

Ventidius, when fighting against the Parthians, would not lead out his soldiers until the Parthians were within five hundred paces. Thus by a rapid advance he came so near them that, meeting them at close quarters, he escaped their arrows, which they shoot from a distance. By this scheme, since he exhibited a certain show of confidence, he quickly subdued the barbarians.[3]

At Numistro, when Hannibal was expecting a battle with Marcellus, he secured a position where his flank was protected by hollows and precipitous roads. By thus making the ground serve as a defence, he won a victory over a most renowned commander.[4]

Again at Cannae, when Hannibal learned that the Volturnus River, at variance with the nature of other streams, sent out high winds in the morning, which carried swirling sand and dust, he so marshalled his line of battle that the entire fury of the elements fell on the rear of his own troops, but struck the Romans in the face and eyes. Since this difficulty

SEXTUS JULIUS FRONTINUS

incommodis mire hosti adversantibus illam memorabilem adeptus est victoriam.

8. Marius adversus Cimbros ac Teutonos constituta die pugnaturus firmatum cibo militem ante castra conlocavit, ut per aliquantum spatii, quo adversarii dirimebantur, exercitus hostium potius[1] labore itineris profligaretur. Fatigationi deinde eorum incommodum aliud obiecit, ita ordinata suorum acie, ut adverso sole et vento et pulvere barbarorum occuparetur exercitus.

9. Cleomenes Lacedaemonius adversus Hippiam Atheniensem, qui equitatu praevalebat, planitiem, in qua dimicaturus erat, arboribus prostratis impediit et inviam fecit equiti.

10. Hiberi in Africa ingenti hostium multitudine excepti timentesque, ne circumirentur, applicuerunt se flumini, quod altis in ea regione ripis praefluebat. Ita a tergo amne defensi et subinde, cum virtute praestarent, incursando in proximos omnem hostium exercitum straverunt.

11. Xanthippus Lacedaemonius sola loci commutatione fortunam Punici belli convertit. Nam cum a desperantibus iam Carthaginiensibus mercede sollicitatus animadvertisset Afros quidem, qui equitatu et elephantis praestabant, colles sectari, a Romanis autem, quorum robur in pedite erat, campestria teneri, Poenos in plana deduxit; ubi per elephantos

[1] *Probably corrupt.*

---

[1] 216 B.C. *Cf.* Val. Max. vii. iv. ext. 2; Plut. *Fab.* 16; Livy xxii. 43, 46. In none of these accounts is a river mentioned as the source of the wind. Livy speaks of the *ventus Volturnus.*

[2] 101 B.C. *Cf.* Plut. *Mar.* 26; Polyaen. VIII. x. 3.

[3] 510 B.C.

was a serious obstacle to the enemy, he won that memorable victory.[1]

After Marius had settled on a day for fighting the Cimbrians and Teutons, he fortified his soldiers with food and stationed them in front of his camp, in order that the army of the enemy might be exhausted by marching over the interval between the opposing armies. Then, when the enemy were thus used up, he confronted them with another embarrassment by so arranging his own line of battle that the barbarians were caught with the sun and wind and dust in their faces.[2]

When Cleomenes, the Spartan, in his battle against Hippias, the Athenian, found that the latter's main strength lay in his cavalry, he thereupon felled trees and cluttered the battlefield with them, thus making it impassable for cavalry.[3]

The Iberians in Africa, upon encountering a great multitude of foes and fearing that they would be surrounded, drew near a river which at that point flowed along between deep banks. Thus, defended by the river in the rear and enabled by their superior prowess to make frequent onsets upon those nearest them, they routed the entire host of their adversaries.

Xanthippus, the Spartan, by merely changing the locality of operations, completely altered the fortunes of the Punic War; for when, summoned as a mercenary by the despairing Carthaginians, he had noticed that the Africans, who were superior in cavalry and elephants, kept to the hills, while the Romans, whose strength was in their infantry, held to the plains, he brought the Carthaginians down to level ground, where he broke the ranks of the

SEXTUS JULIUS FRONTINUS

dissipatis ordinibus Romanorum sparsos milites per Numidas persecutus eorum exercitum fudit, in illam diem terra marique victorem.

12. Epaminondas dux Thebanorum adversus Lacedaemonios directurus aciem, pro fronte eius decurrere equitibus iussis, cum ingentem pulverem hostium oculis obiecisset exspectationemque equestris certaminis praetendisset, circumducto pedite ab ea parte, ex qua decursus in aversam hostium aciem ferebat, inopinantium terga adortus cecidit.

13. Lacedaemonii CCC contra innumerabilem multitudinem Persarum Thermopylas occupaverunt, quarum angustiae non amplius quam parem numerum comminus pugnaturum poterant admittere. Eaque ratione, quantum ad congressus facultatem, aequati numero barbarorum, virtute autem praestantes, magnam eorum partem ceciderunt nec superati forent, nisi per proditorem Ephialten Trachinium circumductus hostis a tergo eos oppressisset.

14. Themistocles dux Atheniensium, cum videret utilissimum Graeciae adversus multitudinem Xerxis navium in angustiis Salaminis decernere idque persuadere civibus non posset, sollertia effecit, ut a barbaris ad utilitates suas Graeci compellerentur. Simulata namque proditione misit ad Xerxen, qui indicaret populares suos de fuga cogitare difficilio-

---

[1] 255 B.C. *Cf.* Polyb. i. 33 ff.; Zonar. viii. 13.
[2] 362 B.C. *Cf.* Polyaen. II. iii. 14.
[3] 480 B.C. *Cf.* Herod. vii. 201 ff.; Polyaen. VII. xv. 5.

Romans with the elephants. Then pursuing their scattered troops with Numidians, he routed their army, which till that day had been victorious on land and sea.[1]

Epaminondas, leader of the Thebans, when about to marshal his troops in battle array against the Spartans, ordered his cavalry to engage in manœuvres along the front. Then, when he had filled the eyes of the enemy with clouds of dust and had caused them to expect an encounter with cavalry, he led his infantry around to one side, where it was possible to attack the enemy's rear from higher ground, and thus, by a surprise attack, cut them to pieces.[2]

Against a countless horde of Persians, three hundred Spartans seized and held the pass of Thermopylae which was capable of admitting only a like number of hand-to-hand opponents. In consequence, the Spartans became numerically equal to the barbarians, so far as opportunity for fighting was concerned, and being superior to them in valour, slew large numbers of them. Nor would they have been overcome, had not the enemy been led around to the rear by the traitor Ephialtes, the Trachinian, and thus been enabled to overwhelm them.[3]

Themistocles, leader of the Athenians, saw that it was most advantageous for Greece to fight in the Straits of Salamis against the vast numbers of Xerxes's vessels, but he was unable to persuade his fellow Athenians of this. He therefore employed a stratagem to make the barbarians force the Greeks to do what was advantageous for the latter; for under pretence of turning traitor, he sent a messenger to Xerxes to inform him that the Greeks were planning flight, and that the situation would be

SEXTUS JULIUS FRONTINUS

remque ei rem futuram, si singulas civitates obsidione adgrederetur. Qua ratione effecit, ut exercitus barbarorum primum inquietaretur, dum tota nocte in statione custodiae est; deinde, ut sui mane integris viribus cum barbaris vigilia marcentibus confligerent, loco ut voluerat arto, in quo Xerxes multitudine qua praestabat uti non posset.

III. DE ACIE ORDINANDA

1. CN. SCIPIO in Hispania adversus Hannonem ad oppidum Indibile, cum animadvertisset Punicam aciem ita directam, ut in dextro cornu Hispani constituerentur, robustus quidem miles, sed qui alienum negotium ageret, in sinistro autem Afri, minus viribus firmi, sed animi constantiores, reducto sinistro latere suorum, dextro cornu, quod validissimis militibus exstruxerat, obliqua acie cum hoste conflixit. Deinde fusis fugatisque Afris Hispanos, qui in recessu spectantium more steterant, facile in deditionem compulit.

2. Philippus Macedonum rex adversus Hyllios[1] gerens bellum, ut animadvertit frontem hostium stipatam electis de toto exercitu viris, latera autem infirmiora, fortissimis suorum in dextro cornu conlocatis, sinistrum latus hostium invasit turbataque tota acie victoriam profligavit.

[1] yllyrios *P.*

---

[1] 480 B.C. *Cf.* Herod. viii. 75; Plut. *Them.* 12; Nep. *Them.* 4.
[2] 218 B.C. Cn. Cornelius Scipio Calvus. *Cf.* Polyb. iii. 76.
[3] Illyrians. [4] 359 B.C. *Cf.* Diodor. xvi. 4.

more difficult for the King if he should besiege each city separately. By this policy, in the first place he caused the host of the barbarians to be kept on the alert doing guard-duty all night; in the second place, he made it possible for his own followers, the next morning, with strength unimpaired, to encounter the barbarians all exhausted with watching, and (precisely as he had wished) in a confined place, where Xerxes could not utilize his superiority in numbers.[1]

III. On the Disposition of Troops for Battle

Gnaeus Scipio, when campaigning in Spain against Hanno, near the town of Indibile, noted that the Carthaginian line of battle was drawn up with the Spaniards posted on the right wing—sturdy soldiers, to be sure, but fighting for others—while on the left were the less powerful, but more resolute, Africans. He accordingly drew back his own left wing, and keeping his battle-line at an angle with the enemy, engaged the enemy with his right wing, which he had formed of his sturdiest soldiers. Then routing the Africans and putting them to flight, he easily forced the surrender of the Spaniards, who had stood apart after the manner of spectators.[2]

When Philip, king of the Macedonians, was waging war against the Hyllians,[3] he noticed that the front of the enemy consisted entirely of men picked from the whole army, while their flanks were weaker. Accordingly he placed the stoutest of his own men on the right wing, attacked the enemy's left, and by throwing their whole line into confusion won a complete victory.[4]

SEXTUS JULIUS FRONTINUS

3. Pammenes Thebanus, conspecta Persarum acie, quae robustissimas copias in dextro cornu conlocatas habebat, simili ratione et ipse suos ordinavit omnemque equitatum et fortissimum quemque peditum in dextro cornu, infirmissimos autem contra fortissimos hostium posuit praecepitque, ut ad primum impetum eorum fuga sibi consulerent et in silvestria confragosaque loca se reciperent. Ita frustrato robore exercitus, ipse optima parte virium suarum dextro cornu totam circumiit aciem hostium et avertit.

4. P. Cornelius Scipio, cui postea Africano cognomen fuit, adversus Hasdrubalem Poenorum ducem in Hispania gerens bellum ita per continuos dies ordinatum produxit exercitum, ut media acies fortissimis fundaretur. Sed cum hostes quoque eadem ratione adsidue ordinati procederent, Scipio eo die, quo statuerat decernere, commutavit instructionis ordinem et firmissimos[1] in cornibus conlocavit ac levem armaturam in media acie, sed retractam. Ita cornibus, quibus ipse praevalebat, infirmissimas hostium partes lunata acie adgressus facile fudit.

5. Metellus in Hispania, eo proelio quo Hirtuleium devicit, cum comperisset cohortes eius, quae validissimae vocabantur, in media acie locatas, ipse mediam suorum aciem reduxit, ne ea parte ante cum hoste confligeret, quam cornibus complicatis medios undique circumvenisset.

---

[1] id est legionarios *of the MSS. is omitted after* firmissimos *as probably a gloss.*

---

[1] 353 B.C. *Cf.* Polyaen. v. xvi. 2.
[2] 206 B.C. *Cf.* Livy xxviii. 14–15; Polyb. xi. 22 ff.
[3] 76 B.C. Q. Caecilius Metellus Pius.

Pammenes, the Theban, having observed the battle-line of the Persians, where the most powerful troops were posted on the right wing, drew up his own men also on the same plan, putting all his cavalry and the bravest of his infantry on the right wing, but stationing opposite the bravest of the enemy his own weakest troops, whom he directed to flee at the first onset of the foe and to retreat to rough, wooded places. When in this way he had made the enemy's strength of no effect, he himself with the best part of his own forces enveloped the whole array of the enemy with his right wing and put them to rout.[1]

Publius Cornelius Scipio, who subsequently received the name Africanus, on one occasion, when waging war in Spain against Hasdrubal, leader of the Carthaginians, led out his troops day after day in such formation that the centre of his battle-line was composed of his best fighting men. But when the enemy also regularly came out marshalled on the same plan, Scipio, on the day when he had determined to fight, altered the scheme of his arrangement and stationed his strongest troops on the wings, having his light-armed troops in the centre, but slightly behind the line. Thus, by attacking the enemy's weakest point in crescent formation from the flank, where he himself was strongest, he easily routed them.[2]

Metellus, in the battle in which he vanquished Hirtuleius in Spain, had discovered that the battalions of Hirtuleius which were deemed strongest were posted in the centre. Accordingly he drew back the centre of his own troops, to avoid encountering the enemy at that part of the line, until by an enveloping movement of his wings he could surround their centre from all sides.[3]

SEXTUS JULIUS FRONTINUS

6. Artaxerxes adversus Graecos, qui Persida intraverant, cum multitudine superaret, latius quam hostes acie instructa in fronte peditem, equitem levemque armaturam in cornibus conlocavit. Atque ita ex industria lentius procedente media acie copias hostium cinxit cecidetque.

7. Contra Hannibal ad Cannas reductis cornibus productaque media acie nostros primo impetu protrusit. Idem conserto proelio, paulatim invicem sinuantibus procedentibusque ad praeceptum cornibus, avide insequentem hostem in mediam aciem suam recepit et ex utraque parte compressum cecidit. Veterano et diu edocto usus exercitu, hoc enim genus ordinationis exsequi nisi peritus et ad omne momentum respondens miles vix potest.[1]

8. Livius Salinator et Claudius Nero, cum Hasdrubal bello Punico secundo decernendi necessitatem evitans in colle confragoso post vineas aciem direxisset, ipsi diductis in latera viribus vacua fronte ex utraque parte circumvenerunt eum atque ita adgressi superarunt.

9. Hannibal, cum frequentibus proeliis a Claudio Marcello superaretur, novissime sic castra metabatur, ut aut montibus aut paludibus aut simili locorum aliqua opportunitate adiutus aciem eo modo conlocaret, ut vincentibus quidem Romanis paene indemnem recipere posset intra munimenta exercitum, cedentibus autem instandi liberum haberet arbitrium.

10. Xanthippus Lacedaemonius in Africa adversus

[1] *The last sentence of 7 is regarded as an interpolation.*

---

[1] 401 B.C. Battle of Cunaxa. *Cf.* Xen. *Anab.* I. viii. 10.
[2] 216 B.C. *Cf.* Livy xxii. 47; Polyb. iii. 115.
[3] 207 B.C. Battle of the Metaurus. *Cf.* Livy xxvii. 48.

Artaxerxes, having superior numbers in his campaign against the Greeks, who had invaded Persia, drew up his line of battle with a wider front than the enemy, placing infantry, cavalry, and light-armed troops on the wings. Then by purposely causing the centre to advance more slowly he enveloped the enemy troops and cut them to pieces.[1]

On the other hand, at Cannae Hannibal, having drawn back his flanks and advanced his centre, drove back our troops at the first assault. Then, when the fighting began, and the flanks gradually worked towards each other moving forward according to instructions, Hannibal enveloped within his own lines the impetuously attacking enemy, forced them towards the centre from both sides, and cut them to pieces, using veteran troops of long training; for hardly anything but a trained army, responsive to every direction, can carry out this sort of tactics.[2]

In the Second Punic War, when Hasdrubal was seeking to avoid the necessity of an engagement, and had drawn up his line on a rough hillside behind protective works, Livius Salinator and Claudius Nero diverted their own forces to the flanks, leaving their centre vacant. Having in this way enveloped Hasdrubal, they attacked and defeated him.[3]

After Hannibal had been defeated in frequent battles by Claudius Marcellus, he finally laid out his camp on this plan: Protected by mountains, marshes, or similar advantages of terrain, he so posted his troops as to be able to withdraw his army, practically without loss, within his fortifications, in case the Romans won, but so as to have free option of pursuit, in case they gave way.

Xanthippus, the Spartan, in the campaign con-

SEXTUS JULIUS FRONTINUS

M. Atilium Regulum levem armaturam in prima acie conlocavit, in subsidio autem robur exercitus praecepitque auxiliaribus, ut emissis telis cederent hosti et, cum se intra suorum ordines recepissent, confestim in latera discurrerent et a cornibus rursus erumperent; exceptumque iam hostem a robustioribus et ipsi circumierunt.

11. Sertorius idem in Hispania adversus Pompeium fecit.[1]

12. Cleandridas Lacedaemonius adversus Lucanos densam instruxit aciem, ut longe minoris exercitus speciem praeberet. Securis deinde hostibus in ipso certamine diduxit ordines et a lateribus circumventos eos fudit.

13. Gastron Lacedaemonius, cum auxilio Aegyptiis adversus Persas venisset et sciret, firmiorem esse Graecum militem magisque a Persis timeri, commutatis armis Graecos in prima posuit acie et, cum illi aequo Marte pugnarent, submisit Aegyptiorum manum. Persae cum Graecis, quos Aegyptios opinabantur, restitissent, superveniente multitudine, quam ut Graecorum expaverant, cesserunt.

14. Cn. Pompeius in Albania, quia hostes et numero et equitatu praevalebant, iuxta collem in angustiis

---

[1] *No. 11 is probably interpolated*

---

[1] These were the "light-armed troops," already mentioned.
[2] 255 B.C.   [3] *Cf.* II. v. 31.
[4] After 443 B.C.   *Cf.* Polyaen. II. x. 4
[5] *Cf.* Polyaen. ii. 16.

ducted in Africa against Marcus Atilius Regulus, placed his light-armed troops in the front line, holding the flower of his army in reserve. Then he directed the auxiliary troops,[1] after hurling their javelins, to give way before the enemy, withdraw within the ranks of their fellow-soldiers, hurry to the flanks, and from there again rush forward to the attack. Thus when the enemy had been met by the stronger troops, they were enveloped also by these light-armed forces.[2]

Sertorius employed the same tactics in Spain in the campaign against Pompey.[3]

Cleandridas, the Spartan, when fighting against the Lucanians, drew up his troops in close array, so as to present the appearance of a much smaller army. Then, when the enemy had thus been put off their guard, at the moment the engagement began he opened up his ranks, enveloped the enemy on the flank, and put them to rout.[4]

Gastron, the Spartan, having come to assist the Egyptians against the Persians, and realizing that the Greek soldiers were more powerful and more dreaded by the Persians, interchanged the arms of the two contingents, placing the Greeks in the front line. When these merely held their own in the encounter, he sent in the Egyptians as reinforcements. Although the Persians had proved equal to the Greeks (deeming them Egyptians), they gave way, so soon as they were set upon by a multitude, of whom (as supposedly consisting of Greeks) they had stood in terror.[5]

When Gnaeus Pompey was fighting in Albania, and the enemy were superior in numbers and in cavalry, he directed his infantry to cover their

SEXTUS JULIUS FRONTINUS

protegere galeas, ne fulgore earum conspicui fierent, iussit pedites, equites deinde in aequum procedere ac velut praetendere peditibus, praecepitque eis, ut ad primum impetum hostium refugerent et, simul ad pedites ventum esset, in latera discederent. Quod ubi explicitum est, patefacto loco subita peditum consurrexit acies invectosque temere hostes inopinato interfusa proelio cecidit.

15. M. Antonius adversus Parthos, qui infinita multitudine sagittarum exercitum eius obruebant, subsidere suos et testudinem facere iussit, supra quam transmissis sagittis sine militum noxa exhaustus est hostis.

16. Hannibal adversus Scipionem in Africa, cum haberet exercitum ex Poenis et auxiliaribus, quorum pars non solum ex diversis partibus, sed etiam ex Italicis constabat, post elephantos LXXX, qui in prima fronte positi hostium turbarent aciem, auxiliares Gallos et Ligures et Baliares Maurosque posuit, ut neque fugere possent Poenis a tergo stantibus et hostem oppositi, si non infestarent, at certe fatigarent. Tum suis et Macedonibus, qui iam fessos Romanos integri exciperent, in secunda acie conlocatis, novissimos Italicos constituit, quorum et timebat fidem et segnitiam verebatur, quoniam plerosque eorum ab Italia invitos extraxerat.

---

[1] 65 B.C. *Cf.* Dio xxxvii. 4.

[2] In the *testudo* the soldiers secured protection by holding their over-lapped shields above their heads, a formation whose appearance suggested the scales of a tortoise.

[3] 36 B.C. *Cf.* Dio xlix. 29–30; Plut. *Anton.* 45.

helmets, in order to avoid being visible in consequence of the reflection, and to take their place in a defile by a hill. Then he commanded his cavalry to advance on the plain and to act as a screen to the infantry, but to withdraw at the first onset of the enemy, and, as soon as they reached the infantry, to disperse to the flanks. When this manœuvre had been executed, suddenly the force of infantry rose up, revealing its position, and pouring with unexpected attack upon the enemy who were heedlessly bent on pursuit, thus cut them to pieces.[1]

When Mark Antony was engaged in battle with the Parthians and these were showering his army with innumerable arrows, he ordered his men to stop and form a *testudo*.[2] The arrows passed over this without harm to the soldiers, and the enemy's supply was soon exhausted.[3]

When Hannibal was contending against Scipio in Africa, having an army of Carthaginians and auxiliaries, part of whom were not only of different nationalities, but actually consisted of Italians, he placed eighty elephants in the forefront, to throw the enemy into confusion. Behind these he stationed auxiliary Gauls, Ligurians, Balearians, and Moors, that these might be unable to run away, since the Carthaginians were standing behind them, and in order that, being placed in front, they might at least harass the enemy, if not do him damage. In the second line he placed his own countrymen and the Macedonians, to be fresh to meet the exhausted Romans; and in the rear the Italians, whose loyalty he distrusted and whose indifference he feared, inasmuch as he had dragged most of them from Italy against their will.

SEXTUS JULIUS FRONTINUS

Scipio adversus hanc formam robur legionis triplici acie in fronte ordinatum per hastatos et principes et triarios opposuit; nec continuas construxit cohortes, sed manipulis inter se distantibus spatium dedit, per quod elephanti ab hostibus acti facile transmitti sine perturbatione ordinum possent. Ea ipsa intervalla expeditis velitibus implevit, ne interluceret acies, dato his praecepto, ut ad impetum elephantorum vel retro vel in latera concederent. Equitatum deinde in cornua divisit et dextro Romanis equitibus Laelium, sinistro Numidis Masinissam praeposuit. Quae tam prudens ordinatio non dubie causa victoriae fuit.

17. Archelaus adversus L. Sullam in fronte ad perturbandum hostem falcatas quadrigas locavit, in secunda acie phalangem Macedonicam, in tertia Romanorum more armatos auxiliares, mixtis fugitivis Italicae gentis, quorum pervicaciae plurimum fidebat; levem armaturam in ultimo statuit; in utroque deinde latere equitatum, cuius amplum numerum habebat, circumeundi hostis causa posuit.

Contra haec Sulla fossas amplae latitudinis utroque latere duxit et capitibus earum castella communiit. Qua ratione, ne circumiretur ab hoste et peditum numero et maxime equitatu superante, consecutus est. Triplicem deinde peditum aciem ordinavit relictis intervallis, per quae levem armaturam et equitem,

---

[1] These names designate three successive lines of battle. The *hastati* were the first line; the two other lines were drawn up behind these.

[2] 202 B.C. Battle of Zama. *Cf.* Livy xxx. 33, 35; Polyb. xv. ix. 6–10, xv. xi. 1–3; Appian *Pun.* 40–41. Livy and Polybius put Laelius on the left wing and Masinissa on the right; Appian puts Laelius on the right and Octavius on the left.

Against this formation Scipio drew up the flower of his legions in three successive front lines, arranged according to *hastati*, *principes*, and *triarii*,[1] not making the cohorts touch, but leaving a space between the detached companies through which the elephants driven by the enemy might easily be allowed to pass without throwing the ranks into confusion. These intervals he filled with light-armed skirmishers, that the line might show no gaps, giving them instructions to withdraw to the rear or the flanks at the first onset of the elephants. The cavalry he distributed on the flanks, placing Laelius in charge of the Roman horsemen on the right, and Masinissa in charge of the Numidians on the left. This shrewd scheme of arrangement was undoubtedly the cause of his victory.[2]

In the battle against Lucius Sulla, Archelaus placed his scythe-bearing chariots in front, for the purpose of throwing the enemy into confusion; in the second line he posted the Macedonian phalanx, and in the third line auxiliaries armed after the Roman way, with a sprinkling of Italian runaway slaves, in whose doggedness he had the greatest confidence. In the last line he stationed the light-armed troops, while on the two flanks, for the purpose of enveloping the enemy, he placed the cavalry, of whom he had a great number.

To meet these dispositions, Sulla constructed trenches of great breadth on each flank, and at their ends built strong redoubts. By this device he avoided the danger of being enveloped by the enemy, who outnumbered him in infantry and especially in cavalry. Next he arranged a triple line of infantry, leaving intervals through which to send, according

SEXTUS JULIUS FRONTINUS

quem in novissimo conlocaverat, cum res exegisset, emitteret. Tum postsignanis qui in secunda acie erant imperavit, ut densos numerososque palos firme in terram defigerent, intraque eos appropinquantibus quadrigis antesignanorum aciem recepit. Tum demum sublato universorum clamore velites et levem armaturam ingerere tela iussit. Quibus factis quadrigae hostium aut implicitae palis aut exterritae clamore telisque in suos conversae sunt turbaveruntque Macedonum structuram. Qua cedente, cum Sulla instaret et Archelaus equitem opposuisset, Romani equites subito emissi averterunt eos consummaveruntque victoriam.

18. C. Caesar Gallorum falcatas quadrigas eadem ratione palis defixis excepit inhibuitque.

19. Alexander ad Arbela, cum hostium multitudinem vereretur, virtuti autem suorum fideret, aciem in omnem partem spectantem ordinavit, ut circumventi undique pugnare possent.

20. Paulus adversus Persen Macedonum regem, cum is phalangem suorum duplicem mediam in partem direxisset eamque levi armatura cinxisset et equitem utroque cornu conlocasset, triplicem aciem cuneis instruxit, inter quos velites subinde emisit. Quo genere cum profligari nihil videret, cedere instituit, ut hac simulatione perduceret hostes in

---

[1] Troops posted behind the standards.
[2] Troops posted in front of the standards and serving for their defence.
[3] 86 B.C.
[4] 331 B.C. *Cf.* Curt. IV. xiii. 30–32; Diodor. XVII. lvii. 5.

to need, the light-armed troops and the cavalry, which he placed in the rear. He then commanded the *postsignani*,[1] who were in the second line, to drive firmly into the ground large numbers of stakes set close together, and as the chariots drew near, he withdrew the line of *antesignani*[2] within these stakes. Then at length he ordered the skirmishers and light-armed troops to raise a general battle-cry and discharge their spears. By these tactics either the chariots of the enemy were caught among the stakes, or their drivers became panic-stricken at the din and were driven by the javelins back upon their own men, throwing the formation of the Macedonians into confusion. As these gave way, Sulla pressed forward, and Archelaus met him with cavalry, whereupon the Roman horsemen suddenly darted forth, drove back the enemy, and achieved victory.[3]

In the same way Gaius Caesar met the scythe-bearing chariots of the Gauls with stakes driven in the ground, and kept them in check.

At Arbela, Alexander, fearing the numbers of the enemy, yet confident in the valour of his own troops, drew up a line of battle facing in all directions, in order that his men, if surrounded, might be able to fight from all sides.[4]

When Perseus, king of the Macedonians, had drawn up a double phalanx of his own troops and had placed them in the centre of his forces, with light-armed troops on each side and cavalry on both flanks, Paulus in the battle against him drew up a triple array in wedge formation, sending out skirmishers every now and then between the wedges. Seeing nothing accomplished by these tactics, he determined to retreat, in order by this feint to lure

SEXTUS JULIUS FRONTINUS

confragosa loca, quae ex industria captaverat. Cum sic quoque, suspecta calliditate recedentium, ordinata sequeretur phalanx, equites a sinistro cornu praeter oram phalangis iussit transcurrere citatis equis, tectos, ut objectis armis ipso impetu praefringerent hostium spicula. Quo genere telorum exarmati Macedones solverunt aciem et terga verterunt.

21. Pyrrhus pro Tarentinis apud Asculum, secundum Homericum versum quo pessimi in medium recipiuntur, dextro cornu Samnites Epirotasque, sinistro Bruttios atque Lucanos cum Sallentinis, in media acie Tarentinos conlocavit, equitatum et elephantos in subsidiis esse iussit.

Contra consules, aptissime divisis in cornua equitibus, legiones in prima acie et in subsidiis conlocaverunt et his immiscuerunt auxilia. XL milia utrimque fuisse constat. Pyrrhi dimidia pars exercitus amissa, apud Romanos V milia desiderata sunt.

22. Cn. Pompeius adversus C. Caesarem Palaepharsali triplicem instruxit aciem, quarum singulae denos ordines in latitudinem habuerunt. Legiones secundum virtutem cuiusque firmissimas in cornibus et in medio conlocavit, spatia his interposita tironibus supplevit. Dextro latere DC equites propter flumen Enipea, qui et alveo suo et alluvie regionem impedierat, reliquum equitatum in sinistro cornu

---

[1] 168 B.C. Battle of Pydna. *Cf.* Livy xliv. 41; Plut. *Aem.* 20.
[2] *Iliad* iv. 299 seems to have become proverbial; *cf.* Ammian. Marc. XXIV. vi. 9.
[3] 279 B.C. *Cf.* Plut. *Pyrrh.* 21.
[4] A town in Thessaly near Pharsalus.

the enemy after him on to rough ground, which he had selected with this in view. When even then the enemy, suspecting his ruse in retiring, followed in good order, he commanded the cavalry on the left wing to ride at full speed past the front of the phalanx, covering themselves with their shields, in order that the points of the enemy's spears might be broken by the shock of their encounter with the shields. When the Macedonians were deprived of their spears, they broke and fled.[1]

Pyrrhus, when fighting in defence of the Tarentines near Asculum, following the Homeric verse,[2] according to which the poorest troops are placed in the centre, stationed Samnites and Epirotes on the right flank, Bruttians, Lucanians, and Sallentines on the left, with the Tarentines in the centre, ordering the cavalry and elephants to be held as reserves.

The consuls, on the other hand, very judiciously distributed their cavalry on the wings, posting legionary soldiers in the first line and in reserve, with auxiliary troops scattered among them. We are informed that there were forty thousand men on each side. Half of Pyrrhus's army was lost; on the Roman side only five thousand.[3]

In the battle against Caesar at Old Pharsalus,[4] Gnaeus Pompey drew up three lines of battle, each one ten men deep, stationing on the wings and in the centre the legions upon whose prowess he could most safely rely, and filling the spaces between these with raw recruits. On the right flank he placed six hundred horsemen, along the Enipeus River, which with its channel and deposits had made the locality impassable; the rest of the cavalry he stationed on the left, together with the auxiliary

SEXTUS JULIUS FRONTINUS

cum auxiliis omnibus locavit, ut inde Iulianum exercitum circumiret.

Adversus hanc ordinationem C. Caesar et ipse triplici acie dispositis in fronte legionibus sinistrum latus, ne circumiri posset, admovit paludibus. In dextro cornu equitem posuit, cui velocissimos miscuit peditum, ad morem equestris pugnae exercitatos. Sex deinde cohortes in subsidio retinuit ad res subitas et dextro latere conversas in obliquum, unde equitatum hostium exspectabat, conlocavit. Nec ulla res eo die plus ad victoriam Caesari contulit; effusum namque Pompei equitatum inopinato excursu averterunt caedendumque tradiderunt.

23. Imperator Caesar Augustus Germanicus, cum subinde Chatti equestre proelium in silvas refugiendo deducerent, iussit suos equites, simulatque ad impedita ventum esset, equis desilire pedestrique pugna confligere; quo genere consecutus est, ne quis iam locus victoriam eius moraretur.

24. C. Duellius, cum videret graves suas naves mobilitate Punicae classis eludi inritamque virtutem militum fieri, excogitavit manus ferreas. Quae ubi hostilem apprenderant navem, superiecto ponte transgrediebatur Romanus et in ipsorum ratibus comminus eos trucidabat.

---

[1] *i.e* in fighting in conjunction with cavalry—doubtless after the method detailed in Caesar's *Gallic War*, i. 48.
[2] 48 B.C. *Cf.* Caes. *B.C.* iii 89, 93, 94; Plut. *Caes.* 44, *Pomp.* 69.   [3] Domitian.
[4] 83 A.D. *Cf.* Suet. *Domit.* 6.

troops, that from this quarter he might envelop the troops of Caesar.

Against these dispositions, Gaius Caesar also drew up a triple line, placing his legions in front and resting his left flank on marshes in order to avoid envelopment. On the right he placed his cavalry, among whom he distributed the fleetest of his foot-soldiers, men trained in cavalry fighting.[1] Then he held in reserve six cohorts for emergencies, placing them obliquely on the right, from which quarter he was expecting an attack of the enemy's cavalry. No circumstance contributed more than this to Caesar's victory on that day; for as soon as Pompey's cavalry poured forth, these cohorts routed it by an unexpected onset, and delivered it up to the rest of the troops for slaughter.[2]

The Emperor Caesar Augustus Germanicus,[3] when the Chatti, by fleeing into the forests, again and again interfered with the course of a cavalry engagement, commanded his men, as soon as they should reach the enemy's baggage-train, to dismount and fight on foot. By this means he made sure that his success should not be blocked by any difficulties of terrain.[4]

When Gaius Duellius saw that his own heavy ships were eluded by the mobile fleet of the Carthaginians and that the valour of his soldiers was thus brought to naught, he devised a kind of grappling-hook. When this caught hold of an enemy ship, the Romans, laying gangways over the bulwarks, went on board and slew the enemy in hand-to-hand combat on their own vessels.[5]

[5] 260 B.C. *Cf.* Flor. II. ii. 8–9; Polyb. i. 22.

SEXTUS JULIUS FRONTINUS

IIII. De Acie Hostium Turbanda

1. Papirius Cursor filius consul, cum aequo Marte adversus obstinatos Samnites concurreret, ignorantibus suis praecepit Spurio Nautio, ut pauci alares et agasones mulis insidentes ramosque per terram trahentes a colle transverso magno tumultu decurrerent. Quibus prospectis proclamavit victorem adesse collegam, occuparent ipsi praesentis proelii gloriam. Quo facto et Romani fiducia concitati proruere et hostes pulvere perculsi terga verterunt.

2. Fabius Rullus Maximus quarto consulatu in Samnio, omni modo frustra conatus aciem hostium perrumpere, novissime hastatos subduxit ordinibus et cum Scipione legato suo circummisit iussitque collem capere, ex quo decurri poterat in hostium terga. Quod ubi factum est, Romanis crevit animus et Samnites perterriti fugam molientes caesi sunt.

3. Minucius Rufus imperator, cum a Scordiscis Dacisque premeretur, quibus impar erat numero, praemisit fratrem et paucos una equites cum aeneatoribus praecepitque, ut, cum vidisset contractum proelium, subitus ex diverso se ostenderet iuberetque concinere aeneatores; resonantibus montium

---

[1] Of the dictator. [2] Spurius Carvilius.
[3] 293 B.C. *Cf.* Livy x. 40–41.
[4] First line troops.
[5] 297 B.C. *Cf.* Livy x. 14.

STRATAGEMS, II. iv. 1-3

IV. On Creating Panic in the Enemy's Ranks

When Papirius Cursor, the son,[1] in his consulship failed to win any advantage in his battle against the stubbornly resisting Samnites, he gave no intimation of his purpose to his men, but commanded Spurius Nautius to arrange to have a few auxiliary horsemen and grooms, mounted on mules and trailing branches over the ground, race down in great commotion from a hill running at an angle with the field. As soon as these came in sight, he proclaimed that his colleague[2] was at hand, crowned with victory, and urged his men to secure for themselves the glory of the present battle before he should arrive. At this the Romans rushed forward, kindling with confidence, while the enemy, disheartened at the sight of the dust, turned and fled.[3]

Fabius Rullus Maximus, when in Samnium in his fourth consulship, having vainly essayed in every way to break through the line of the enemy, finally withdrew the *hastati*[4] from the ranks and sent them round with his lieutenant Scipio, under instructions to seize a hill from which they could rush down upon the rear of the enemy. When this had been done, the courage of the Romans rose, and the Samnites, fleeing in terror, were cut to pieces.[5]

The general Minucius Rufus, hard pressed by the Scordiscans and Dacians, for whom he was no match in numbers, sent his brother and a small squadron of cavalry on ahead, along with a detachment of trumpeters, directing him, as soon as he should see the battle begin, to show himself suddenly from the opposite quarter and to order the trumpeters to blow their horns. Then, when the hill-tops re-echoed with

SEXTUS JULIUS FRONTINUS

iugis species ingentis multitudinis offusa est hostibus, qua perterriti dedere terga.

4. Acilius Glabrio consul adversus Antiochi regis aciem, quam is in Achaia pro angustiis Thermopylarum direxerat, iniquitatibus loci non inritus tantum, sed cum iactura quoque repulsus esset, nisi circummissus ab eo Porcius Cato, qui tum, iam consularis, tribunus militum a populo factus in exercitu erat, deiectis iugis Callidromi montis Aetolis, qui praesidio ea tenebant, super imminentem castris regiis collem a tergo subitus apparuisset. Quo facto perturbatis Antiochi copiis utrimque inrupere Romani et fusis fugatisque castra ceperunt.

5. C. Sulpicius Peticus consul contra Gallos dimicaturus iussit muliones clam in montes proximos cum mulis abire et indidem conserto iam proelio velut equis insidentes ostentare se pugnantibus; quare Galli existimantes adventare auxilia Romanis cessere iam paene victores.

6. Marius circa Aquas Sextias, cum in animo haberet postera die depugnare adversus Teutonos, Marcellum cum parva manu equitum peditumque nocte post terga hostium misit et ad implendam multitudinis speciem agasones lixasque armatos simul ire iussit iumentorumque magnam partem instra-

---

[1] 109 B.C.
[2] 191 B.C. *Cf.* Livy xxxvi. 14–19; Plut. *Cat. Maj.* 12 ff.; Appian *Syr.* 17 ff.
[3] 358 B.C. Peticus was dictator in this year, having been consul in 364 and 361. *Cf.* Livy vii. 14–15. Appian *Gall.* 1 gives a different stratagem.

the sound, the impression of a huge multitude was borne in upon the enemy, who fled in terror.[1]

The consul Acilius Glabrio, when confronted by the army of King Antiochus, which the latter had drawn up in front of the Pass of Thermopylae in Greece, was not only hampered by the difficulties of terrain, but would have been repulsed with loss besides, had not Porcius Cato prevented this. Cato, although an ex-consul, was in the army as a tribune of the soldiers, elected to this office by the people. [Having been sent by Glabrio to make a détour], he dislodged the Aetolians, who were guarding the crest of Mt. Callidromus, and then suddenly appeared from the rear on the summit of a hill commanding the camp of the king. The forces of Antiochus were thus thrown into panic, whereupon the Romans attacked them from front and rear, repulsed and scattered the enemy, and captured their camp.[2]

The consul Gaius Sulpicius Peticus, when about to fight against the Gauls, ordered certain muleteers secretly to withdraw with their mules to the hills near by, and then, after the engagement began, to exhibit themselves repeatedly to the combatants, as though mounted on horses. The Gauls, therefore, imagining that reinforcements were coming, fell back before the Romans, though already almost victorious.[3]

At Aquae Sextiae, Marius, purposing to fight a decisive battle with the Teutons on the morrow, sent Marcellus by night with a small detachment of horse and foot to the rear of the enemy, and, to complete the illusion of a large force, ordered armed grooms and camp-followers to go along with them, and also a large part of the pack-animals, wearing

SEXTUS JULIUS FRONTINUS

torum centunculis, ut per hoc facies equitatus obiceretur, praecepitque, ut, cum animadvertissent committi proelium, ipsi in terga hostium descenderent. Qui apparatus tantum terroris intulit, ut asperrimi hostes in fugam versi sint.

7. Licinius Crassus fugitivorum bello apud Camalatrum educturus militem adversus Castum et Cannicum duces Gallorum XII cohortes cum C. Pomptinio et Q. Marcio Rufo legatis post montem circummisit; quae cum commisso iam proelio a tergo clamore sublato decucurrissent, ita fuderunt hostes, ut ubique fuga, nusquam pugna capesseretur.

8. M. Marcellus, cum vereretur, ne paucitatem militum eius clamor detegeret, simul lixas calonesque et omnis generis sequellas conclamare iussit atque hostem magni exercitus specie exterruit.

9. Valerius Laevinus adversus Pyrrhum, occiso quodam gregali tenens gladium cruentum, utrique exercitui persuasit Pyrrhum interemptum; quamobrem hostes destitutos se ducis morte credentes, consternati a mendacio in castra se pavidi receperunt.

10. Iugurtha in Numidia adversus C. Marium, cum Latinae quoque linguae usum ei conversatio pristina castrorum dedisset, in primam aciem procucurrit et occisum a se C. Marium clare praedicavit atque ita multos nostrorum avertit.

---

[1] 102 B.C.   *Cf.* Plut. *Mar.* 20; Polyaen. VIII. x. 2.
[2] 71 B.C.   [3] 216 B.C.   *Cf.* Livy XXIII. xvi. 13–14.
[4] 280 B.C.   *Cf.* Plut. *Pyrrh.* 17.
[5] 107 B.C.   *Cf.* Sall. *Jug.* ci. 6–8.

saddle-cloths, in order by this means to present the appearance of cavalry. He commanded these men to fall upon the enemy from the rear, as soon as they should notice that the engagement had begun. This scheme struck such terror into the enemy that despite their great ferocity they turned and fled.[1]

Licinius Crassus in the Slave War, when about to lead forth his troops at Camalatrum against Castus and Cannicus, the leaders of the Gauls, sent twelve cohorts around behind the mountain with Gaius Pomptinius and Quintus Marcius Rufus, his lieutenants. When the engagement began, these troops, raising a shout, poured down from the mountain in the rear, and so routed the enemy that they fled in all directions with no attempt at battle.[2]

Marcus Marcellus on one occasion, fearing that a feeble battle-cry would reveal the small number of his forces, commanded that sutlers, servants, and camp-followers of every sort should join in the cry. He thus threw the enemy into panic by giving the appearance of having a large army.[3]

Valerius Laevinus, in the battle against Pyrrhus, killed a common soldier, and, holding up his dripping sword, made both armies believe that Pyrrhus had been slain. The enemy, therefore, panic-stricken at the falsehood, and thinking that they had been rendered helpless by the death of their commander, betook themselves in terror back to camp.[4]

In his struggle against Gaius Marius in Numidia, Jugurtha, having acquired facility in the use of the Latin language as a result of his early association with Roman camps, ran forward to the front line and shouted that he had slain Gaius Marius, thus causing many of our men to flee.[5]

SEXTUS JULIUS FRONTINUS

11. Myronides Atheniensis dubio proelio adversus Thebanos rem gerens repente in dextrum suorum cornu prosiluit et exclamavit sinistro iam se vicisse; qua re et suis alacritate et hostibus iniecto metu vicit.

12. Croesus praevalido hostium equitatui camelorum gregem opposuit, quorum novitate et odore consternati equi non solum insidentes praecipitaverunt, sed peditum quoque suorum ordines protriverunt vincendosque hosti praebuerunt.

13. Pyrrhus Epirotarum rex pro Tarentinis adversus Romanos eodem modo elephantis ad perturbandam aciem usus est.

14. Poeni quoque adversus Romanos frequenter idem fecerunt.[1]

15. Volscorum castra cum prope a virgultis silvaque posita essent, Camillus ea omnia, quae conceptum ignem usque in vallum perferre poterant, incendit et sic adversarios exuit castris.[2]

16. P. Crassus bello sociali eodem modo prope cum copiis omnibus interceptus est.[3]

17. Hispani contra Hamilcarem boves vehiculis adiunctos in prima fronte constituerunt vehiculaque tedae et sebi et sulphuris plena, signo pugnae dato, incenderunt; actis deinde in hostem bubus consternatam aciem perruperunt.

---

[1] *No. 14 is regarded as an interpolation.*
[2] *Identical with* IV. vii. 40 *and regarded as an interpolation in this place.*
[3] *Identical with* IV. vii. 41 *and regarded as an interpolation in this place.*

---

[1] 457 B.C. *Cf.* Polyaen. I. xxxv. 1.
[2] 546 B.C. *Cf.* Herod. i. 80; Polyaen. VII. vi. 6; they attribute this stratagem to Cyrus.
[3] 280 B.C. *Cf.* Flor. I. xviii. 8; Plut. *Pyrrh.* 17.

STRATAGEMS II. iv. 11–17

Myronides, the Athenian, in an indecisive battle which he was waging against the Thebans, suddenly darted forward to the right flank of his own troops and shouted that he had already won victory on the left. Thus, by inspiring courage in his own men and fear in the enemy, he gained the day.[1]

Against overwhelming forces of the enemy's cavalry Croesus once opposed a troop of camels. At the strange appearance and smell of these beasts, the horses were thrown into panic, and not merely threw their riders, but also trampled the ranks of their own infantry under foot, thus delivering them into the hands of the enemy to defeat.[2]

Pyrrhus, king of the Epirotes, fighting on behalf of the Tarentines against the Romans, employed elephants in the same way, in order to throw the Roman army into confusion.[3]

The Carthaginians also often did the same thing in their battles against the Romans.[4]

The Volscians having on one occasion pitched their camp near some brush and woods, Camillus set fire to everything which could carry the conflagration up to their entrenchments, and thus deprived his adversaries of their camp.

In the same way, Publius Crassus in the Social War narrowly escaped being cut off with all his forces.

The Spaniards, when fighting against Hamilcar, hitched steers to carts and placed them in the front line. These carts they filled with pitch, tallow, and sulphur, and when the signal for battle was given, set them afire. Then, driving the steers against the enemy, they threw the line into panic and broke through.[5]

---

[4] *Cf.* II. v. 4.   [5] 229 B.C. *Cf.* Appian *Hisp.* 5.

SEXTUS JULIUS FRONTINUS

18. Falisci et Tarquinienses, compluribus suorum in habitum sacerdotum subornatis, faces et angues furiali habitu praeferentibus, aciem Romanorum turbaverunt.

19. Idem Veientes et Fidenates facibus adreptis fecerunt.[1]

20. Atheas rex Scytharum, cum adversus ampliorem Triballorum exercitum confligeret, iussit a feminis puerisque et omni imbelli turba greges asinorum ac boum ad postremam hostium aciem admoveri et erectas hastas praeferri; famam deinde diffudit, tamquam auxilia sibi ab ulterioribus Scythis adventarent. Qua adseveratione avertit hostem.

V. De Insidiis

1. Romulus, per latebras copiarum parte disposita, cum ad Fidenas accessisset, simulata fuga temere hostes insecutos eo perduxit, ubi occultos milites habebat, qui undique adorti effusos et incautos ceciderunt.

2. Q. Fabius Maximus consul, auxilio Sutrinis missus adversus Etruscos, omnes hostium copias in se convertit; deinde simulato timore in superiora loca velut fugiens recessit effuseque subeuntes adgressus non acie tantum superavit, sed etiam castris exuit.

---

[1] *No. 19 is probably an interpolation.*

---

[1] 356 B.C. *Cf.* Livy VII. xvii. 3.
[2] 426 B.C. *Cf.* Livy iv. 33; Flor. I. xii. 7.
[3] *Cf* Polyaen. VII. xliv. 1.
[4] *Cf.* Livy i. 14; Polyaen. VIII. iii. 2.
[5] 310 B.C. Q. Fabius Maximus Rullianus. *Cf.* Livy ix. 35.

The Faliscans and Tarquinians disguised a number of their men as priests, and had them hold torches and snakes in front of them, like Furies. Thus they threw the army of the Romans into panic.[1]

On one occasion the men of Veii and Fidenae snatched up torches and did the same thing.[2]

When Atheas, king of the Scythians, was contending against the more numerous tribe of the Triballi, he commanded that herds of asses and cattle should be brought up in the rear of the enemy's forces by women, children, and all the non-combatant population, and that spears, held aloft, should be carried in front of these. Then he spread abroad the rumour that reinforcements were coming to him from the more distant Scythian tribes. By this declaration he forced the enemy to withdraw.[3]

V. On Ambushes

Romulus, when he had drawn near to Fidenae, distributed a portion of his troops in ambush, and pretended to flee. When the enemy recklessly followed, he led them on to the point where he was holding his men in hiding, whereupon the latter, attacking from all sides, and taking the enemy off their guard, cut them to pieces in their onward rush.[4]

The consul Quintus Fabius Maximus, having been sent to aid the Sutrians against the Etruscans, caused the full brunt of the enemy's attack to fall upon himself. Then, feigning fear, he retired to higher ground, as though in retreat, and when the enemy rushed upon him pell-mell he attacked, and not merely defeated them in battle but captured their camp.[5]

SEXTUS JULIUS FRONTINUS

3. Sempronius Gracchus adversus Celtiberos metu simulato continuit exercitum; emissa deinde armatura levi, quae hostem lacesseret ac statim pedem referret, evocavit hostes. Deinde inordinatos adgressus usque eo cecidit, ut etiam castra caperet.

4. L. Metellus consul in Sicilia bellum adversus Hasdrubalem gerens, ob ingentem eius exercitum et CXXX elephantos intentior, simulata diffidentia intra Panormum copias tenuit fossamque ingentis magnitudinis ante se duxit. Conspecto deinde exercitu Hasdrubalis, qui in prima acie elephantos habebat, praecepit hastatis, tela in beluas iacerent protinusque se intra munimenta reciperent. Ea ludificatione rectores elephantorum concitati in ipsam fossam elephantos egerunt. Quo ut primum inlati sunt, partim multitudine telorum confecti, partim retro in suos acti totam aciem turbaverunt. Tunc Metellus, hanc opperiens occasionem, cum toto exercitu erupit et adgressus a latere Poenos cecidit ipsisque elephantis potitus est.

5. Thamyris Scytharum regina Cyrum Persarum ducem aequo Marte certantem simulato metu elicuit ad notas militi suo angustias atque ibi, repente converso agmine, natura loci adiuta devicit.

6. Aegyptii conflicturi acie in eis campis, quibus

---

[1] 179 B C.   Cf. Livy xl. 48.   [2] The modern Palermo.
[3] 251 B C.   Cf. Polyb. i. 40.
[4] 529 B.C.   Cf. Justin. i. 8; Herod. i. 204 ff.

Sempronius Gracchus, when waging war against the Celtiberians, feigned fear and kept his army in camp. Then, by sending out light-armed troops to harass the enemy and retreat forthwith, he caused the enemy to come out; whereupon he attacked them before they could form, and crushed them so completely that he also captured their camp.[1]

When the consul Lucius Metellus was waging war in Sicily against Hasdrubal—and with all the more alertness because of Hasdrubal's immense army and his one hundred and thirty elephants—he withdrew his troops, under pretence of fear, inside Panormus[2] and constructed in front a trench of huge proportions. Then, observing Hasdrubal's army, with the elephants in the front rank, he ordered the *hastati* to hurl their javelins at the beasts and straightway to retire within their defences. The drivers of the elephants, enraged at such derisive treatment, drove the elephants straight towards the trench. As soon as the beasts were brought up to this, part were dispatched by a shower of darts, part were driven back to their own side, and threw the entire host into confusion. Then Metellus, who was biding his time, burst forth with his whole force, attacked the Carthaginians on the flank, and cut them to pieces. Besides this, he captured the elephants themselves.[3]

When Thamyris, queen of the Scythians, and Cyrus, king of the Persians, became engaged in an indecisive combat, the queen, feigning fear, lured Cyrus into a defile well-known to her own troops, and there, suddenly facing about, and aided by the nature of the locality, won a complete victory.[4]

The Egyptians, when about to engage in battle on

SEXTUS JULIUS FRONTINUS

iunctae paludes erant, alga eas contexerunt commissoque proelio fugam simulantes in insidias hostes evocaverunt, qui rapidius per ignota invecti loca limo inhaeserunt circumventique sunt.

7. Viriathus, ex latrone dux Celtiberorum, cedere se Romanis equitibus simulans usque ad locum voraginosum et praealtum eos perduxit et, cum ipse per solidos ac notos sibi transitus evaderet, Romanos ignaros locorum immersosque limo cecidit.

8. Fulvius imperator Cimbrico bello conlatis cum hoste castris equites suos iussit succedere ad munitiones eorum lacessitisque barbaris simulata fuga regredi. Hoc cum per aliquot dies fecisset, avide insequentibus Cimbris, animadvertit castra eorum solita nudari. Itaque per partem exercitus custodita consuetudine ipse cum expeditis post castra hostium consedit occultus effusisque eis ex more repente adortus et desertum proruit vallum et castra cepit.

9. Cn. Fulvius, cum in finibus nostris exercitus Faliscorum longe nostro maior castra posuisset, per suos milites quaedam procul a castris aedificia succendit, ut Falisci suos id fecisse credentes spe praedae diffunderentur.

10. Alexander Epirotes adversus Illyrios conlocata in insidiis manu quosdam ex suis habitu Illyriorum

---

[1] 147–139 B.C. In ii. xiii. 4, Viriathus is *dux Lusitanorum*.
[2] Livy xl. 30–32 says that Q. Fulvius Flaccus used this stratagem with the Celtiberians in 181 B.C There is no account of Fulvius's warring with the Cimbrians.

a plain near a marsh, covered the marsh with seaweed, and then, when the battle began, feigning flight, drew the enemy into a trap; for the latter, while advancing too swiftly over the unfamiliar ground, were caught in the mire and surrounded.

Viriathus, who from being a bandit became leader of the Celtiberians, on one occasion, while pretending to give way before the Roman cavalry, led them on to a place full of deep holes. There, while he himself made his way out by familiar paths that afforded good footing, the Romans, ignorant of the locality, sank in the mire and were slain.[1]

Fulvius, commander in the Cimbrian war, having pitched his camp near the enemy, ordered his cavalry to approach the fortifications of the barbarians and to withdraw in pretended flight, after making an attack. When he had done this for several days, with the Cimbrians in hot pursuit, he noticed that their camp was regularly left exposed. Accordingly, maintaining his usual practice with part of his force, he himself, with light-armed troops, secretly took a position behind the camp of the enemy, and as they poured forth according to their custom, he suddenly attacked and demolished the unguarded rampart and captured their camp.[2]

Gnaeus Fulvius, when a force of Faliscans far superior to ours had encamped on our territory, had his soldiers set fire to certain buildings at a distance from the camp, in order that the Faliscans, thinking their own men had done this, might scatter in hope of plunder.

Alexander, the Epirote, when waging war against the Illyrians, first placed a force in ambush, and then dressed up some of his own men in Illyrian garb,

SEXTUS JULIUS FRONTINUS

instruxit et iussit vastare suam, id est Epiroticam regionem. Quod cum Illyrii viderent fieri, ipsi passim praedari coeperunt eo securius, quod praecedentes veluti pro exploratoribus habebant; a quibus ex industria in loca iniqua deducti caesi fugatique sunt.

11. Leptines quoque Syracusanus adversus Carthaginienses vastari suos agros et incendi villas castellaque quaedam imperavit. Carthaginienses, a suis id fieri rati, et ipsi tamquam in adiutorium exierunt exceptique ab insidiatoribus fusi sunt.

12. Maharbal, missus a Carthaginiensibus adversus Afros rebellantes, cum sciret gentem avidam esse vini, magnum eius modum mandragora permiscuit, cuius inter venenum ac soporem media vis est. Tum proelio levi commisso ex industria cessit. Nocte deinde intempesta relictis intra castra quibusdam sarcinis et omni vino infecto fugam simulavit; cumque barbari occupatis castris in gaudium effusi avide medicatum merum hausissent et in modum defunctorum strati iacerent, reversus aut cepit eos aut trucidavit.

13. Hannibal, cum sciret sua et Romanorum castra in eis locis esse, quae lignis deficiebantur, ex industria in regione deserta plurimos armentorum greges intra vallum reliquit, qua velut praeda Romani potiti in summis lignationis angustiis insalu-

---

[1] 397–396 B.C. *Cf.* Polyaen. v. viii. 1.
[2] A Carthaginian officer under Hannibal.
[3] Polyaen. v. x. 1 attributes this stratagem to Himilco, 396 B.C.

ordering them to lay waste his own, that is to say, Epirote territory. When the Illyrians saw that this was being done, they themselves began to pillage right and left—the more confidently since they thought that those who led the way were scouts. But when they had been designedly brought by the latter into a disadvantageous position, they were routed and killed.

Leptines, the Syracusan, also, when waging war against the Carthaginians, ordered his own lands to be laid waste and certain farm-houses and forts to be set on fire. The Carthaginians, thinking this was done by their own men, went out themselves also to help; whereupon they were set upon by men lying in wait, and were put to rout.[1]

Maharbal,[2] sent by the Carthaginians against the rebellious Africans, knowing that the tribe was passionately fond of wine, mixed a large quantity of wine with mandragora, which in potency is something between a poison and a soporific. Then after an insignificant skirmish he deliberately withdrew. At dead of night, leaving in the camp some of his baggage and all the drugged wine, he feigned flight. When the barbarians captured the camp and in a frenzy of delight greedily drank the drugged wine, Maharbal returned, and either took them prisoners or slaughtered them while they lay stretched out as if dead.[3]

Hannibal, on one occasion, aware that both his own camp and that of the Romans were in places deficient in wood, deliberately abandoned the district, leaving many herds of cattle within his camp. The Romans, securing possession of these as booty, gorged themselves with flesh, which, owing to

bribus se cibis oneraverunt. Hannibal, reducto nocte exercitu, securos eos et semicruda graves carne maiorem in modum vexavit.

14. Ti. Gracchus in Hispania, certior factus hostem inopem commercio laborare, instructissima castra omnibus esculentis deseruit; quae adeptum hostem et repertis intemperanter repletum gravemque reducto exercitu subito oppressit.

15. Chii, qui adversus Erythraeos bellum gerebant, speculatorem eorum in loco edito deprehensum occiderunt et vestem eius suo militi dederunt, qui ex eodem iugo Erythraeos signo dato in insidias evocavit.

16. Arabes, cum esset nota consuetudo eorum, qua de adventu hostium interdiu fumo, nocte igne significare instituerant, ut sine intermissione ea fierent, praeceperunt, adventantibus autem adversariis intermitterentur; qui cum cessantibus luminibus existimarent ignorari adventum suum, avidius ingressi oppressique sunt.

17. Alexander Macedo, cum hostis in saltu editiore castra communisset, subducta parte copiarum praecepit his quos relinquebat, ut ex more ignes excitarent speciemque praeberent totius exercitus; ipse per avias regiones circumducta manu hostem superiore adgressus loco depulit.

---

[1] 179–178 B.C.
[2] 327 B.C. Polyaen. IV. iii. 29 and Curt. VII. xi. 1 tell of the employment of a different stratagem.

the scarcity of firewood, was raw and indigestible. Hannibal, returning by night with his army, finding them off their guard and gorged with raw meat, inflicted great loss upon them.

Tiberius Gracchus, when in Spain, upon learning that the enemy were suffering from lack of provisions, provided his camp with an elaborate supply of eatables of all kinds and then abandoned it. When the enemy had got possession of the camp and had gorged themselves to repletion with the food they found, Gracchus brought back his army and suddenly crushed them.[1]

The Chians, when waging war against the Erythreans, caught an Erythrean spy on a lofty eminence and put him to death. They then gave his clothes to one of their own soldiers, who, by giving a signal from the same eminence, lured the Erythreans into an ambush.

The Arabians, since their custom of giving notice of the arrival of an enemy by means of smoke by day, and by fire at night, was well known, issued orders on one occasion that these practices should continue without interruption until the enemy actually approached, when they should be discontinued. The enemy, imagining from the absence of the fires that their approach was unknown, advanced too eagerly and were overwhelmed.

Alexander of Macedon, when the enemy had fortified their camp on a lofty wooded eminence, withdrew a portion of his forces, and commanded those whom he left to kindle fires as usual, and thus to give the impression of the complete army. He himself, leading his forces around through untravelled regions, attacked the enemy and dislodged them from their commanding position.[2]

18. Memnon Rhodius, cum equitatu praevaleret et hostem in collibus se continentem in campos vellet deducere, quosdam ex militibus suis sub specie perfugarum misit in hostium castra, qui adfirmarent exercitum Memnonis tam perniciosa seditione furere, ut subinde aliqua pars eius dilaberetur. Huic adfirmationi ut fidem faceret, passim in conspectu hostium iussit parva castella muniri, velut in ea se recepturi essent qui dissidebant. Hac persuasione sollicitati, qui in montibus se continuerant, in plana descenderunt et, dum castella temptant, ab equitatu circumventi sunt.

19. Harrybas rex Molossorum, bello petitus a Bardyli Illyrio, maiorem aliquanto exercitum habente, amolitus[1] imbelles suorum in vicinam regionem Aetoliae famam sparsit, tamquam urbes ac res suas Aetolis concederet. Ipse cum his, qui arma ferre poterant, insidias in montibus et locis confragosis distribuit. Illyrii timentes, ne quae Molossorum erant ab Aetolis occuparentur, velut ad praedam festinantes neglectis ordinibus accelerare coeperunt; quos dissipatos, nihil tale exspectantes, Harrybas ex insidiis fudit fugavitque.

20. T. Labienus C. Caesaris legatus adversus Gallos ante adventum Germanorum, quos auxilio his venturos sciebat, confligere cupiens diffidentiam simulavit posi-

[1] molitus *HP. corr. d.*

---

[1] Polyaen. v. xliv. 2 has a slightly different version.

Memnon, the Rhodian, being superior in cavalry, and wishing to draw down to the plains an enemy who clung to the hills, sent certain of his soldiers under the guise of deserters to the camp of the enemy, to say that the army of Memnon was inspired with such a serious spirit of mutiny that some portion of it was constantly deserting. To lend credit to this assertion, Memnon ordered small redoubts to be fortified here and there in view of the enemy, as though the disaffected were about to retire to these. Inveigled by these representations, those who had been keeping themselves on the hills came down to level ground, and, as they attacked the redoubts, were surrounded by the cavalry.[1]

When Harrybas, king of the Molossians, was attacked in war by Bardylis, the Illyrian, who commanded a considerably larger army, he dispatched the non-combatant portion of his subjects to the neighbouring district of Aetolia, and spread the report that he was yielding up his towns and possessions to the Aetolians. He himself, with those who could bear arms, placed ambuscades here and there on the mountains and in other inaccessible places. The Illyrians, fearful lest the possessions of the Molossians should be seized by the Aetolians, began to race along in disorder, in their eagerness for plunder. As soon as they became scattered, Harrybas, emerging from his concealment and taking them unawares, routed them and put them to flight.

Titus Labienus, lieutenant of Gaius Caesar, eager to engage in battle with the Gauls before the arrival of the Germans, who, he knew, were coming to their aid, pretended discouragement, and, pitching his

SEXTUS JULIUS FRONTINUS

tisque in diversa ripa castris profectionem edixit in posterum diem. Galli credentes eum fugere flumen, quod medium erat, instituerunt transmittere. Labienus circumacto exercitu inter ipsas superandi amnis difficultates eos cecidit.

21. Hannibal, cum explorasset neglegenter castra Fulvi Romani ducis munita, ipsum praeterea multa temere audere, prima luce, cum densiores nebulae praestarent obscuritatem, paucos equites munitionum nostrarum vigilibus ostendit; quo Fulvius repente movit exercitum. Hannibal per diversam partem castra eius occupavit et ita[1] in tergum Romanorum effusus octo milia fortissimorum militum cum ipso duce trucidavit.

22. Idem Hannibal, cum inter Fabium dictatorem et Minucium magistrum equitum divisus esset exercitus et Fabius occasionibus immineret, Minucius pugnandi cupiditate flagraret, castra in campo, qui medius inter hostes erat, posuit et, cum partem peditum in confragosis rupibus celasset, ipse ad evocandum hostem misit, qui proximum tumulum occuparent. Ad quos opprimendos cum eduxisset copias Minucius, insidiatores ab Hannibale dispositi subito consurrexerunt et delessent Minuci exercitum, nisi Fabius periclitantibus subvenisset.

23. Idem Hannibal, cum ad Trebiam in conspectu haberet Semproni Longi consulis castra, medio amne

---

[1] illa, *Gund.*; ita, *most edd.*

---

[1] 53 B.C. *Cf.* Caes. *B.G.* vi. 7–8; Dio xl. 31.
[2] 210 B.C. At Herdonia. *Cf.* Livy xxvii. 1; Appian *Hann.* 48.
[3] 217 B.C. *Cf.* Livy xxii. 28; Polyb. iii. 104–105.

camp across the stream, announced his departure for the following day. The Gauls, imagining that he was in flight, began to cross the intervening river. Labienus, facing about with his troops, cut the Gauls to pieces in the very midst of their difficulties of crossing.[1]

Hannibal, on one occasion, learned that the camp of Fulvius, the Roman commander, was carelessly fortified and that Fulvius himself was taking many rash chances besides. Accordingly, at daybreak, when dense mists afforded cover, he permitted a few of his horsemen to show themselves to the sentries of our fortifications; whereupon Fulvius suddenly advanced. Meanwhile, Hannibal, at a different point, entered Fulvius's camp, and overwhelming the Roman rear, slew eight thousand of the bravest soldiers along with their commander himself.[2]

Once, when the Roman army had been divided between the dictator Fabius and Minucius, master of the horse, and Fabius was watching for a favourable opportunity, while Minucius was burning with eagerness for battle, the same Hannibal pitched his camp on the plain between the hostile armies, and having concealed a portion of his troops among rough rocks, sent others to seize a neighbouring hillock, as a challenge to the foe. When Minucius had led out his forces to crush these, the men placed here and there in ambush by Hannibal suddenly sprang up, and would have annihilated Minucius's army, had not Fabius come to help them in their distress.[3]

When the same Hannibal was encamped in the depths of winter at the Trebia, with the camp of the consul, Sempronius Longus, in plain view and

SEXTUS JULIUS FRONTINUS

interfluente, saevissima hieme Magonem et electos in insidiis posuit. Deinde Numidas equites ad eliciendam Semproni credulitatem adequitare vallo eius iussit, quibus praeceperat, ut ad primum nostrorum incursum per nota refugerent vada. Hos consul et adortus temere et secutus ieiunum exercitum in maximo frigore transitu fluminis rigefecit. Mox torpore et inedia adfectis Hannibal suum militem opposuit, quem ad id ignibus oleoque et cibo foverat; nec defuit partibus Mago, quin terga hostium in hoc ordinatus caederet.

24. Idem ad Trasumennum, cum arta quaedam via inter lacum et radices montis in campos patentes duceret, simulata fuga per angustias ad patentia evasit ibique castra posuit ac nocte dispositis militibus et per collem, qui imminebat, et in lateribus angustiarum prima luce, nebula quoque adiutus, aciem direxit. Flaminius velut fugientem insequens, cum angustias esset ingressus, non ante providit insidias, quam simul a fronte, lateribus, tergo circumfusus ad internecionem cum exercitu caederetur.

25. Idem Hannibal adversus Iunium dictatorem nocte intempesta DC equitibus imperavit, ut in plures turmas segregati per vices sine intermissione circa castra hostium se ostentarent. Ita tota nocte Romanis

---

[1] 218 B.C. *Cf.* Livy xxi. 54 ff; Polyb. iii. 71.
[2] Generally considered to be the narrow passage between the lake and Monte Gualandro, near Borghetto. Livy's description suits this locality; that of Polybius does not.
[3] 217 B.C. *Cf.* Livy xxii. 4; Polyb. iii. 83 ff.
[4] M. Junius Pera.

only the river flowing between, he placed Mago and picked men in ambush. Then he commanded Numidian cavalry to advance up to Sempronius's fortifications, in order to lure forth the simple-minded Roman. At the same time, he ordered these troops to retire by familiar fords at our first onset. By heedlessly attacking and pursuing the Numidians, the consul gave his troops a chill, as a result of fording the stream in the bitter cold and without breakfast. Then, when our men were suffering from numbness and hunger, Hannibal led against them his own troops, whom he had got in condition for that purpose by warm fires, food, and rubbing down with oil. Mago also did his part, and cut to pieces the rear of his enemy at the point where he had been posted for the purpose.[1]

At Trasimenus, where a narrow way,[2] running between the lake and the base of the hills, led out to the open plain, the same Hannibal, feigning flight, made his way through the narrow road to the open districts and pitched his camp there. Then, posting soldiers by night at various points over the rising ground of the hill and at the ends of the defile, at daybreak, under cover of a fog, he marshalled his line of battle. Flaminius, pursuing the enemy, who seemed to be retreating, entered the defile and did not see the ambush until he was surrounded in front, flank, and rear, and was annihilated with his army.[3]

The same Hannibal, when contending against the dictator Junius,[4] ordered six hundred cavalrymen to break up into a number of squadrons, and at dead of night to appear in successive detachments without intermission around the camp of the enemy. Thus all night long the Romans were harassed and

SEXTUS JULIUS FRONTINUS

in vallo statione ac pluvia, quae forte continua fuerat, inquietatis confectisque, cum receptui signum mane Iunius dedisset, Hannibal suos requietos eduxit et castra eius invasit.

26. Epaminondas Thebanus in eundem modum, cum Lacedaemonii ad Isthmon vallo ducto Peloponneson tuerentur, paucorum opera levis armaturae tota nocte inquietavit hostem. Ac deinde prima luce revocatis suis, cum Lacedaemonii se recepissent, subito universum exercitum, quem quietum habuerat, admovit et per ipsa munimenta destituta propugnatoribus inrupit.

27. Hannibal directa acie ad Cannas DC equites Numidas transfugere iussit, qui ad fidem faciendam gladios et scuta nostris tradiderunt et in ultimum agmen recepti, ubi primum concurri coepit, strictis minoribus quos occultaverant gladiis, scutis iacentium adsumptis, Romanorum aciem ceciderunt.

28. Iapydes P. Licinio proconsuli paganos quoque sub specie deditionis obtulerunt, qui recepti et in postrema acie conlocati terga Romanorum ceciderunt.

29. Scipio Africanus, cum adversa haberet bina hostium castra, Syphacis et Carthaginiensium, statuit Syphacis, ubi multa incendii alimenta erant, adgredi nocte ignemque inicere, ut ea re Numidas quidem

---

[1] 216 B.C. At Capua. *Cf.* Polyaen. VI. xxxviii. 6; Zonar. ix. 3.
[2] 369 B.C. *Cf.* Polyaen. II. iii. 9.
[3] 216 B.C. *Cf.* Livy xxii. 48; Val. Max. VII. iv. *ext.* 2; Appian *Hann.* 20 ff. Livy and Appian give the number as five hundred.

worn out by sentry duty on the rampart and by the rain, which happened to fall continuously, so that in the morning, when Junius gave the signal for recall, Hannibal led out his own troops, who had been well rested, and took Junius's camp by assault.[1]

In the same way, when the Spartans had drawn entrenchments across the Isthmus and were defending the Peloponnesus, Epaminondas, the Theban, with the help of a few light-armed troops, harassed the enemy all night long. Then at daybreak, after he had recalled his own men and the Spartans had also retired, he suddenly moved forward the entire force which he had kept at rest, and burst directly through the ramparts, which had been left without defenders.[2]

At the battle of Cannae, Hannibal, having drawn up his line of battle, ordered six hundred Numidian cavalry to go over to the enemy. To prove their sincerity, these surrendered their swords and shields to our men, and were dispatched to the rear. Then, as soon as the engagement began, drawing out small swords, which they had secreted, and picking up the shields of the fallen, they slaughtered the troops of the Romans.[3]

Under pretence of surrender, the Iapydes handed over some of their tribesmen to Publius Licinius, the Roman proconsul. These were received and placed in the last line, whereupon they cut to pieces the Romans who were bringing up the rear.

Scipio Africanus, when facing the two hostile camps of Syphax and the Carthaginians, decided to make a night attack on that of Syphax, where there was a large supply of inflammable material, and to set fire to it, in order thus to cut down the Numidians

SEXTUS JULIUS FRONTINUS

ex suis castris trepidantes caederet, Poenos autem, quos certum erat ad succurrendum sociis procursuros, insidiis dispositis exciperet. Utrumque ex sententia cessit, nam tamquam ad fortuitum incendium sine armis procurrentis adortus cecidit.

30. Mithridates, a Lucullo virtute frequenter superatus, insidiis eum appetiit, Adathante quodam eminente viribus subornato, ut transfugeret et fide parta hosti facinus perpetraret; quod is strenue quidem, sed sine eventu conatus est. Receptus enim a Lucullo in gregem equitum non sine tacita custodia habitus est, quia nec credi subito transfugae nec inhiberi reliquos oportebat. Cum deinde frequentibus excursionibus promptam et enixam operam exhiberet, fide adquisita tempus elegit, quo missa principia quietem omnibus castrensibus dabant praetoriumque secretius praestabant. Casus adiuvit Lucullum. Nam qui ad vigilantem usque admitteretur, fatigatum nocturnis cogitationibus illo tempore quiescentem invenit. Cum deinde, tamquam nuntiaturus subitum aliquid ac necessarium, intrare vellet et pertinaciter a servis valetudini domini consulentibus excluderetur, veritus, ne suspectus

---

[1] 203 B.C. *Cf.* Livy xxx. 5–6; Polyb. xiv. 4.

as they scurried in terror from their camp, and also, by laying ambuscades, to catch the Carthaginians, who, he knew, would rush forward to assist their allies. Both plans succeeded. For when the enemy rushed forward unarmed, thinking the conflagration accidental, Scipio fell upon them and cut them to pieces.[1]

Mithridates, after repeated defeats in battle at the hands of Lucullus, made an attempt against his life by treachery, hiring a certain Adathas, a man of extraordinary strength, to desert and to perpetrate the deed, so soon as he should gain the confidence of the enemy. This plan the deserter did his best to execute, but his efforts failed. For, though admitted by Lucullus to the cavalry troop, he was quietly kept under surveillance, since it was neither well to put trust at once in a deserter, nor to prevent other deserters from coming. After this fellow had exhibited a ready and earnest devotion on repeated raids, and had won confidence, he chose a time when the dismissal of the staff-officers brought with it repose throughout the camp, and caused the general's headquarters to be less frequented. Chance favoured Lucullus; for whereas the deserter expected to find Lucullus awake, in which case he would have been at once admitted to his presence, he actually found him at that time fast asleep, exhausted with revolving plans in his mind the night before. Then when Adathas pleaded to be admitted, on the ground that he had an unexpected and imperative message to deliver, he was kept out by the determined efforts of the slaves, who were concerned for their master's health. Fearing consequently that he was an object of

SEXTUS JULIUS FRONTINUS

esset, equis quos ante portam paratos habebat ad
Mithridaten refugit inritus.

31. Sertorius in Hispania, cum apud Lauronem
oppidum vicina castra Pompei castris haberet et duae
tantummodo regiones essent, ex quibus pabulum peti
posset, una in propinquo, altera longius sita, eam
quae in propinquo erat subinde a levi armatura in-
festari, ulteriorem autem vetuit ab ullo armato adiri,
donec persuasit adversariis, tutiorem esse quae erat
remotior. Quam cum petissent Pompeiani, Octavium
Graecinum[1] cum decem cohortibus in morem Romanorum armatis et decem Hispanorum levis armaturae et Tarquitium Priscum cum duobus milibus
equitum ire iubet ad insidias tendendas pabulatori-
bus. Illi strenue imperata faciunt. Explorata enim
locorum natura, in vicina silva nocte praedictas
copias abscondunt ita, ut in prima parte leves
Hispanos, aptissimos ad furta bellorum, ponerent,
paulo interius scutatos, in remotissimo equites, ne
fremitu eorum cogitata proderentur; quiescere omnes
silentio servato in horam tertiam diei iubent. Cum
deinde Pompeiani securi oneratique pabulo de reditu
cogitarent et hi quoque, qui in statione fuerant,
quiete invitati ad pabulum conligendum dilaberentur,
emissi primum Hispani velocitate gentili in palantes
effunduntur et convulnerant confunduntque nihil

[1] graecinium *HP*. Graecinum *Oud*.

---

[1] 72 B.C. *Cf.* Appian *Mithr.* 79; Plut. *Lucull.* 16. The
name of the deserter varies in the different accounts.

[2] About 9 A.M.

suspicion, he mounted the horse which he held in readiness outside the gate, and fled to Mithridates without accomplishing his purpose.[1]

When Sertorius was encamped next to Pompey near the town of Lauron in Spain, there were only two tracts from which forage could be gathered, one near by, the other farther off. Sertorius gave orders that the one near by should be continually raided by light-armed troops, but that the remoter one should not be visited by any troops. Thus, he finally convinced his adversaries that the more distant tract was safer. When, on one occasion, Pompey's troops had gone to this region, Sertorius ordered Octavius Graecinus, with ten cohorts armed after the Roman fashion, and ten cohorts of light-armed Spaniards along with Tarquitius Priscus and two thousand cavalry, set forth to lay an ambush against the foragers. These men executed their instructions with energy; for after examining the ground, they hid the above-mentioned forces by night in a neighbouring wood, posting the light-armed Spaniards in front, as best suited to stealthy warfare, the shield-bearing soldiers a little further back, and the cavalry in the rear, in order that the plan might not be betrayed by the neighing of the horses. Then they ordered all to repose in silence till the third hour of the following day.[2] When Pompey's men, entertaining no suspicion and loaded down with forage, thought of returning, and those who had been on guard, lured on by the situation, were slipping away to forage, suddenly the Spaniards, darting out with the swiftness characteristic of their race, poured forth upon the stragglers, inflicted many wounds upon them, and put them to rout, to their great

SEXTUS JULIUS FRONTINUS

tale exspectantes. Prius deinde quam resisti his inciperet, scutati erumpunt e saltu et redeuntes in ordinem consternant avertuntque; fugientibus equites immissi toto eos spatio, quo redibatur in castra, persecuti caedunt. Curatum quoque, ne quis effugeret, nam reliqui CCL equites praemissi facile per compendia itinerum effusis habenis, antequam ad castra Pompei perveniretur, conversi occurrerunt eis, qui primi fugerant. Ad cuius rei sensum Pompeio emittente legionem cum D. Laelio in praesidium suorum, subducti in dextrum latus velut cesserunt equites, deinde circumita legione hanc quoque a tergo infestaverunt, cum iam et a fronte qui pabulatores persecuti erant incursarent; sic legio quoque inter duas acies hostium cum legato suo elisa est. Ad cuius praesidium Pompeio totum educente exercitum, Sertorius quoque e collibus suos instructos ostendit effecitque, ne Pompeio expediret; ita praeter duplex damnum, eadem sollertia inlatum, spectatorem quoque eum cladis suorum continuit. Hoc primum proelium inter Sertorium et Pompeium fuit; X milia hominum de Pompei exercitu amissa et omnia impedimenta Livius auctor est.[1]

32. Pompeius in Hispania, dispositis ante qui ex occulto adgrederentur, simulato metu deduxit instantem hostem in loca insessa; deinde, ubi res poposcit,

---

[1] *The last sentence is considered by some an interpolation.*

[1] 76 B.C. *Cf.* Plut. *Sert.* 18; Appian *B.C.* i. 109. The allusion to Livy cannot be identified.

amazement. Then, before resistance to this first assault could be organised, the shield-bearing troops, bursting forth from the forest, overthrew and routed the Romans who were returning to the ranks, while the cavalry, dispatched after those in flight, followed them all the way back to camp, cutting them to pieces. Provision was also made that no one should escape. For two hundred and fifty reserve horsemen, sent ahead for the purpose, found it a simple matter to race forward by short cuts, and then to turn back and meet those who had first fled, before they reached Pompey's camp. On learning of this, Pompey sent out a legion under Decimus Laelius to reinforce his men, whereupon the cavalry of the enemy, withdrawing to the right flank, pretended to give way, and then, passing round the legion, assaulted it from the rear, while those who had followed up the foragers attacked it from the front also. Thus the legion with its commander was crushed between the two lines of the enemy. When Pompey led out his entire army to help the legion, Sertorius exhibited his forces drawn up on the hillside, and thus baulked Pompey's purpose. Thus, in addition to inflicting a twofold disaster, as a result of the same strategy, Sertorius forced Pompey to be the helpless witness of the destruction of his own troops. This was the first battle between Sertorius and Pompey. According to Livy, ten thousand men were lost in Pompey's army, along with the entire transport.[1]

Pompey, when warring in Spain, having first posted troops here and there to attack from ambush, by feigning fear, drew the enemy on in pursuit, till they reached the place of the ambuscade. Then

SEXTUS JULIUS FRONTINUS

conversus et in fronte et utrisque lateribus ad internecionem cecidit, capto etiam duce eorum Perperna.

33. Idem adversus Mithridaten in Armenia, numero et genere equitum praevalentem, tria milia levis armaturae et D equites nocte in valle sub virgultis, quae inter bina castra erant, disposuit. Prima deinde luce in stationem hostium emisit equites ita formatos, ut, cum universus cum exercitu hostium equitatus proelium inisset, servatis ordinibus paulatim cederent, donec spatium darent consurgendi a tergo ob hoc dispositis. Quod postquam ex sententia contigit, conversis qui terga dedisse videbantur, medium hostem trepidantem cecidit, ipsos etiam equos pedite comminus accedente confodit. Eoque proelio fiduciam regi, quam in equestribus copiis habebat, detraxit.

34. Crassus bello fugitivorum apud Cantennam bina castra comminus cum hostium castris vallavit. Nocte deinde commotis copiis, manente praetorio in maioribus castris, ut fallerentur hostes, ipse omnes copias eduxit et in radicibus praedicti montis constituit; divisoque equitatu praecepit L. Quintio, partem Spartaco obiceret pugnaque eum frustraretur, parte alia Gallos Germanosque ex factione Casti et Cannici eliceret ad pugnam et fuga simulata deduceret, ubi

---

[1] 72 B.C. Cf. Livy Per. 96; Oros. v. xxiii. 13; Appian B.C. i. 115.
[2] 66 B.C. Cf. Appian Mithr. 98; Dio xxxvi. xlvii. 3–4.

when the opportune moment arrived, wheeling about, he slaughtered the foe in front and on both flanks, and likewise captured their general, Perperna.[1]

The same Pompey, in Armenia, when Mithridates was superior to him in the number and quality of his cavalry, stationed three thousand light-armed men and five hundred cavalry by night in a valley under cover of bushes lying between the two camps. Then at daybreak he sent forth his cavalry against the position of the enemy, planning that, as soon as the full force of the enemy, cavalry and infantry, became engaged in battle, the Romans should gradually fall back, still keeping ranks, until they should afford room to those who had been stationed for the purpose of attacking from the rear to arise and do so. When this design turned out successfully, those who had seemed to flee turned about, enabling Pompey to cut to pieces the enemy thus caught in panic between his two lines. Our infantry also, engaging in hand-to-hand encounter, stabbed the horses of the enemy. That battle destroyed the faith which the king had reposed in his cavalry.[2]

In the Slave War, Crassus fortified two camps close beside the camp of the enemy, near Mt. Cantenna. Then, one night, he moved his forces, leading them all out and posting them at the base of the mountain above mentioned, leaving his headquarters tent in the larger camp in order to deceive the enemy. Dividing the cavalry into two detachments, he directed Lucius Quintius to oppose Spartacus with one division and fool him with a mock encounter; with the other to lure to combat the Germans and Gauls, of the faction of Castus and Cannicus, and, by feigning flight, to draw them on to the spot

SEXTUS JULIUS FRONTINUS

ipse aciem instruxerat. Quos cum barbari insecuti essent, equite recedente in cornua, subito acies Romana adaperta cum clamore procurrit. XXXV milia armatorum eo proelio interfecta cum ipsis ducibus Livius tradit, receptas quinque Romanas aquilas, signa sex et XX, multa spolia, inter quae quinque fasces cum securibus.[1]

35. C. Cassius in Syria adversus Parthos ducemque Osacen equitem ostendit a fronte, cum a tergo peditem in confragoso loco occultasset. Dein cedente equitatu et per nota se recipiente, in praeparatas insidias perduxit exercitum Parthorum et cecidit.

36. Ventidius Parthos et Labienum, alacres successibus victoriarum, dum suos ipse per simulationem metus continet, evocavit et in loca iniqua deductos adgressus per abreptionem adeo debellavit, ut destituto Labieno provincia excederent Parthi.

37. Idem adversus Pharnastanis Parthos, cum ipse exiguum numerum militum haberet, illis autem fiduciam ex multitudine videret increscere, ad latus castrorum XVIII cohortes in obscura valle posuit, equitatu post terga peditum conlocato. Tum paucos admodum milites in hostem misit; qui ubi simulata fuga hostem effuse sequentem ultra locum insidiarum

[1] *The last sentence is considered by some an interpolation.*

---

[1] 71 B.C.   *Cf.* Livy *Per.* 97; Plut. *Crass.* 11; Oros. v. xxiv. 6.   [2] 51 B.C.   *Cf.* Dio xl. 29.
[3] 39 B.C.   *Cf.* Dio xlviii. 39–40; Justin. XLII. iv. 7–8.

where Crassus himself had drawn up his troops in battle array. When the barbarians followed, the cavalry fell back to the flanks, and suddenly the Roman force disclosed itself and rushed forward with a shout. In that battle Livy tells us that thirty-five thousand armed men, with their commanders, were slain; five Roman eagles and twenty-six standards were recaptured, along with much other booty, including five sets of rods and axes.[1]

Gaius Cassius, when fighting in Syria against the Parthians and their leader Osaces, exhibited only cavalry in front, but had posted infantry in hiding on rough ground in the rear. Then, when his cavalry fell back and retreated over familiar roads, he drew the army of the Parthians into the ambush prepared for them and cut them to pieces.[2]

Ventidius, keeping his own men in camp on pretence of fear, caused the Parthians and Labienus, who were elated with victorious successes, to come out for battle. Having lured them into an unfavourable situation, he attacked them by surprise and so overwhelmed them that the Parthians refused to follow Labienus and evacuated the province.[3]

The same Ventidius, having himself only a small force available for use against the Parthians under Pharnastanes, but observing that the confidence of the enemy was growing in consequence of their numbers, posted eighteen cohorts at the side of the camp in a hidden valley, with cavalry stationed behind the infantry. Then he sent a very small detachment against the enemy. When these by feigning flight had drawn the enemy in hot pursuit beyond the place of ambush, the force at the side rose up, whereupon Ventidius drove the Parthians

SEXTUS JULIUS FRONTINUS

perduxere, coorta a latere acie praecipitatos in fugam, in his Pharnastanem, interfecit.

38. C. Caesar, suis et Afranii castris contrarias tenentibus planitias, cum utriusque partis plurimum interesset colles proximos occupare idque propter saxorum asperitatem esset difficile, tamquam Ilerdam repetiturum retro agmen ordinavit, faciente inopia fidem destinationi. Intra brevissimum deinde spatium exiguo circuitu flexit repente ad montis occupandos. Quo visu perturbati Afraniani velut captis castris et ipsi effuso cursu eosdem montes petiere. Quod futurum cum praedivinasset Caesar, partim peditatu quem praemiserat, partim a tergo summissis equitibus inordinatos est adortus.

39. Antonius apud Forum Gallorum, cum Pansam consulem adventare comperisset, insidiis per silvestria Aemiliae viae dispositis agmen eius excepit fuditque et ipsum eo vulnere adfecit, quo intra paucos dies exanimaretur.

40. Iuba rex in Africa bello civili Curionis animum simulato regressu impulit in vanam alacritatem; cuius spei vanitate deceptus Curio, dum tamquam fugientem Sabboram regium praefectum persequitur, devenit in patentes campos, ubi Numidarum circumventus equitatu, perdito exercitu, cecidit.

41. Melanthus dux Atheniensium, cum provocatus a rege hostium Xantho Boeotio descendisset ad pug-

---

[1] 39 B.C. *Cf.* Dio XLVIII. xli. 1–4; Plut. *Ant.* 33.
[2] 49 B.C. *Cf.* Caes. *B.C.* i. 65–72.
[3] 43 B.C. Mutina. *Cf.* Appian *B.C.* iii. 66 ff.; Cic. *ad Fam.* x. 30.
[4] 49 B.C. *Cf.* Caes. *B.C.* ii. 40–42; Dio xli. 41–42; Appian *B.C.* ii. 45.

in precipitate flight and slaughtered them, Pharnastanes among them.[1]

On one occasion when the camps of Gaius Caesar and Afranius were pitched in opposite plains, it was the special ambition of each side to secure possession of the neighbouring hills—a task of extreme difficulty on account of the jagged rocks. In these circumstances, Caesar marshalled his army as though to march back again to Ilerda, a move supported by his deficiency of supplies. Then, within a short time, making a small detour, he suddenly started to seize the hills. The followers of Afranius, alarmed at sight of this, just as though their camp had been captured, started out themselves at top speed to gain the same hills. Caesar, having forecast this turn of affairs, fell upon Afranius's men, before they could form—partly with infantry, which he had sent ahead, partly with cavalry sent up in the rear.[2]

Antonius, near Forum Gallorum, having heard that the consul Pansa was approaching, met his army by means of ambuscades, set here and there in the woodland stretches along the Aemilian Way, thus routing his troops and inflicting on Pansa himself a wound from which he died in a few days.[3]

Juba, king in Africa at the time of the Civil War, by feigning a retirement, once roused unwarranted elation in the heart of Curio. Under the influence of this mistaken hope, Curio, pursuing Sabboras, the king's general, who, he thought, was in flight, came to open plains, where, surrounded by the cavalry of the Numidians, he lost his army and perished himself.[4]

Melanthus, the Athenian general, on one occasion came out for combat, in response to the challenge of the king of the enemy, Xanthus, the Boeotian. As

nam, ut primum comminus stetit, "inique," inquit, "Xanthe, et contra pactum facis; adversus solum enim cum altero processisti." Cumque admiratus ille, quisnam se comitaretur, respexisset, aversum uno ictu confecit.

42. Iphicrates Atheniensis ad Cherronessum, cum sciret Lacedaemoniorum ducem Anaxibium exercitum pedestri itinere ducere, firmissimam manum militum eduxit e navibus et in insidiis conlocavit, naves autem omni tamquam onustas milite palam transnavigare iussit; ita securos et nihil exspectantes Lacedaemonios a tergo ingressus itinere oppressit fuditque.

43. Liburni, cum vadosa loca obsedissent, capitibus tantum eminentibus fidem fecerunt hosti alti maris ac triremem, quae eos persequebatur, implicatam vado ceperunt.

44. Alcibiades dux Atheniensium in Hellesponto adversus Mindarum Lacedaemoniorum ducem, cum amplum exercitum et plures naves haberet, nocte expositis in terram quibusdam militum suorum, parte quoque navium post quaedam promunturia occultata, ipse cum paucis profectus ita, ut contemptu sui hostem invitaret, eundem insequentem fugit, donec in praeparatas insidias perduceret. Aversum deinde et egredientem in terram per eos, quos ad hoc ipsum exposuerat, cecidit.

---

[1] *Cf.* Polyaen. i. 19.
[2] 389-388 B.C. *Cf.* Xen. *Hell.* IV. viii. 32 ff.
[3] 410 B.C. *Cf.* Xen. *Hell.* I. i. 11 ff.; Polyaen. I. xl. 9; Diodor. xiii. 50.

soon as they stood face to face, Melanthus exclaimed: "Your conduct is unfair, Xanthus, and contrary to agreement. I am alone, but you have come out with a companion against me." When Xanthus wondered who was following him and looked behind, Melanthus dispatched him with a single stroke, as his head was turned away.[1]

Iphicrates, the Athenian, on one occasion in the Chersonesus, aware that Anaxibius, commander of the Spartans, was proceeding with his troops by land, disembarked a large force of men from his vessels and placed them in ambush, but directed his ships to sail in full view of the enemy, as though loaded with all his forces. When the Spartans were thus thrown off their guard and apprehended no danger, Iphicrates, attacking them by land from the rear as they marched along, crushed and routed them.[2]

The Liburnians on one occasion, when they had taken a position among some shallows, by allowing only their heads to appear above the surface of the water, caused the enemy to believe that the water was deep. In this way a galley which followed them became stranded on the shoal, and was captured.

Alcibiades, commander of the Athenians at the Hellespont against Mindarus, leader of the Spartans, having a large army and numerous vessels, landed some of his soldiers by night, and hid part of his ships behind certain headlands. He himself, advancing with a few troops, so as to lure the enemy on in scorn of his small force, fled when pursued, until he finally drew the foe into the trap which had been laid. Then attacking the enemy in the rear, as he disembarked, he cut him to pieces with the aid of the troops which he had landed for this very purpose.[3]

SEXTUS JULIUS FRONTINUS

45. Idem, navali proelio decertaturus, constituit malos quosdam in promunturio praecepitque his quos ibi relinquebat, ut, cum commissum proelium sensissent, panderent vela. Quo facto consecutus est, ut hostes aliam classem in auxilium ei supervenire arbitrati verterentur.

46. Memnon Rhodius navali proelio, cum haberet ducentarum navium classem et hostium naves elicere ad proelium vellet, ita ordinavit suos, ut paucarum navium malos erigeret easque primas agi iuberet; hostes procul conspicati numerum arborum et ex eo navium quoque coniectantes obtulerunt se certamini et a pluribus occupati superatique sunt.

47. Timotheus dux Atheniensium adversus Lacedaemonios navali acie decertaturus, cum instructa classis eorum ad pugnam processisset, ex velocissimis navibus viginti praemisit, quae omni arte varioque flexu eluderent hostem; ut primum deinde sensit minus agiliter moveri adversam partem, progressus praelassatos facile superavit.

VI. DE EMITTENDO HOSTE, NE CLAUSUS PROELIUM EX DESPERATIONE REDINTEGRET

1. GALLOS eo proelio, quod Camilli ductu gestum est, desiderantes navigia, quibus Tiberim transirent, senatus censuit transvehendos et commeatibus quoque prosequendos.

Eiusdem generis hominibus postea per Pompti-

---

[1] 375 B.C. *Cf.* Polyaen. III. x. 6, 12, 16.

The same Alcibiades, on one occasion, when about to engage in a naval combat, erected a number of masts on a headland, and commanded the men whom he left there to spread sails on these as soon as they noticed that the engagement had begun. By this means he caused the enemy to retreat, since they imagined another fleet was coming to his assistance.

Memnon, the Rhodian, in a naval encounter, possessing a fleet of two hundred ships, and wishing to lure the vessels of the enemy out to battle, made arrangements for raising the masts of only a few of his ships, ordering these to proceed first. When the enemy from a distance saw the number of masts, and from that inferred the number of vessels, they offered battle, but were fallen upon by a larger number of ships and defeated.

Timotheus, leader of the Athenians, when about to engage in a naval encounter with the Spartans, as soon as the Spartan fleet came out arrayed in line of battle, sent ahead twenty of his swiftest vessels, to baulk the enemy in every way by various tactics. Then as soon as he observed that the enemy were growing less active in their manœuvres, he moved forward and easily defeated them, since they were already worn out.[1]

VI. On Letting the Enemy Escape, lest, Brought to Bay, He Renew the Battle in Desperation

When the Gauls, after the battle fought under Camillus's generalship, desired boats to cross the Tiber, the Senate voted to set them across and to supply them with provisions as well.

On a subsequent occasion also a free passage was

SEXTUS JULIUS FRONTINUS

num agrum fugientibus via data est, quae Gallica appellatur.

2. T. Marcius eques Romanus, cui duobus Scipionibus occisis exercitus imperium detulit, cum circumventi ab eo Poeni, ne inulti morerentur, acrius pugnarent, laxatis manipulis et concesso fugae spatio dissipatos sine periculo suorum trucidavit.

3. C. Caesar Germanos inclusos, ex desperatione fortius pugnantis, emitti iussit fugientesque adgressus est.

4. Hannibal, cum ad Trasumennum inclusi Romani acerrime pugnarent, diductis ordinibus fecit eis abeundi potestatem euntesque sine suorum sanguine stravit.

5. Antigonus rex Macedonum Aetolis, qui in obsidionem ab eo compulsi fame urguebantur statuerantque eruptione facta commori, viam fugae dedit; atque ita infracto impetu eorum insecutus aversos cecidit.

6. Agesilaus Lacedaemonius adversus Thebanos, cum acie confligeret intellexissetque hostes locorum condicione clausos ob desperationem fortius dimicare, laxatis suorum ordinibus apertaque Thebanis ad evadendum via, rursus in abeuntis contraxit aciem et sine iactura suorum cecidit aversos.

---

[1] 349 B.C. L. Furius Camillus, son of the great Camillus.
[2] 212 B.C. Livy xxv. 37 gives his praenomen as Lucius.
[3] 217 B.C. [4] 223–221 B.C. Antigonus Doson.
[5] 394 B.C. *Cf.* Polyaen. II. i. 19; Plut. *Agesil.* 18. Xen. *Hell.* iv. 3 notes the failure of Agesilaus to employ this stratagem.

afforded to the people of the same race when retreating through the Pomptine district. This road goes by the name of the "Gallic Way."[1]

Titus Marcius, a Roman knight, on whom the army conferred the supreme command after the two Scipios were slain, succeeded in enveloping the Carthaginians. When the latter, in order not to die unavenged, fought with increased fury, Marcius opened up the maniples, afforded room for escape, and as the enemy became separated, slaughtered them without danger to his own men.[2]

When certain Germans whom Gaius Caesar had penned in fought the more fiercely from desperation, he ordered them to be allowed to escape, and then attacked them as they fled.

At Trasimenus, when the Romans had been enveloped and were fighting with the greatest fury, Hannibal opened up his ranks and gave them an opportunity of escape, whereupon, as they fled, he overwhelmed them without loss of his own troops.[3]

When the Aetolians, blockaded by Antigonus, king of the Macedonians, were suffering from famine and had resolved to make a sally in face of certain death, Antigonus afforded them an avenue of flight. Thus having cooled their ardour, he attacked them from the rear and cut them to pieces.[4]

Agesilaus, the Spartan, when engaged in battle with the Thebans, noticed that the enemy, hemmed in by the character of the terrain, were fighting with greater fury on account of their desperation. Accordingly he opened up his ranks and afforded the Thebans a way of escape. But when they tried to retreat, he again enveloped them, and cut them down from behind without loss of his own troops.[5]

SEXTUS JULIUS FRONTINUS

7. Cn. Manlius consul, cum ex acie reversus capta ab Etruscis Romanorum castra invenisset, omnibus portis statione circumdatis inclusos hostes in eam rabiem efferavit, ut ipse in proelio caderet. Quod ut animadverterunt legati eius, ab una porta remota statione exitum Etruscis dederunt et effusos persecuti, occurrente altero consule Fabio, ceciderunt.

8. Themistocles victo Xerxe volentes suos pontem rumpere prohibuit, cum docuisset aptius esse eum expelli Europa, quam cogi ex desperatione pugnare. Idem misit ad eum, qui indicaret, in quo periculo esset, nisi fugam maturaret.

9. Pyrrhus Epirotarum rex, cum quandam civitatem cepisset clausisque portis ex ultima necessitate fortiter dimicantes eos, qui inclusi erant, animadvertisset, locum illis ad fugam dedit.

10. Idem inter praecepta imperatoria memoriae tradidit, non esse pertinaciter instandum hosti fugienti, ne non solum ea re fortius ex necessitate resisteret, sed ut postea quoque facilius acie cederet, cum sciret non usque ad perniciem fugientibus instaturos victores.

---

[1] 480 B.C. *Cf.* Livy ii. 47.
[2] *i.e.* the bridge which Xerxes had constructed across the Hellespont at the time of his invasion of Greece.
[3] 480 B.C. *Cf.* Justin. II. xiii. 5 ff.; Polyaen. I. xxx. 4. Plut. *Them.* 16 and Herod. viii. 108 attribute the advice against destroying the bridge to Aristides and Eurybiades respectively.

Gnaeus Manlius, the consul, on returning from battle found the camp of the Romans in possession of the Etruscans. He therefore posted guards at all the gates and roused the enemy, thus shut up within, to such a pitch of fury that he himself was slain in the fighting. When his lieutenants realized the situation, they withdrew the guards from one gate and afforded the Etruscans an opportunity of escape. But when the latter poured forth, the Romans pursued them and cut them to pieces, with the help of the other consul, Fabius, who happened to come up.[1]

When Xerxes had been defeated and the Athenians wished to destroy his bridge,[2] Themistocles prevented this, showing that it was better for them that Xerxes should be expelled from Europe rather than be forced to fight in desperation. He also sent to the king a messenger to tell him in what danger he would be, in case he failed to make a hasty retreat.[3]

When Pyrrhus, king of the Epirotes, had captured a certain city and had noticed that the inhabitants, shut up inside, had closed the gates and were fighting valiantly from dire necessity, he gave them an opportunity to escape.

The same Pyrrhus, among many other precepts on the art of war, recommended never to press relentlessly on the heels of an enemy in flight—not merely in order to prevent the enemy from resisting too furiously in consequence of necessity, but also to make him more inclined to withdraw another time, knowing that the victor would not strive to destroy him when in flight.

SEXTUS JULIUS FRONTINUS

VII. De Dissimulandis Adversis

1. Tullus Hostilius rex Romanorum commisso adversus Veientes proelio, cum Albani deserto exercitu Romanorum proximos peterent tumulos eaque res turbasset nostrorum animos, clare pronuntiavit iussu suo Albanos id fecisse, ut hostem circumveniret. Quae res et terrorem Veientibus et Romanis fiduciam attulit remque inclinatam consilio restituit.

2. L. Sulla, cum praefectus eius, comitante non exigua equitum manu, commisso iam proelio ad hostis transfugisset, pronuntiavit iussu suo id factum; eaque ratione militum animos non tantum a confusione retraxit, sed quadam etiam spe utilitatis, quae id consilium secutura esset, confirmavit.

3. Idem, cum auxiliares eius missi ab ipso circumventi ab hostibus et interfecti essent verereturque, ne propter hoc damnum universus trepidaret exercitus, pronuntiavit auxiliares, qui ad defectionem conspirassent, consilio a se in loca iniqua deductos. Ita manifestissimam cladem ultionis simulatione velavit et militum animos hac persuasione confirmavit.

4. Scipio, cum Syphacis legati nuntiarent ei regis sui nomine, ne fiducia societatis eius ex Sicilia in Africam transiret, veritus, ne confunderentur animi suorum abscisa spe peregrinae societatis, dimisit pro-

---

[1] *Cf.* Livy i. 27; Val. Max. vii. iv. 1; Dionys. iii. 24. Nep. *Ages.* 6 and *Datam.* 6 attributes like stratagems to Agesilaus and Datames.

STRATAGEMS, II. vii. 1-4

VII. On Concealing Reverses

Tullus Hostilius, king of the Romans, on one occasion had engaged in battle with the Veientines, when the Albans, deserting the army of the Romans, made for the neighbouring hills. Since this action disconcerted our troops, Tullus shouted in a loud voice that the Albans had done that by his instructions, with the object of enveloping the foe. This declaration struck terror into the hearts of the Veientines and lent confidence to the Romans. By this device he turned the tide of battle.[1]

When a lieutenant of Lucius Sulla had gone over to the enemy at the beginning of an engagement, accompanied by a considerable force of cavalry, Sulla announced that this had been done by his own instructions. He thereby not merely saved his men from panic, but encouraged them by a certain expectation of advantage to result from this plan.

The same Sulla, when certain auxiliary troops dispatched by him had been surrounded and cut to pieces by the enemy, fearing that his entire army would be in a panic on account of this disaster, announced that he had purposely placed the auxiliaries in a place of danger, since they had plotted to desert. In this way he veiled a very palpable reverse under the guise of discipline, and encouraged his soldiers by convincing them that he had done this.

When the envoys of King Syphax told Scipio in the name of their king not to cross over to Africa from Sicily in expectation of an alliance, Scipio, fearing that the spirits of his men would receive a shock, if the hope of a foreign alliance were cut off,

pere[1] legatos et famam diffudit, tamquam ultro a Syphace accerseretur.

5. Q. Sertorius, cum acie decertaret, barbarum, qui ei nuntiaverat Hirtuleium perisse, pugione traiecit, ne et in aliorum id notitiam perferret et ob hoc animi suorum infirmarentur.

6. Alcibiades Atheniensis, cum ab Abydenis proelio urgueretur subitoque magno cursu tristem adventare animadvertisset tabellarium, prohibuit palam dicere, quid adferret. Dehinc secreto sciscitatus, a Pharnabazo regio praefecto classem suam oppugnari celatis et hostibus et militibus proelium finiit ac protinus ad eripiendam classem ducto exercitu opem tulit suis.

7. Hannibalem venientem in Italiam tria milia Carpetanorum reliquerunt; quos ille, exemplo ne et ceteri moverentur, edixit a se esse dimissos et insuper in fidem eius rei paucos levissimae operae domos remisit.

8. L. Lucullus, cum animadvertisset Macedonas equites, quos in auxilio habebat, subito consensu ad hostem transfugere, signa canere iussit et turmas, quae eos sequerentur, immisit. Hostis committi proelium ratus transfugientes telis excepit. Macedones, qui viderent neque recipi se ab adversariis et premi ab his, quos deserebant, necessario ad iustum proelium conversi hostem invaserunt.

[1] prorelegatos *HP* ; propere legatos *Mod. Cf. Livy* xxix. 24.

[1] 204 B.C. *Cf.* Livy xxix. 23–24; Polyaen. VIII. xvi. 7.
[2] 75 B.C. [3] 409 B.C.
[4] 218 B.C. *Cf.* Livy xxi. 23. [5] 74–66 B.C.

summarily dismissed the envoys, and spread abroad the report that he was expressly sent for by Syphax.[1]

Once when Quintus Sertorius was engaged in battle, he plunged a dagger into the barbarian who had reported to him that Hirtuleius had fallen, for fear the messenger might bring this news to the knowledge of others and in this way the spirit of his own troops should be broken.[2]

When Alcibiades, the Athenian, was hard pressed in battle by the Abydenes and suddenly noticed a courier approaching at great speed and with dejected countenance, he prevented the courier from telling openly what tidings he brought. Having privately learned that his fleet was beset by Pharnabazus, the commander of the king, he concealed the fact both from the enemy and from his own soldiers, and finished the battle. Then straightway marching to rescue his fleet, he bore aid to his friends.[3]

When Hannibal entered Italy, three thousand Carpetani deserted him. Fearing that the rest of his troops might be affected by their example, he proclaimed that they had been discharged by him, and as further proof of that, he sent home a few others whose services were of very little importance.[4]

When Lucius Lucullus noticed that the Macedonian cavalry, whom he had as auxiliaries, were suddenly deserting to the enemy in a body, he ordered the trumpets to sound and sent out squadrons to pursue the deserters. The enemy, thinking that an engagement was beginning, received the deserters with javelins, whereupon the Macedonians, seeing that they were not welcomed by the enemy and were attacked by those whom they were deserting, were forced to resort to a genuine battle and assaulted the enemy.[5]

SEXTUS JULIUS FRONTINUS

9. Datames dux Persarum adversum Autophradaten in Cappadocia, cum partem equitum suorum transfugere comperisset, ceteros omnes venire secum iussit adsecutusque transfugas conlaudavit, quod eum alacriter praecessissent, hortatusque est eos etiam, ut fortiter hostem adorirentur. Pudor transfugis attulit paenitentiam et consilium suum, quia non putabant deprehensum, mutaverunt.

10. T. Quintius Capitolinus consul cedentibus Romanis ementitus est in altero cornu hostes fugatos et ita confirmatis suis victoriam rettulit.

11. Cn. Manlius adversus Etruscos, vulnerato collega M. Fabio, qui sinistrum cornu ducebat, et ob id ea parte cedente, quod etiam occisum crederent consulem, cum turmis equitum occurrit, clamitans et collegam vivere et se dextro cornu vicisse; qua constantia redintegratis animis suorum vicit.

12. Marius adversus Cimbros et Teutonos, cum metatores eius per imprudentiam ita castris locum cepissent, ut sub potestate barbarorum esset aqua, flagitantibus eam suis, digito hostem ostendens "illinc," inquit, "petenda est"; quo instinctu adsecutus est, ut protinus barbari tollerentur.

13. T. Labienus post Pharsalicam pugnam, cum

---

[1] 362 B.C. Nep. *Datam.* 6, Diodor. xv. 91 and Polyaen. VII. xxi. 7 give a slightly different version of this episode.

[2] 468 B.C. *Cf.* Livy II. lxiv. 5–7; Dionys. ix. 57.

[3] 480 B.C. According to Livy ii. 46–47 and Dionys ix. 11, Q. Fabius and Manlius were wounded, and Marcus Fabius was the author of this stratagem.

[4] 102 B.C. *Cf.* Flor. III. iii. 7–10; Plut. *Mar.* 18.

# STRATAGEMS, II. vii. 9-13

Datames, commander of the Persians against Autophradates in Cappadocia, learning that part of his cavalry were deserting, ordered the rest of his troops to follow with him. Upon coming up with the deserters, he commended them for outstripping him in their eagerness, and also urged them to attack the enemy courageously. Seized with shame and penitence, the deserters changed their purpose, imagining that it had not been detected.[1]

The consul Titus Quinctius Capitolinus, when the Romans yielded ground in battle, falsely claimed that the enemy had been routed on the other flank. By thus lending courage to his men, he won a victory.[2]

When Gnaeus Manlius was fighting against the Etruscans, his colleague Marcus Fabius, commander of the left flank, was wounded, and that section of the army therefore gave way, imagining that the consul had been slain. Thereupon Manlius confronted the broken line with squadrons of horse, shouting that his colleague was alive and that he himself had been victorious on the right flank. By this dauntless spirit, he restored the courage of his men and won the victory.[3]

When Marius was fighting against the Cimbrians and Teutons, his engineers on one occasion had heedlessly chosen such a site for the camp that the barbarians controlled the water supply. In response to the soldiers' demand for water, Marius pointed with his finger toward the enemy and said: "There is where you must get it." Thus inspired, the Romans straightway drove the barbarians from the place.[4]

Titus Labienus, after the Battle of Pharsalia,

SEXTUS JULIUS FRONTINUS

victis partibus Dyrrhachium refugisset, miscuit vera falsis et, non celato exitu pugnae, aequatam partium fortunam gravi vulnere Caesaris finxit; et hac adsimulatione reliquis Pompeianarum partium fiduciam fecit.

14. M. Cato, cum Ambraciam eo tempore, quo sociae naves ab Aetolis oppugnabantur, imprudens uno lembo appulisset, quamquam nihil secum praesidii haberet, coepit signum voce gestuque dare, quo videretur subsequentis suorum navis vocare, eaque adseveratione hostem terruit, tamquam plane appropinquarent, qui quasi ex proximo citabantur. Aetoli, ne adventu Romanae classis opprimerentur, reliquerunt oppugnationem.

VIII. De Restituenda per Constantiam Acie

1. Servius Tullius adulescens proelio, quo rex Tarquinius adversus Sabinos conflixit, signiferis segnius dimicantibus raptum signum in hostem misit; cuius recuperandi gratia Romani ita ardenter pugnaverunt, ut et signum et victoriam rettulerint.

2. Furius Agrippa consul cedente cornu signum militare ereptum signifero in hostes Hernicos et Aequos misit. Quo facto eius proelium restitutum

---

[1] 48 B.C. [2] 191 B.C.
[3] This type of stratagem was very frequently resorted to. *Cf.* Florus i. 11; Val. Max. iii. ii. 20; Caes. *B.G.* iv. 25; Livy iv. xxix. 3 and *passim*.

when his side had been defeated and he himself had fled to Dyrrhachium, combined falsehood with truth, and while not concealing the outcome of the battle, pretended that the fortunes of the two sides had been equalized in consequence of a severe wound received by Caesar. By this pretence, he created confidence in the other followers of Pompey's party.[1]

Marcus Cato, having inadvertently landed with a single galley in Ambracia at a time when the allied fleet was blockaded by the Aetolians, although he had no troops with him, began nevertheless to make signals by voice and gesture, in order to give the impression that he was summoning the approaching ships of his own forces. By this earnestness he alarmed the enemy, just as though the troops, whom he pretended to be summoning from near at hand, were visibly approaching. The Aetolians, accordingly, fearing that they would be crushed by the arrival of the Roman fleet, abandoned the blockade.[2]

VIII. On Restoring Morale by Firmness

In the battle in which King Tarquinius encountered the Sabines, Servius Tullius, then a young man, noticing that the standard-bearers fought half-heartedly, seized a standard and hurled it into the ranks of the enemy. To recover it, the Romans fought so furiously that they not only regained the standard, but also won the day.[3]

The consul Furius Agrippa, when on one occasion his flank gave way, snatched a military standard from a standard-bearer and hurled it into the hostile ranks of the Hernici and Aequi. By this act the

SEXTUS JULIUS FRONTINUS

est; summa enim alacritate Romani ad recipiendum signum incubuerunt.

3. T. Quintius Capitolinus consul signum in hostes Faliscos eiecit militesque id repetere iussit.

4. M. Furius Camillus tribunus militum consulari potestate, cunctante exercitu, arreptum manu signiferum in hostes Volscos et Latinos traxit; ceteros puduit non sequi.

5. Salvius Pelignus bello Persico idem fecit.[1]

6. M. Furius averso exercitu, cum occurrisset, adfirmavit non recepturum se in castra quemquam nisi victorem, reductisque in aciem victoria potitus est.

7. Scipio apud Numantiam, cum aversum suum videret exercitum, pronuntiavit pro hoste sibi futurum, quisquis in castra redisset.

8. Servilius Priscus dictator, cum signa legionum ferri in hostis Faliscos iussisset, signiferum cunctantem occidi imperavit; quo exemplo perterriti hostem invaserunt.

9. Cossus Cornelius magister equitum adversus Fidenates idem fecit.[2]

---

[1] *No. 5 is probably an interpolation.*
[2] *No. 9 is probably an interpolation.*

---

[1] 446 B.C. *Cf.* Livy III. lxx. 2–11.
[2] There is no other record of Titus Quinctius Capitolinus's warring with the Faliscans. Livy iv. 29 attributes this stratagem to Titus Quinctius Cincinnatus in the war with the Volscians, 431 B.C.
[3] 386 B.C. *Cf.* Livy vi. 7–8.
[4] Plut. *Aem.* 20 attributes to Salvius a stratagem similar to the first three of this chapter. 168 B.C.

STRATAGEMS, II. VIII. 2–9

day was saved, for the Romans with the greatest eagerness pressed forward to recapture the standard.[1]

The consul Titus Quinctius Capitolinus hurled a standard into the midst of the hostile ranks of the Faliscans and commanded his troops to regain it.[2]

Marcus Furius Camillus, military tribune with consular power, on one occasion when his troops held back, seized a standard-bearer by the hand and dragged him into the hostile ranks of the Volscians and Latins, whereupon the rest were shamed into following.[3]

Salvius, the Pelignian, did the same in the Persian War.[4]

Marcus Furius, meeting his army in retreat, declared he would receive in camp no one who was not victorious. Thereupon he led them back to battle and won the day.[5]

Scipio, at Numantia, seeing his forces in retreat, proclaimed that he would treat as an enemy whoever should return to camp.[6]

The dictator Servilius Priscus, having given the command to carry the standards of the legions against the hostile Faliscans, ordered the standard-bearer to be executed for hesitating to obey. The rest, cowed by this example, advanced against the foe.[7]

Cornelius Cossus, master of the horse, did the same in an engagement with the people of Fidenae.[8]

---

[5] 381 B.C. Camillus. *Cf.* Livy vi. 24.
[6] 133 B.C.
[7] 418 B.C. According to Livy iv. 46–47, the battle was with the Aequi, not the Faliscans.
[8] 426 B.C. *Cf.* Livy IV. xxxiii. 7; Flor. I. xi. 2–3. This stratagem is similar to number 10, rather than number 8.

SEXTUS JULIUS FRONTINUS

10. Tarquinius adversus Sabinos cunctantes equites detractis frenis concitatisque equis perrumpere aciem iussit.

11. M. Atilius consul bello Samnitico ex acie refugientibus in castra militibus aciem suorum opposuit, adfirmans secum et bonis civibus dimicaturos eos, nisi cum hostibus maluissent; ea ratione universos in aciem reduxit.

12. L. Sulla, cedentibus iam legionibus exercitui Mithridatico ductu Archelai, stricto gladio in primam aciem procucurrit appellansque milites dixit, si quis quaesisset, ubi imperatorem reliquissent, responderent pugnantem in Boeotia; cuius rei pudore universi eum secuti sunt.

13. Divus Iulius ad Mundam referentibus suis pedem equum suum abduci a conspectu iussit et in primam aciem pedes prosiluit; milites, dum destituere imperatorem erubescunt, redintegraverunt proelium.

14. Philippus veritus, ne impetum Scytharum sui non sustinerent, fidelissimos equitum a tergo posuit praecepitque, ne quem commilitonum ex acie fugere paterentur, perseverantius abeuntes trucidarent. Qua denuntiatione cum effecisset, ut etiam timidissimi malent ab hostibus quam ab suis interfici, victoriam adquisivit.

---

[1] 294 B.C. *Cf.* Livy x. 36. A slightly different version of this story is told in IV. i. 29.

[2] 85 B.C. *Cf.* Plut. *Sulla* 21; Polyaen. VIII. ix. 2; Appian *Mithr.* 49.

[3] 45 B.C. *Cf.* Plut. *Caes.* 56; Polyaen. VIII. xxiii. 16; Vell. ii. 55.

Tarquinius, when his cavalry showed hesitation in the battle against the Sabines, ordered them to fling away their bridles, put spurs to their horses, and break through the enemy's line.

In the Samnite War, the consul Marcus Atilius, seeing his troops quitting the battle and taking refuge in camp, met them with his own command and declared that they would have to fight against him and all loyal citizens, unless they preferred to fight against the enemy. In this way he marched them back in a body to the battle.[1]

When Sulla's legions broke before the hosts of Mithridates led by Archelaus, Sulla advanced with drawn sword into the first line and, addressing his troops, told them, in case anybody asked where they had left their general, to answer: "Fighting in Boeotia." Shamed by these words, they followed him to a man.[2]

The deified Julius, when his troops gave way at Munda, ordered his horse to be removed from sight, and strode forward as a foot-soldier to the front line. His men, ashamed to desert their commander, thereupon renewed the fight.[3]

Philip, on one occasion, fearing that his troops would not withstand the onset of the Scythians, stationed the trustiest of his cavalry in the rear, and commanded them to permit no one of their comrades to quit the battle, but to kill them if they persisted in retreating. This proclamation induced even the most timid to prefer to be killed by the enemy rather than by their own comrades, and enabled Philip to win the day.[4]

[4] Justin. I. vi. 10-13 attributes a similar stratagem to Astyages.

SEXTUS JULIUS FRONTINUS

De His Quae Post Proelium Fiunt

VIIII. Si Res Prospere Cesserit, de Consummandis Reliquiis Belli

1. C. Marius, victis proelio Teutonis, reliquias eorum, quia nox intervenerat, circumsedens, sublatis subinde clamoribus per paucos suorum territavit insomnemque hostem detinuit, ex eo adsecutus, ut postero die inrequietum facilius debellaret.

2. Claudius Nero, victis Poenis, quos Hasdrubale duce in Italiam ex Hispania traicientes exceperat, caput Hasdrubalis in castra Hannibalis eiecit; quo factum est, ut et Hannibal luctu et exercitus desperatione adventantis praesidii adfligerentur.

3. L. Sulla his, qui Praeneste obsidebantur, occisorum in proelio ducum capita hastis praefixa ostendit atque ita obstinatorum pervicaciam fregit.

4. Arminius dux Germanorum capita eorum, quos occiderat, similiter praefixa ad vallum hostium admoveri iussit.

5. Domitius Corbulo, cum Tigranocertam obsideret et Armenii pertinaciter viderentur toleraturi obsidionem, in Vadandum unum[1] ex megistanis, quos ceperat, animadvertit caputque eius ballista excussum intra munimenta hostium misit. Id forte decidit in

[1] *I have supplied this word in accordance with Gundermann's suggestion.*

---

[1] 102 B.C.
[2] 207 B.C. *Cf.* Livy xxvii. 51 ; Zonar. ix. 9.
[3] 82 B.C. *Cf.* Appian *B.C.* i. 93–94.
[4] 9 A.D.

On Measures taken after Battle

IX. On Bringing the War to a Close after a Successful Engagement

After Gaius Marius had defeated the Teutons in battle, and night had put an end to the conflict, he encamped round about the remnants of his opponents. By causing a small group of his own men to raise loud cries from time to time, he kept the enemy in a state of alarm and prevented them from securing rest. He thus succeeded more easily in crushing them on the following day, since they had had no sleep.[1]

Claudius Nero, having met the Carthaginians on their way from Spain to Italy under the command of Hasdrubal, defeated them and threw Hasdrubal's head into Hannibal's camp. As a result, Hannibal was overwhelmed with grief and the army gave up hope of receiving reinforcements.[2]

When Lucius Sulla was besieging Praeneste, he fastened on spears the heads of Praenestine generals who had been slain in battle, and exhibited them to the besieged inhabitants, thus breaking their stubborn resistance.[3]

Arminius, leader of the Germans, likewise fastened on spears the heads of those he had slain, and ordered them to be brought up to the fortifications of the enemy.[4]

When Domitius Corbulo was besieging Tigranocerta and the Armenians seemed likely to make an obstinate defence, Corbulo executed Vadandus, one of the nobles he had captured, shot his head out of a ballista, and sent it flying within the fortifications of the enemy. It happened to fall in the

SEXTUS JULIUS FRONTINUS

medium concilium, quod cum maxime habebant barbari; ad cuius conspectum velut ostento consternati ad deditionem festinaverunt.

6. Hermocrates Syracusanus superatis acie Carthaginiensibus veritus, ne captivi, quorum ingentem manum in potestatem redegerat, parum diligenter custodirentur, quia eventus dimicationis in epulas et securitatem compellere victores poterat, finxit proxima nocte equitatum hostilem venturum. Qua exspectatione adsecutus est, ut solito attentius vigiliae agerentur.

7. Idem, rebus prospere gestis et ob id resolutis suis in nimiam securitatem somnoque et mero pressis, in castra hostium transfugam misit, qui prohiberet eos a fuga: dispositas enim ubique a Syracusanis insidias. Quarum metu illi continuerunt se intra castra. Hermocrates detentos eos postero die habilioribus iam suis tradidit bellumque confecit.

8. Miltiades, cum ingentem Persarum multitudinem apud Marathona fudisset, Athenienses circa gratulationem morantis compulit, ut festinarent ad opem urbi ferendam, quam classis Persarum petebat; cumque praecucurrisset implessetque moenia armatis, Persae, rati ingentem numerum esse Atheniensium et alio milite apud Marathona pugnatum, alium pro muris

---

[1] 60 A.D.
[2] 413 B.C. *Cf.* Polyaen. I. xliii. 2; Thuc. vii. 73; Plut. *Nic.* 26.

midst of a council which the barbarians were holding at that very moment, and the sight of it (as though it were some portent) so filled them with consternation that they made haste to surrender.[1]

When Hermocrates, the Syracusan, had defeated the Carthaginians in battle, and was afraid that the prisoners, of whom he had taken an enormous number, would be carelessly guarded, since the successful issue of the struggle might prompt the victors to revelry and neglect, he pretended that the cavalry of the enemy were planning an attack on the following night. By instilling this fear, he succeeded in having the guard over the prisoners maintained even more carefully than usual.

When the same Hermocrates had achieved certain successes, and for that reason his men, through a spirit of over-confidence, had abandoned all restraint and were sunk in a drunken stupor, he sent a deserter into the camp of the enemy to prevent their flight by declaring to them that ambuscades of Syracusans had been posted everywhere. From fear of these, the enemy remained in camp. Having thus detained them, Hermocrates, on the following day, when his own men were more fit, gave the enemy over to their mercy and ended the war.[2]

When Miltiades had defeated a huge host of Persians at Marathon, and the Athenians were losing time in rejoicing over the victory, he forced them to hurry to bear aid to the city, at which the Persian fleet was aiming. Having thus got ahead of the enemy, he filled the walls with warriors, so that the Persians, thinking that the number of the Athenians was enormous and that they themselves had met one army at Marathon while another was now

SEXTUS JULIUS FRONTINUS

suis opponi, circumactis extemplo navibus Asiam repetierunt.

9. Pisistratus Atheniensis, cum excepisset Megarensium classem, qua illi ad Eleusin noctu applicuerant, ut operatas Cereris sacro feminas Atheniensium raperent, magnaque edita caede eorum ultus esset suos, eadem quae ceperat navigia Atheniensi milite complevit, quibusdam matronis habitu captivarum in conspectu locatis. Qua facie decepti Megarenses tamquam suis et cum successu renavigantibus effuse obvii inermesque rursus oppressi sunt.

10. Cimon dux Atheniensium, victa classe Persarum apud insulam Cypron, milites suos captivis armis induit et eisdem barbarorum navibus ad hostem navigavit in Pamphyliam apud flumen Eurymedonta. Persae, qui et navigia et habitum superstantium adgnoscerent, nihil caverunt. Subito itaque oppressi eodem die et navali et pedestri proelio victi sunt.

X. Si Res Durius Cesserit, de Adversis Emendandis

1. T. Didius in Hispania, cum acerrimo proelio conflixisset, quod nox diremerat, magno numero utrimque caeso complura suorum corpora intra noctem sepelienda curavit. Hispani postero die ad simile officium progressi, quia plures ex ipsorum numero quam ex Romanis caesos reppererant, victos se esse

---

[1] 490 B.C. *Cf.* Herod. vi. 115 ff.
[2] 604 B.C. *Cf.* Justin. ii. 8; Polyaen. I. xx. 2; Plut. *Solon* 8.
[3] 466 B.C. *Cf.* Thuc. i. 100; Polyaen. I. xxxiv. 1; Diodor. xi. 61.

confronting them on the walls, straightway turned their vessels about and laid their course for Asia.[1]

When the fleet of the Megarians approached Eleusis at night with the object of kidnapping the Athenian matrons who had made sacrifice to Ceres, Pisistratus, the Athenian, engaged it in battle and, by ruthlessly slaughtering the enemy, avenged his own countrymen. Then he filled these same captured ships with Athenian soldiers, placing in full view certain matrons dressed as captives. The Megarians, deceived by these appearances, thinking their own people were sailing back, and that, too, crowned with victory, rushed out to meet them, in disorder and without weapons, whereupon they were a second time overwhelmed.[2]

Cimon, the Athenian general, having defeated the fleet of the Persians near the island of Cyprus, fitted out his men with the weapons of the prisoners and in the barbarians' own ships set sail to meet the enemy in Pamphylia, near the Eurymedon River. The Persians, recognizing the vessels and the garb of those standing on deck, were quite off their guard. Thus on the same day they were suddenly crushed in two battles, one on sea and one on land.[3]

X. On Repairing One's Losses after a Reverse

When Titus Didius was warring in Spain and had fought an extremely bitter engagement, to which darkness put an end, leaving a large number of slain on both sides, he provided for the burial by night of many bodies of his own men. On the following day, the Spaniards, coming out to perform a like duty, found more of their men slain than of

secundum eam dinumerationem argumentati, ad condiciones imperatoris descenderunt.

2. T. Marcius eques Romanus, qui reliquiis exercitus praefuit, cum in propinquo bina castra Poenorum paucis milibus passuum distarent, cohortatus milites proxima castra intempesta nocte adortus est; et cum hostem victoriae fiducia incompositum adgressus ne nuntios quidem cladis reliquisset, brevissimo tempore militi ad requiem dato, eadem nocte raptim famam rei gestae praegressus altera eorundem castra invasit. Ita bis simili usus eventu, deletis utrubique Poenis, amissas populo Romano Hispanias restituit.

XI. De Dubiorum Animis in Fide Retinendis

1. P. Valerius Epidauri timens oppidanorum perfidiam, quia parum praesidii habebat, gymnicos ludos procul ab urbe apparavit. Quo cum omnis fere multitudo spectandi causa exisset, clausit portas nec ante admisit Epidaurios, quam obsides a principibus acciperet.

2. Cn. Pompeius, cum suspectos haberet Chaucenses et vereretur, ne praesidium non reciperent, petiit ab eis, ut aegros interim apud se refici paterentur; fortissimis deinde habitu languentium missis civitatem occupavit continuitque.

---

[1] 98–93 b.c.
[2] 212 b.c. Livy xxv. 37 gives his praenomen as Lucius.

the Romans, and arguing according to this calculation that they had been beaten, came to terms with the Roman commander.[1]

Titus Marcius, a Roman knight, who had charge of the remnants of the army [of the Scipios] in Spain, seeing near at hand two camps of the Carthaginians a few miles distant from each other, urged on his men and attacked the nearer camp at dead of night. Since the enemy, being flushed with victory, were without organization, Marcius by his attack did not leave even so much as a single man to report the disaster. Granting his troops merely the briefest time for rest, and outstripping the news of his exploit, he attacked the second camp the same night. Thus, by a double success, he destroyed the Carthaginians in both places and restored to the Roman people the lost provinces of Spain.[2]

XI. ON ENSURING THE LOYALTY OF THOSE WHOM ONE MISTRUSTS

WHEN Publius Valerius had an insufficient garrison at Epidaurus and therefore feared perfidy on the part of the townspeople, he prepared to celebrate athletic contests at some distance from the city. When nearly all the population had gone there to see the show, he closed the gates and refused to admit the Epidaurians until he had taken hostages from their chief men.

Gnaeus Pompey, suspecting the Chaucensians and fearing that they would not admit a garrison, asked that they would meanwhile permit his invalid soldiers to recover among them. Then, sending his strongest men in the guise of invalids, he seized the city and held it.

SEXTUS JULIUS FRONTINUS

3. Alexander devicta perdomitaque Thracia petens Asiam, veritus, ne post ipsius discessum sumerent arma, reges eorum praefectosque et omnis, quibus videbatur inesse cura detractae libertatis, secum velut honoris causa traxit, ignobiles autem relictis plebeiosque praefecit, consecutus, uti principes beneficiis eius obstricti nihil novare vellent, plebs vero ne posset quidem, spoliata principibus.

4. Antipater, conspecto Peloponesiorum exercitu, qui audita morte Alexandri ad infestandum imperium eius confluxerant, dissimulans scire se, qua mente venissent, gratias his egit, quod ad auxilium ferendum Alexandro adversus Lacedaemonios convenissent, adiecitque id se regi scripturum; ceterum ipsos, quia sibi opera eorum in praesentia non esset necessaria, abirent domos, hortatus est, et hac adseveratione periculum, quod ex novitate rerum imminebat, discussit.

5. Scipio Africanus in Hispania, cum inter captivas eximiae formae virgo et nobilis [1] ad eum perducta esset omniumque oculos in se converteret, summa custodia habitam sponso nomine Alicio reddidit insuperque aurum, quod parentes eius redempturi captivam donum Scipioni adtulerant, eidem sponso pro nuptiali munere dedit. Qua multiplici magnificentia universa gens victa imperio populi Romani accessit.

---

[1] *The text is confused and uncertain here. Oud. conjectures* eximiae formae virgo nubilis.

---

[1] 334 B.C. *Cf.* Justin. XI v. 1–3.
[2] *i.e.* he intentionally acted on the assumption that they had not heard of the death of Alexander, though he knew this assumption to be false.
[3] 331–330 B.C.

When Alexander had conquered and subdued Thrace and was setting out for Asia, fearing that after his departure the Thracians would take up arms, he took with him, as though by way of conferring honour, their kings and officials—all in fact who seemed to take to heart the loss of freedom. In charge of those left behind he placed common and ordinary persons, thus preventing the officials from wishing to make any change, as being bound to him by favours, and the common people from even being able to do so, since they had been deprived of their leaders.[1]

When Antipater beheld the army of the Peloponnesians, who had assembled to assail his authority on hearing of the death of Alexander, he pretended not to understand with what purpose they had come, and thanked them for having gathered to aid Alexander against the Spartans, adding that he would write to the king about this.[2] But inasmuch as he did not need their assistance at present, he urged them to go home, and by this statement dispelled the danger which threatened him from the new order of affairs.[3]

When Scipio Africanus was warring in Spain, there was brought before him among the captive women a noble maiden of surpassing beauty who attracted the gaze of everyone. Scipio guarded her with the greatest pains and restored her to her betrothed, Alicius by name, presenting to him likewise, as a marriage gift, the gold which her parents had brought to Scipio as a ransom. Overcome by this manifold generosity, the whole tribe leagued itself with the government of Rome.[4]

[4] 210 B.C. *Cf.* Livy xxvi. 50; Val. Max. IV. iii. 1; Gell. vii. 8; Polyb. x. 19.

SEXTUS JULIUS FRONTINUS

6. Alexandrum quoque Macedonem traditum est eximiae pulchritudinis virgini captivae, cum finitimae gentis principi fuisset desponsa, summa abstinentia ita pepercisse, ut illam ne adspexerit quidem. Qua mox ad sponsum remissa, universae gentis per hoc beneficium animos conciliavit sibi.[1]

7. Imperator Caesar Augustus Germanicus eo bello, quo victis hostibus cognomen Germanici meruit, cum in finibus Cubiorum castella poneret, pro fructibus locorum, quae vallo comprehendebat, pretium solvi iussit; atque ita iustitiae fama omnium fidem adstrinxit.

XII. Quae Facienda Sint pro Castrorum Defensione, si Satis Fiduciae in Praesentibus Copiis non Habeamus

1. T. Quintius consul, cum Volsci castra eius adgressuri forent, cohortem tantummodo in statione detinuit, reliquum exercitum ad quiescendum dimisit. Aeneatoribus praecepit, ut vallum insidentes equis circumirent concinerentque. Qua facie et simulatione cum et propulsasset et detinuisset per totam noctem hostes, ad lucis exortum fessos vigilia repente facta eruptione facile superavit.

2. Q. Sertorius in Hispania hostium equitatui maxime impar, qui usque ad ipsas munitiones nimia fiducia succedebat, nocte scrobes aperuit et ante eos

---

[1] *No. 6 is apparently interpolated.*

[1] *Cf.* Gell. vii. 8; Ammian. Marc. xxiv. iv. 27.
[2] Domitian. [3] 83 A.D.
[4] 468 B.C. *Cf.* Livy ii. 64–65.

The story goes that Alexander of Macedon likewise, having taken captive a maiden of exceeding beauty betrothed to the chief of a neighbouring tribe, treated her with such extreme consideration that he refrained even from gazing at her. When the maiden was later returned to her lover, Alexander, as a result of this kindness, secured the attachment of the entire tribe.[1]

When the Emperor Caesar Augustus Germanicus,[2] in the war in which he earned his title by conquering the Germans, was building forts in the territory of the Cubii, he ordered compensation to be made for the crops which he had included within his fortifications. Thus the renown of his justice won the allegiance of all.[3]

XII. What to do for the Defence of the Camp, in case a Commander lacks Confidence in His Present Forces

The consul, Titus Quinctius, as the Volscians were about to attack his camp, kept only one cohort on duty, and dismissed the remainder of the army to take their rest, directing the trumpeters to mount their horses and make the round of the camp sounding their trumpets. By exhibiting this semblance of strength, he kept the enemy off and held them throughout the night. Then at daybreak, attacking them by a sudden sortie when they were exhausted with watching, he easily defeated them.[4]

Quintus Sertorius, when in Spain, was completely outmatched by the cavalry of the enemy, who in their excessive confidence advanced up to his very fortifications. Accordingly during the night he con-

SEXTUS JULIUS FRONTINUS

aciem direxit. Cum deinde turmales secundum consuetudinem adventarent, recepit aciem; persecuti aciem in fossas deciderunt et eo modo victi sunt.

3. Chares dux Atheniensium, cum exspectaret auxilia et vereretur, ne interea contemptu praesentis paucitatis hostes castra eius oppugnarent, complures ex eis quos habebat per aversam partem nocte emissos iussit, qua praecipue conspicui forent hostibus, redire in castra et accedentium novarum virium speciem praebere; atque ita simulatis auxiliis tutus est, donec instrueretur exspectatis.

4. Iphicrates Atheniensis, cum campestribus locis castra haberet explorassetque Thracas ex collibus, per quos unus erat descensus, nocte ad diripienda castra venturos, clam eduxit exercitum et in utraque viae latera, per quam transituri Thraces erant, distributum conlocavit; hostemque decurrentem in castra, in quibus multi ignes per paucorum curam instituti speciem manentis ibi multitudinis servabant, a lateribus adortus oppressit.

XIII. De Effugiendo

1. Galli pugnaturi cum Attalo aurum omne et argentum certis custodibus tradiderunt, a quibus, si

---

[1] 80–72 B.C. [2] 366–338 B.C.
[3] 389 B.C. This same story is told in I. v. 24. *Cf.* also Polyaen. III. ix. 41, 46, 50.

structed trenches and drew up his line of battle in front of them. Then when the cavalry approached, as was their wont, he drew back his line. The enemy following close on his heels, fell into the trenches and thus were defeated.[1]

Chares, the Athenian commander, on one occasion was expecting reinforcements, but feared that meanwhile the enemy, despising his small force, would attack his camp. He therefore ordered that a number of the soldiers under his command should pass out at night by the rear of the camp, and should return by a route where they would be clearly observed by the enemy, thus creating the impression that fresh forces were arriving. In this way, he defended himself by pretended reinforcements, until he was equipped with those he was expecting.[2]

Iphicrates, the Athenian, being encamped on one occasion on level ground, happened to learn that the Thracians were intending to come down from the hills, over which there was but a single line of descent, with the purpose of plundering his camp by night. He therefore secretly led forth his troops and posted them on both sides of the road over which the Thracians were to pass. Then when the enemy descended upon the camp, in which a large number of watch-fires, built by the hands of a few men, produced the impression that a mighty host was still there, Iphicrates was enabled to attack them on the flank and crush them.[3]

XIII. On Retreating

When the Gauls were about to fight with Attalus, they handed over all their gold and silver to trusty guards, with instructions to scatter it, in case their

SEXTUS JULIUS FRONTINUS

acie fusi essent, spargeretur, quo facilius conligenda praeda hostem impeditum effugerent.

2. Tryphon Syriae rex victus per totum iter fugiens pecuniam sparsit et sectanda ea Antiochi equites moratos[1] effugit.

3. Q. Sertorius, pulsus acie a Q. Metello Pio, ne fugam quidem sibi tutam arbitratus, abire dispersos milites iussit, admonitos in quem locum vellet convenire.

4. Viriathus dux Lusitanorum copias nostras locorumque iniquitatem evasit eadem qua Sertorius ratione, sparso exercitu, dein reconlecto.

5. Horatius Cocles, urguente Porsennae exercitu, iussit suos per pontem redire in urbem eumque, ne eos insequeretur hostis, intercidere. Quod dum efficitur, in capite eius propugnator ipse insequentes detinuit. Audito deinde fragore pontis abrupti, deiecit se in alveum eumque in armis et vulneribus oneratus tranavit.

6. Afranius in Hispania ad Ilerdam, cum Caesarem fugeret, instante eo castra posuit; cum idem Caesar fecisset et pabulatum suos dimisisset, ille signum repente itineri dedit.

7. Antonius, cum ex Parthis instantibus reciperet

---

[1] moratus *H, Gund.*; moratos *d*, Wachs.

---

[1] Cic. *Pro Lege Manil.* ix. 22 and Flor. III. v. 18 attribute a similar stratagem to Mithridates.
[2] 134 B.C.   [3] 75 B.C.   *Cf.* Plut. *Pomp.* 19.
[4] 147–139 B.C.   *Cf.* Appian *Hisp.* 62   *Cf.* note to ii. v. 7.
[5] 507 B.C.   *Cf.* Livy ii. 10; Dionys. v. 23–25; Plut. *Publ.* 16.   [6] 49 B.C.   *Cf.* Caes. *B.C.* i. 80.

forces should be routed in battle, in order that thereby the enemy might be occupied in picking up the spoils and they themselves might more easily escape.[1]

Tryphon, king of Syria, when defeated, scattered money along the whole line of his retreat. While the cavalry of Antiochus delayed to pick this up, he effected his escape.[2]

Quintus Sertorius, when defeated in battle by Quintus Metellus Pius, being convinced that not even an organized retreat was safe, commanded his soldiers to disband and retire, informing them at what point he desired them to reassemble.[3]

Viriathus, leader of the Lusitanians, extricated himself from an awkward position, and from the menace of our troops, by the same method as Sertorius, disbanding his forces and then reassembling.[4]

Horatius Cocles, when Porsenna's army was pressing hard upon him, commanded his supporters to return over the bridge to the City, and then to destroy the bridge in order that the foe might not follow them. While this was being done, he himself, as defender of the bridgehead, held up the oncoming enemy. Then, when the crash told him that the bridge had been destroyed, he threw himself into the stream, and swam across it in his armour, exhausted though he was by wounds.[5]

Afranius, when fleeing from Caesar near Ilerda in Spain, pitched camp, while Caesar was pressing close upon him. When Caesar did the same and sent his men off to gather forage, Afranius suddenly gave the signal to continue the retreat.[6]

When Anthony was retreating, hard pressed by

SEXTUS JULIUS FRONTINUS

exercitum et, quotiens prima luce moveret, totiens urguentibus barbarorum sagittis infestaretur abeuntium agmen, in quintam horam continuit suos fidemque stativorum fecit. Qua persuasione digressis inde Parthis, iustum iter reliquo die sine interpellatione confecit.

8. Philippus in Epiro victus, ne fugientem eum Romani premerent, indutias ad sepeliendos qui caesi erant impetravit et ob id remissioribus custodibus evasit.

9. P. Claudius, navali proelio superatus a Poenis, cum per hostium praesidia necesse haberet erumpere, reliquas viginti naves tamquam victrices iussit ornari; atque ita Poenis existimantibus superiores fuisse acie nostros terribilis excessit.

10. Poeni classe superati, quia instantem avertere Romanum studebant, simulaverunt in vada naves suas incidisse haerentisque imitati effecerunt, ut victor eorum timens casum spatium ad evadendum daret.

11. Commius Atrebas, cum victus a Divo Iulio ex Gallia in Britanniam fugeret et forte ad Oceanum

---

[1] 36 B.C.
[2] 198 B.C.
[3] 249 B.C. Battle of Drepanum. Eutrop. ii. 26, Polyb. I. li. 11–12, and Oros. IV. x. 3, give the number of ships as thirty.

the Parthians, as often as he broke camp at daybreak, his retiring troops were assailed by volleys of arrows from the barbarians. Accordingly one day he kept his men back till nearly noon, thus producing the impression that he had made a permanent camp. As soon as the Parthians had become persuaded of this and had withdrawn, he accomplished his regular march for the remainder of the day without interference.[1]

When Philip had suffered defeat in Epirus, in order that the Romans might not overwhelm him in flight, he secured the grant of a truce to bury the dead. In consequence of this, the guards relaxed their vigilance, so that Philip slipped away.[2]

Publius Claudius, defeated by the Carthaginians in a naval engagement and thinking it necessary to break through the forces of the enemy, ordered his twenty remaining vessels to be dressed out as though victorious. The Carthaginians, therefore, thought our men had proved themselves superior in the encounter, so that Claudius became an object of fear to the enemy and thus made his escape.[3]

The Carthaginians, on one occasion, when defeated in a naval battle, desiring to shake off the Romans who were close upon them, pretended that their vessels had caught on shoals and imitated the movements of stranded galleys. In this way they caused the victors, in fear of meeting a like disaster, to afford them an opportunity of escape.

Commius, the Atrebatian, when defeated by the deified Julius, fled from Gaul to Britain, and happened to reach the Channel at a time when the wind was

vento quidem secundo, sed aestu recedente venisset, quamvis naves in siccis litoribus haererent, pandi nihilominus vela iussit. Quae cum persequens eum Caesar ex longinquo tumentia et flatu plena vidisset, ratus prospero sibi eripi cursu recessit.

fair, but the tide was out. Although the vessels were stranded on the flats, he nevertheless ordered the sails to be spread. Caesar, who was following from a distance, seeing the sails swelling with the full breeze, and imagining Commius to be escaping from his hands and to be proceeding on a prosperous voyage, abandoned the pursuit.

BOOK III

LIBER TERTIUS

Si priores libri responderunt titulis suis et lectorem hucusque cum attentione perduxerunt, edam nunc circa oppugnationes urbium defensionesque στρατηγήματα. Nec morabor ulla praelocutione, prius traditurus quae oppugnandis urbibus usui sunt, tum quae obsessos instruere possint. Depositis autem operibus et machinamentis, quorum expleta iam pridem inventione nullam video ultra artium materiam, has circa expugnationem species στρατηγημάτων fecimus:

    I. De repentino impetu.
   II. De fallendis his qui obsidebuntur.
  III. De eliciendis ad proditionem.
 IIII. Per quae hostes ad inopiam redigantur.
    V. Quemadmodum persuadeatur obsidionem permansuram.
  VI. De districtione praesidiorum hostilium.
 VII. De fluminum derivatione et vitiatione aquarum.
VIII. De iniciendo obsessis pavore.
VIIII. De inruptione ex diversa parte quam exspectabimur.
    X. De insidiis, per quas eliciantur obsessi.
  XI. De simulatione regressus.

---

[1] A curious illustration of the rashness of prophecy.

BOOK III

If the preceding books have corresponded to their titles, and I have held the attention of the reader up to this point, I will now treat of ruses that deal with the siege and defence of towns. Waiving any preface, I will first submit those which are useful in the siege of cities, then those which offer suggestions to the besieged. Laying aside also all considerations of works and engines of war, the invention of which has long since reached its limit,[1] and for the improvement of which I see no further hope in the applied arts, I shall recognize the following types of stratagems connected with siege operations:

    I. On surprise attacks.
   II. On deceiving the besieged.
  III. On inducing treachery.
  IV. By what means the enemy may be reduced to want.
   V. How to persuade the enemy that the siege will be maintained.
  VI. On distracting the attention of a hostile garrison.
 VII. On diverting streams and contaminating waters.
VIII. On terrorizing the besieged.
  IX. On attacks from an unexpected quarter.
   X. On setting traps to draw out the besieged.
  XI. On pretended retirements.

SEXTUS JULIUS FRONTINUS

Ex contrario circa tutelam obsessorum :

    XII. De excitanda cura suorum.
   XIII. De emittendo et recipiendo nuntio.
  XIIII. De introducendis auxiliis et commeatibus suggerendis.
    XV. Quemadmodum efficiatur ut abundare videantur quae deerunt.
   XVI. Qua ratione proditoribus et transfugis occurratur.
  XVII. De eruptionibus.
 XVIII. De constantia obsessorum.

I. De Repentino Impetu

1. T. Quintius consul, victis acie Aequis et Volscis, cum Antium oppidum expugnare statuisset, ad contionem vocato exercitu exposuit, quam id necessarium et facile esset, si non differretur; eoque impetu, quem exhortatio concitaverat, adgressus est urbem.

2. M. Cato in Hispania animadvertit potiri se quodam oppido posse, si inopinatos invaderet. Quadridui itaque iter biduo per confragosa et deserta emensus, nihil tale metuentes oppressit hostes. Victoribus deinde suis causam tam facilis eventus requirentibus dixit, tum illos victoriam adeptos, cum quadridui iter biduo corripuerint.

---

[1] 468 B.C.   [2] 195 B.C.

On the other hand, stratagems connected with the protection of the besieged:

   XII. On stimulating the vigilance of one's own troops.
  XIII. On sending and receiving messages.
  XIV. On introducing reinforcements and supplying provisions.
   XV. How to produce the impression of abundance of what is lacking.
  XVI. How to meet the menace of treason and desertion.
 XVII. On sorties.
XVIII. Concerning steadfastness on the part of the besieged.

I. On Surprise Attacks

The consul Titus Quinctius, having conquered the Aequians and Volscians in an engagement, decided to storm the walled town of Antium. Accordingly he called an assembly of the soldiers and explained how necessary this project was and how easy, if only it were not postponed. Then, having roused enthusiasm by his address, he assaulted the town.[1]

Marcus Cato, when in Spain, saw that he could gain possession of a certain town, if only he could assault the enemy by surprise. Accordingly, having in two days accomplished a four days' march through rough and barren districts, he crushed his foes, who were fearing no such event. Then, when his men asked the reason of so easy a success, he told them that they had won the victory as soon as they had accomplished the four days' march in two.[2]

SEXTUS JULIUS FRONTINUS

II. De Fallendis His Qui Obsidebuntur

1. Domitius Calvinus, cum obsideret Lueriam, oppidum Ligurum, non tantum situ et operibus, verum etiam propugnatorum praestantia tutum, circumire muros frequenter omnibus copiis instituit easdemque reducere in castra. Qua consuetudine inductis ita oppidanis, ut crederent exercitationis id gratia facere Romanum, et ob hoc nihil ab eo conatu caventibus, morem illum obambulandi in subitum direxit impetum, occupatisque moenibus expressit, ut se ipsos dederent oppidani.

2. C. Duellius consul subinde exercendo milites remigesque consecutus est, ut securis Carthaginiensibus usque in id tempus innoxiae consuetudinis subito admota classe murum occuparet.

3. Hannibal in Italia multas urbes cepit, cum Romanorum habitu quosdam suorum, ex longo belli usu latine quoque loquentis, praemitteret.

4. Arcades Messeniorum castellum obsidentes, factis quibusdam armis ad similitudinem hostilium, eo tempore quo successura alia praesidia his exploraverunt, instructi eorum qui exspectabantur ornatu admissique per hunc errorem ut socii, possessionem loci cum strage hostium adepti sunt.

5. Cimon dux Atheniensium in Caria insidiatus cuidam civitati religiosum incolis templum Dianae

---

[1] 260 b.c.
[2] 216–203 b.c.

II. On Deceiving the Besieged

When Domitius Calvinus was besieging Lueria, a town of the Ligurians, protected not only by its location and siege-works, but also by the superiority of its defenders, he instituted the practice of marching frequently around the walls with all his forces, and then marching back to camp. When the townspeople had been induced by this routine to believe that the Roman commander did this for the purpose of drill, and consequently took no precautions against his efforts, he transformed this practice of parading into a sudden attack, and gaining possession of the walls, forced the inhabitants to surrender.

The consul Gaius Duellius, by frequently exercising his soldiers and sailors, succeeded in preventing the Carthaginians from taking notice of a practice which was innocent enough, until suddenly he brought up his fleet and seized their fortifications.[1]

Hannibal captured many cities in Italy by sending ahead certain of his own men, dressed in the garb of Romans and speaking Latin, which they had acquired as a result of long experience in the war.[2]

The Arcadians, when besieging a stronghold of the Messenians, fabricated certain weapons to resemble those of the enemy. Then, at the time when they learned that another force was to relieve the first, they dressed themselves in the uniform of those who were expected, and being admitted as comrades in consequence of this confusion, they secured possession of the place and wrought havoc among the foe.

Cimon, the Athenian general, having designs on a certain city in Caria, under cover of night set fire to a temple of Diana, held in high reverence by the in-

SEXTUS JULIUS FRONTINUS

lucumque, qui extra muros erat, noctu improvisus incendit; effusisque oppidanis ad opem adversus ignes ferendam vacuam defensoribus cepit urbem.

6. Alcibiades dux Atheniensium, cum civitatem Agrigentinorum egregie munitam obsideret, petito ab eis concilio diu tamquam de rebus ad commune pertinentibus disseruit in theatro, ubi ex more Graecorum locus consultationi praebebatur; dumque consilii specie tenet multitudinem, Athenienses, quos ad id praeparaverat, incustoditam urbem ceperunt.

7. Epaminondas Thebanus in Arcadia die festo effuse extra moenia vagantibus hostium feminis plerosque ex militibus suis muliebri ornatu immiscuit. Qua simulatione illi intra portas sub noctem recepti ceperunt oppidum et suis aperuerunt.

8. Aristippus Lacedaemonius festo die Tegeatarum, quo omnis multitudo ad celebrandum Minervae sacrum urbe egressa erat, iumenta saccis frumentariis palea refertis onusta Tegeam misit, agentibus ea militibus, qui negotiatorum specie inobservati portas aperuerunt suis.

9. Antiochus in Cappadocia ex castello Suenda, quod obsidebat, iumenta frumentatum egressa intercepit occisisque calonibus, eorundem vestitu milites suos tamquam frumentum reportantes summisit. Quo

---

[1] About 470 B.C.

[2] 415 B.C. Thuc. vi. 51, Polyaen. I. xl. 4 and Diodor. XIII. iv. 4, make Catana, and not Agrigentum, the scene of this stratagem.

[3] 379 B.C. Polyaen. II. iii. 1 has a different version of this story, but in II. iv. 3 attributes a somewhat similar stratagem to Pelopidas.

habitants, and also to a grove outside the walls. Then, when the townspeople poured out to fight the conflagration, Cimon captured the city, since it was left without defenders.[1]

Alcibiades, the Athenian commander, while besieging the strongly fortified city of the Agrigentines, requested a conference of the citizens, and, as though discussing matters of common concern, addressed them at length in the theatre, where according to the custom of the Greeks it was usual to afford a place for consultation. Then, while he held the crowd on the pretence of deliberation, the Athenians, whom he had previously prepared for this move, captured the city, thus left unguarded.[2]

When Epaminondas, the Theban, was campaigning in Arcadia, and on a certain holiday the women of the enemy strolled in large numbers outside the walls, he sent among them a number of his own troops dressed in women's attire. In consequence of this disguise, the men were admitted towards nightfall to the town, whereupon they seized it and threw it open to their companions.[3]

Aristippus, the Spartan, on a holiday of the Tegeans, when the whole population had gone out of the city to celebrate the rites of Minerva, sent to Tegea a number of mules laden with grain-bags filled with chaff. The mules were driven by soldiers disguised as traders, who, escaping notice, threw open the gates of the town to their comrades.

When Antiochus was besieging the fortified town of Suenda in Cappadocia, he intercepted some beasts of burden which had gone out to procure grain. Then, killing their attendants, he dressed his own soldiers in their clothes and sent them in as though bringing

errore illi custodibus deceptis castellum intraverunt admiseruntque milites Antiochi.

10. Thebani, cum portum Sicyoniorum nulla vi redigere in potestatem suam possent, navem ingentem armatis compleverunt, exposita super merce, ut negotiatorum specie fallerent. Ab ea deinde parte murorum, quae longissime remota erat a mari, paucos disposuerunt, cum quibus e nave quidam egressi inermes simulata rixa concurrerent. Sicyoniis ad dirimendum id iurgium advocatis, Thebanae naves et portum vacantem et urbem occupaverunt.

11. Timarchus Aetolus, occiso Charmade Ptolomaei regis praefecto, clamide interempti et galeari ad Macedonicum ornatus est habitum; per hunc errorem pro Charmade in Saniorum portum receptus occupavit.

III. De Eliciendis ad Proditionem

1. Papirius Cursor consul apud Tarentum Miloni, qui cum praesidio Epirotarum urbem obtinebat, salutem ipsi et popularibus, si per illum oppido potiretur, pollicitus est. Quibus praemiis ille corruptus persuasit Tarentinis, ut se legatum ad consulem mitterent, a quo plena promissa ex pacto referens in securita-

---

[1] Polyaen. v. xvi. 3 makes Pammenes the author of this stratagem, 369 B.C.

[2] Ptolemy Ceraunus, king of Macedonia, 280 B.C.

back the grain. The sentinels fell into the trap and, mistaking the soldiers for teamsters, let the troops of Antiochus enter the fortifications.

When the Thebans were unable by the utmost exertions to gain possession of the harbour of the Sicyonians, they filled a large vessel with armed men, exhibiting a cargo in full view on deck, in order, under the guise of traders, to deceive their enemies. Then at a point of the fortifications remote from the sea they stationed a few men, with whom certain unarmed members of the crew upon disembarking were to engage in a fracas, on the pretence of a quarrel. When the Sicyonians were summoned to stop the altercation, the Theban crews seized both the unguarded harbour and the town.[1]

Timarchus, the Aetolian, having killed Charmades, general of King Ptolemy,[2] arrayed himself in Macedonian fashion in the cloak and casque of the slain commander. Through this disguise he was admitted as Charmades into the harbour of the Sanii and secured possession of it.

III. On Inducing Treachery

When the consul Papirius Cursor was before Tarentum, and Milo was holding the town with a force of Epirotes, Papirius promised safety to Milo and the townspeople if he should secure possession of the town through Milo's agency. Bribed by these inducements, Milo persuaded the Tarentines to send him as ambassador to the consul, from whom, in conformity with their understanding, he brought back liberal promises by means of which he caused

SEXTUS JULIUS FRONTINUS

tem oppidanos resolvit atque ita incustoditam urbem Cursori tradidit.

2. M. Marcellus, cum Syracusanum quendam Sosistratum ad proditionem sollicitasset, ex eo cognovit remissiores custodias fore die festo, quo Epicydes praebiturus esset vini epularumque copiam. Insidiatus igitur hilaritati et quae eam sequebatur socordiae munimenta conscendit vigilibusque caesis aperuit exercitui Romano urbem nobilibus victoriis claram.

3. Tarquinius Superbus, cum Gabios in deditionem accipere non posset, filium suum Sextum Tarquinium caesum virgis ad hostem misit. Is incusata patris saevitia persuasit Gabinis, odio suo adversus regem uterentur, et dux ad bellum electus tradidit patri Gabios.

4. Cyrus Persarum rex comitem suum Zopyrum, explorata eius fide, truncata de industria facie, ad hostes dimisit. Ille adsentante iniuriarum fide creditus inimicissimus Cyro, cum hanc persuasionem adiuvaret procurrendo propius, quotiens acie decertaretur, et in eum tela dirigendo, commissam sibi Babyloniorum urbem tradidit Cyro.

5. Philippus, oppido Saniorum exclusus, Apollonidi praefecto eorum ad proditionem corrupto persuasit, ut plaustrum lapide quadrato oneratum in ipso aditu

---

[1] 272 B.C. *Cf.* Zonar. viii. 6.
[2] 212 B.C. Livy xxv. 23 ff., Plut. *Marc.* 18, and Polyaen. viii. 11 name Damippus a Spartan, rather than Sosistratus, as the source of the information.
[3] *Cf.* Livy i. 53; Val. Max. VII. iv. 2; Polyaen. viii. 6.
[4] 518 B.C. Herod. iii. 153, Justin. I. x. 15, and Polyaen. vii. 13, represent Zopyrus as mutilating himself, and make Darius rather than Cyrus the monarch at the time.

the citizens to relapse into a feeling of security, and was thus enabled to hand the city over to Cursor, since it was left unguarded.[1]

Marcus Marcellus, having tempted a certain Sosistratus of Syracuse to turn traitor, learned from him that the guards would be less strict on a holiday when a certain citizen named Epicydes was to make a generous distribution of wine and food. So, taking advantage of the gaiety and the consequent laxness of discipline, he scaled the walls, slew the sentinels, and threw open to the Roman army a city already made famous as the scene of noted victories.[2]

When Tarquinius Superbus was unable to induce Gabii to surrender, he scourged his son Sextus with rods and sent him among the enemy, where he arraigned the cruelty of his father and persuaded the Gabians to utilize his hatred against the king. Accordingly he was chosen leader in the war, and delivered Gabii over to his father.[3]

Cyrus, king of the Persians, having proved the loyalty of his attendant Zopyrus, deliberately mutilated his face and sent him among the enemy. In consequence of their belief in his wrongs, he was regarded as implacably hostile to Cyrus, and promoted this belief by running up and discharging his weapons against Cyrus, whenever an engagement took place, till finally the city of the Babylonians was entrusted to him and by him delivered into the hands of Cyrus.[4]

Philip, when prevented from gaining possession of the town of the Sanians, bribed one of their generals, Apollonides, to turn traitor, inducing him to plant a cart laden with dressed stone at the

SEXTUS JULIUS FRONTINUS

portae poneret. Confestim deinde signo dato insecutus oppidanos circa impedita portae claustra trepidantis oppressit.

6. Hannibal apud Tarentum, quae a praesidio Romano duce Livio tenebatur, Cononeum quendam Tarentinum, quem ad proditionem sollicitaverat, eiusmodi fallacia instruxit, ut ille per causam venandi noctu procederet, quasi id per hostem interdiu non liceret. Egresso ipsi apros subministrabant, quos ille tamquam ex captura Livio offerret; idque cum saepius factum esset et ideo minus observaretur, quadam nocte Hannibal venatorum habitu Poenos comitibus eius immiscuit. Qui cum onusti venatione quam ferebant recepti essent a custodibus, protinus eos adorti occiderunt. Tum fracta porta admissus cum exercitu Hannibal omnes Romanos interfecit, exceptis his qui in arcem profugerant.

7. Lysimachus rex Macedonum, cum Ephesios oppugnaret et illi in auxilio haberent Mandronem archipiratam, qui plerumque oneratas praeda naves Ephesum appellabat, corrupto ei ad proditionem iunxit fortissimos Macedonum, quos ille restrictis manibus pro captivis Ephesum introduceret. Postea raptis ex arce armis urbem Lysimacho tradiderunt.

---

[1] 359–336 B.C.
[2] 212 B.C. *Cf.* Appian *Hann.* 32; Livy xxv. 8–9; Polyb. viii 26. The name of the traitor is variously given.
[3] 287 B.C. *Cf.* Polyaen. v. 19.

very entrance to the gate. Then straightway giving the signal, he followed after the townspeople, who were huddled in panic around the blocked entrance of the gate, and succeeded in overwhelming them.[1]

When Hannibal was before Tarentum, and this town was held by a Roman garrison under the command of Livius, he induced a certain Cononeus of Tarentum to turn traitor, and concerted with him a stratagem whereby he was to go out at night for the purpose of hunting, on the ground that the enemy rendered this impossible by day. When he went forth, Hannibal supplied him with boars to present to Livius as trophies of the chase. When this had repeatedly been done, and for that reason was less noticed, Hannibal one night dressed a number of Carthaginians in the garb of hunters and introduced them among Cononeus's attendants. When these men, loaded with the game they were carrying, were admitted by the guards, they straightway attacked and slew the latter. Then breaking down the gate, they admitted Hannibal with his troops, who slew all the Romans, save those who had fled for refuge to the citadel.[2]

When Lysimachus, king of the Macedonians, was besieging the Ephesians, these were assisted by the pirate chief Mandro, who was in the habit of bringing into Ephesus galleys laden with booty. Accordingly Lysimachus bribed Mandro to turn traitor, and attached to him a number of dauntless Macedonians to be taken into the city as captives, with hands pinioned behind their backs. These men subsequently snatched weapons from the citadel and delivered the town into the hands of Lysimachus.[3]

SEXTUS JULIUS FRONTINUS

IIII. Per quae Hostes ad Inopiam Redigantur

1. Fabius Maximus vastatis Campanorum agris, ne quid eis ad fiduciam obsidionis superesset, recessit sementis tempore, ut frumentum quod reliquum habebant in sationes conferrent; reversus deinde renata protrivit et ad famem redactis potitus est.

2. Antigonus adversus Athenienses idem fecit.[1]

3. Dionysius, multis urbibus captis, cum Reginos adgredi vellet, qui copiis abundabant, simulabat pacem petiitque ab eis, ut commeatus exercitui ipsius subministrarent. Quod cum impetrasset exhausto oppidanorum frumento adgressus urbem destitutam alimentis superavit.

4. Idem et adversus Himeraeos fecisse dicitur.[2]

5. Alexander oppugnaturus Leucadiam commeatibus abundantem prius castella, quae in confinio erant, cepit omnesque ex his Leucadiam passus est confugere, ut alimenta inter multos celerius absumerentur.

6. Phalaris Agrigentinus, cum quaedam loca munitione tuta in Sicilia oppugnaret, simulato foedere frumenta, quae residua habere se dicebat, apud eos deposuit; deinde data opera, ut camerae tectorum,

---

[1] *No. 2 is probably interpolated.*
[2] *No. 4 is probably interpolated.*

---

[1] Apparently a confusion of two occasions. *Cf.* Livy XXIII. xlviii. 1–2 and XXV. xiii. 215 or 211 B.C.

[2] 263 B.C. *Cf.* Polyaen. IV. vi. 20.

[3] 391 B.C. The version of Diodor. xiv. 108 is slightly different.

[4] 387 B.C. *Cf.* Polyaen. v. ii. 10.

[5] Son of Pyrrhus. 266–263 B.C.

IV. By what Means the Enemy may be Reduced to Want

Fabius Maximus, having laid waste the lands of the Campanians, in order that they might have nothing left to warrant the confidence that a siege could be sustained, withdrew at the time of the sowing, that the inhabitants might plant what seed they had remaining. Then, returning, he destroyed the new crop and thus made himself master of the Campanians, whom he had reduced to famine.[1]

Antigonus employed the same device against the Athenians.[2]

Dionysius, having captured many cities and wishing to attack the Rhegians, who were well provided with supplies, pretended to desire peace, and begged of them to furnish provisions for his army. When he had secured his request and had consumed the grain of the inhabitants, he attacked their town, now stripped of food, and conquered it.[3]

He is said to have employed the same device also against the people of Himera.[4]

When Alexander[5] was about to besiege Leucadia, a town well-supplied with provisions, he first captured the fortresses on the border and allowed all the people from these to flee for refuge to Leucadia, in order that the food-supplies might be consumed with greater rapidity when shared by many.

Phalaris of Agrigentum, when besieging certain places in Sicily protected by fortifications, pretended to make a treaty and deposited with the Sicilians all the wheat which he said he had remaining, taking pains, however, that the chambers of the buildings in which the grain was stored should have

SEXTUS JULIUS FRONTINUS

in quibus id conferebatur, rescissae pluviam reciperent, eos fiducia conditi commeatus proprio tritico abusos initio aestatis adgressus inopia compulit ad deditionem.

V. Quemadmodum Persuadeatur Obsidionem Permansuram

1. Clearchus Lacedaemonius, exploratum habens Thracas omnia victui necessaria in montes comportasse unaque spe sustentari, quod crederent eum commeatus inopia recessurum, per id tempus, quo legatos eorum venturos opinabatur, aliquem ex captivis in conspectu iussit occidi et membratim tamquam alimenti causa in contubernia distribui. Thraces nihil non facturum perseverantiae causa eum credentes, qui tam detestabiles epulas sustinuisset experiri, in deditionem venerunt.

2. Ti. Gracchus, Lusitanis dicentibus in X annos cibaria se habere et ideo obsidionem non expavescere, "undecimo," inquit, "anno vos capiam." Qua voce perterriti Lusitani, quamvis instructi commeatibus, statim se dediderunt.

3. A. Torquato Graecam urbem oppugnanti cum diceretur, iuventutem ibi studiose iaculis et sagittis exerceri, "pluris eam," inquit, "propediem vendam."

---

[1] 570–554 b.c. *Cf.* Polyaen. v. i. 3. According to Polyaenus the Sicilians were to return to Phalaris not the grain he had left with them but the crops from their later harvests.

[2] 402–401 b.c. *Cf.* Polyaen. ii. ii. 8.

[3] 179–178 b.c.

leaky roofs. Then when the Sicilians, relying on the wheat which Phalaris had deposited with them, had used up their own supplies, Phalaris attacked them at the beginning of summer and as a result of their lack of provisions forced them to surrender.[1]

V. How to Persuade the Enemy that the Siege will be Maintained

When Clearchus, the Spartan, had learned that the Thracians had conveyed to the mountains all things necessary for their subsistence and were buoyed up by the sole hope that he would withdraw in consequence of lack of supplies, at the time when he surmised their envoys would come, he ordered one of the prisoners to be put to death in full view and his body to be distributed in pieces among the tents, as though for the mess. The Thracians, believing that Clearchus would stick at nothing in order to hold out, since he brought himself to try such loathsome food, delivered themselves up.[2]

When the Lusitanians told Tiberius Gracchus that they had supplies for ten years and for that reason stood in no fear of a siege, he answered: "Then I'll capture you in the eleventh year." Terror-stricken by this language, the Lusitanians, though well supplied with provisions, at once surrendered.[3]

When Aulus Torquatus was besieging a Greek city and was told that the young men of the city were engaged in earnest practice with the javelin and bow, he replied: "Then the price at which I shall presently sell them shall be higher."

SEXTUS JULIUS FRONTINUS

VI. De Districtione Praesidiorum Hostium

1. Scipio, Hannibale in Africam reverso, cum plura oppida, quae ratio illi in potestatem redigenda dictabat, firmis praesidiis diversae partis obtinerentur, subinde aliquam manum submittebat ad infestanda ea. Novissime etiam tamquam direpturus civitates aderat, deinde simulato metu refugiebat. Hannibal, ratus veram esse eius trepidationem, deductis undique praesidiis, tamquam de summa rerum decertaturus insequi coepit. Ita consecutus Scipio, quod petierat, nudatas propugnatoribus urbis per Masinissam et Numidas cepit.

2. P. Cornelius Scipio, intellecta difficultate expugnandi Delminum, quia concursu omnium defendebatur, adgredi alia oppida coepit et revocatis[1] ad sua defendenda singulis vacuatam auxiliis Delminum cepit.

3. Pyrrhus Epirotarum rex adversus Illyrios, cum civitatem quae caput gentis erat redigere in potestatem suam vellet, eius desperatione ceteras urbes petere coepit consecutusque est, ut hostes fiducia velut satis munitae urbis eius ad tutelam aliarum dilaberentur; quo facto revocatis ipse rursus omnibus suis vacuam eam defensoribus cepit.

4. Cornelius Rufinus consul, cum aliquanto tem-

---

[1] revocatis *Bennett*; evocatis *MSS.*, *edd.*

[1] 202 B.C.   [2] 155 B.C.
[3] 296–280 B.C.

STRATAGEMS, III. vi. 1-4

VI. On Distracting the Attention of a Hostile Garrison

When Hannibal had returned to Africa, many towns were still held by strong forces of the Carthaginians. Scipio's policy demanded that these towns should be reduced. Accordingly he often sent troops to assault them. Finally he would appear before the towns as though bent on sacking them, and would then retire, feigning fear. Hannibal, thinking his alarm real, withdrew the garrison from all points, and began to follow, as though determined to fight a decisive battle. Scipio, having thus accomplished what he intended, with the assistance of Masinissa and the Numidians, captured the towns, which had thus been stripped of their defenders.[1]

Publius Cornelius Scipio, appreciating the difficulty of capturing Delminus, because it was defended by the concerted efforts of the population of the district, began to assault other towns. Then, when the inhabitants of the various towns had been called back to defend their homes, Scipio took Delminus, which had been left without support.[2]

Pyrrhus, king of Epirus, in his war against the Illyrians, aimed to reduce their capital, but despairing of this, began to attack the other towns, and succeeded in making the enemy disperse to protect their other cities, since they had confidence in the apparently adequate fortification of the capital. When he had accomplished this, he recalled his own forces and captured the town, now left without defenders.[3]

The consul Cornelius Rufinus for some time be-

SEXTUS JULIUS FRONTINUS

pore Crotona oppidum frustra obsedisset, quod inexpugnabile faciebat adsumpta in praesidium Lucanorum manus, simulavit se coepto desistere. Captivum deinde magno praemio sollicitatum misit Crotona, qui, tamquam ex custodia effugisset, persuasit discessisse Romanos. Id verum Crotonienses arbitrati dimisere auxilia destitutique propugnatoribus inopinati et invalidi capti sunt.

5. Mago dux Carthaginiensium, victo Cn. Pisone et in quadam turre circumsesso, suspicatus ventura ei subsidia perfugam misit, qui persuaderet appropinquantibus captum iam Pisonem; qua ratione deterritis eis reliqua victoriae consummavit.

6. Alcibiades in Sicilia, cum Syracusanos capere vellet, ex Catiniensibus, apud quos tum exercitum continebat, quendam exploratae sollertiae submisit ad Syracusanos. Is in publicum concilium introductus persuasit infestissimos esse Catinienses Atheniensibus et, si adiuvarentur a Syracusanis, futurum, ut opprimerent eos et Alcibiadem. Qua re adducti Syracusani universis viribus Catinam petituri processerunt, relicta ipsorum urbe, quam a tergo adortus Alcibiades desolatam, ut speraverat, adflixit.

7. Cleonymus Atheniensis Troezenios, qui praesi-

---

[1] 277 B.C. *Cf.* Zonar. viii. 6.   [2] 216–203 B C.
[3] 415 B.C. This account agrees with that of Polyaen. I. xl. 5. Thuc. vi. 64 ff., and Diodor. XIII. vi. 2 ff., attribute the stratagem to Nicias and Lamachus, and give a different version of its result.

sieged the city of Crotona, without success, since it had been made impregnable by the arrival of a band of Lucanian reinforcements. He therefore pretended to desist from his undertaking, and by offers of great rewards induced a certain prisoner to go to Crotona. This emissary, by feigning to have escaped from custody, persuaded the inhabitants to believe his report that the Romans had withdrawn. The people of Crotona, thinking this to be true, dismissed their allies. Then, weakened by being stripped of their defenders, they were surprised and captured.[1]

Mago, general of the Carthaginians, having defeated Gnaeus Piso and having blockaded the tower wherein he had taken refuge, suspecting that reinforcements would come to his relief, sent a deserter to persuade the approaching troops that Piso was already captured. Having thus scared them off, Mago made his victory complete.[2]

Alcibiades, wishing to capture the city of Syracuse in Sicily, chose from among the people of Catana, where he was encamped, a certain man of tested shrewdness and sent him to the Syracusans. This man, when brought before the public assembly of the Syracusans, persuaded them that the people of Catana were very hostile to the Athenians, and that, if assisted by the Syracusans, they would crush the Athenians and Alcibiades along with them. Induced by these representations, the Syracusans left their own city and set out in full force to join the people of Catana, whereupon Alcibiades attacked Syracuse from the rear, and finding it unprotected, as he had hoped, brought it under subjection.[3]

When the people of Troezen were held in subjec-

SEXTUS JULIUS FRONTINUS

dio Crateri tum tenebantur, adgressus tela quaedam, in quibus scriptum erat venisse se ad liberandam eorum rem publicam, intra muros iecit et eodem tempore captivos quosdam conciliatos sibi remisit, qui Craterum detractarent. Per hoc consilium seditione intestina apud obsessos concitata admoto exercitu potitus est civitate.

VII. DE FLUMINUM DERIVATIONE ET VITIATIONE AQUARUM

1. P. SERVILIUS Isauram oppidum, flumine ex quo hostes aquabantur averso, ad deditionem siti compulit.

2. C. Caesar in Gallia Cadurcorum civitatem amne cinctam et fontibus abundantem ad inopiam aquae redegit, cum fontes cuniculis avertisset et fluminis usum per sagittarios arcuisset.

3. L. Metellus in Hispania citeriore in castra hostium humili loco posita fluvium ex superiore parte immisit et subita inundatione turbatos per dispositos in hoc ipsum insidiatores cecidit.

4. Alexander apud Babylona, quae media flumine Euphrate dividebatur, fossam pariter et aggerem instituit, ut in usum eius existimarent hostes egeri terram; atque ita subito flumine averso per alveum

---

[1] 277–276 B.C. Polyaen. II. xxix 1 calls Cleonymus a king of Sparta.

[2] 78–76 B.C.

[3] 51 B.C. *Cf.* Hirt. *B.G.* viii. 40 ff.

[4] 143–142 B.C. Quintus Caecilius Metellus Macedonicus, although the better manuscript readings give *L.* as the praenomen.

tion by troops under the command of Craterus, the Athenian Cleonymus made an assault on the town and hurled within its walls missiles inscribed with messages stating that Cleonymus had come to liberate their state. At the same time certain prisoners whom he had won over to his side were sent back to disparage Craterus. By this plan he stirred up internal strife among the besieged and, bringing up his troops, gained possession of the city.[1]

VII. On Diverting Streams and Contaminating Waters

Publius Servilius diverted the stream from which the inhabitants of Isaura drew their water, and thus forced them to surrender in consequence of thirst.[2]

Gaius Caesar, in one of his Gallic campaigns, deprived the city of the Cadurci of water, although it was surrounded by a river and abounded in springs; for he diverted the springs by subterranean channels, while his archers shut off all access to the river.[3]

Lucius Metellus, when fighting in Hither Spain, diverted the course of a river and directed it from a higher level against the camp of the enemy, which was located on low ground. Then, when the enemy were in panic from the sudden flood, he had them slain by men whom he had stationed in ambush for this very purpose.[4]

At Babylon, which is divided into two parts by the river Euphrates, Alexander constructed both a ditch and an embankment, the enemy supposing that the earth was being taken out merely to form the embankment. Alexander, accordingly, suddenly

veterem, qui siccatus ingressum praebebat, urbem intravit.

5. Semiramis adversus eosdem Babylonios eodem Euphrate averso idem fecisse dicitur.[1]

6. Clisthenes Sicyonius ductum aquarum in oppidum Crisaeorum ferentem rupit; mox adfectis siti restituit aquam elleboro corruptam, qua usos profluvio ventris deficientes cepit.

VIII. De Iniciendo Obsessis Pavore

1. Philippus, cum Prinassum castellum nulla vi capere posset, terram ante ipsos muros egessit simulavitque agi cuniculum; castellani, quia subrutos se existimarant, dediderunt.

2. Pelopidas Thebanus Magnetum duo oppida simul oppugnaturus non ita longo spatio distantia, quo tempore ad alterum eorum exercitum admovebat, praecepit, ut ex composito ab aliis castris quattuor equites coronati notabili alacritate velut victoriam nuntiantes venirent. Ad cuius simulationem curavit, ut silva quae in medio erat incenderetur, praebitura speciem urbis ardentis; praeterea quosdam capti-

[1] *No. 5 is probably an interpolation.*

---

[1] Herod. i. 191, Xen. *Cyrop.* vii. 5, and Polyaen. VII. vi. 5, attribute this stratagem to Cyrus rather than Alexander.

[2] 595–585 B.C. Polyaen. vi. 13 attributes this stratagem to Eurylochus; Pausan. x. xxxvii. 7, to Solon.

[3] 201 B.C. *Cf.* Polyb. xvi. 11; Polyaen. IV. xviii. 1.

diverting the stream, entered the town along the former river bed, which had dried up and thus afforded an entrance to the town.[1]

Semiramis is said to have done the same thing in the war against the Babylonians, by diverting the same Euphrates.

Clisthenes of Sicyon cut the water-pipes leading into the town of the Crisaeans. Then when the townspeople were suffering from thirst, he turned on the water again, now poisoned with hellebore. When the inhabitants used this, they were so weakened by diarrhoea that Clisthenes overcame them.[2]

VIII. On Terrorizing the Besieged

When Philip was unable by the utmost exertions to capture the fortress of Prinassus, he made excavations of earth directly in front of the walls and pretended to be constructing a tunnel. The men within the fortress, imagining that they were being undermined, surrendered.[3]

Pelopidas, the Theban, on one occasion planned to make a simultaneous attack on two towns of the Magnetes, not very far distant from each other. As he advanced against one of these towns, he gave orders that, in accordance with preconcerted arrangements, four horsemen should come from the other camp with garlands on their heads and with the marked eagerness of those who announce a victory. To complete the illusion, he arranged to have a forest between the two cities set on fire, to give the appearance of a burning town. Besides this, he ordered certain prisoners to be led along, dressed in

vorum habitu eodem iussit perduci. Qua adseveratione perterriti qui obsidebantur, dum in parte iam se superatos existimant, defecerunt.

3. Cyrus Persarum rex, incluso Sardibus Croeso, qua praeruptus mons nullum aditum praestabat, ad moenia malos exaequantis altitudinem iugi subrexit, quibus simulacra hominum armata Persici habitus imposuerat, noctuque eos monti admovit. Tum prima luce ex altera parte muros adgressus est. Ubi orto sole simulacra illa armatorum referentia habitum refulserunt, oppidani captam urbem a tergo credentes et ob hoc in fugam dilapsi victoriam hostibus concesserunt.

VIIII. De Inruptione ex Diversa Parte quam Exspectabimur

1. Scipio apud Carthaginem sub decessum[1] aestus maritimi, secutus deum ut dicebat ducem, ad muros urbis accessit et cedente stagno, qua non exspectabatur, inrupit.

2. Fabius Maximus Cunctatoris filius apud Arpos praesidio Hannibalis occupatos, considerato situ urbis, sescentos milites obscura nocte misit, qui per munitam eoque minus frequentem oppidi partem scalis evecti in murum portas revellerent. Hi adiuti decidentium aquarum sono, qui operis strepitum obscura-

[1] discessum *MSS., edd.; Gund. suggests the above reading.*

[1] 369–364 b.c. *Cf.* Polyaen. ii. iv. 1.
[2] 546 b c. *Cf.* Polyaen. vii. vi. 10; Ctes. *ed. Müller*, pp. 46–60. Herod. i. 84 makes no mention of this stratagem.
[3] 210 b.c. *Cf.* Livy xxvi. 45–46; Polyb. x. 10 ff.; Appian *Hisp.* 21 ff.

the costume of the townspeople. When the besieged had been terrified by these demonstrations, deeming themselves already defeated in one quarter, they ceased to offer resistance.[1]

Cyrus, king of the Persians, at one time forced Croesus to take refuge in Sardis. On one side a steep hill prevented access to the town. Here near the walls Cyrus erected masts equal to the height of the ridge of the hill, and on them placed dummies of armed men dressed in Persian uniforms. At night he brought these to the hill. Then at dawn he attacked the walls from the other side. As soon as the sun rose and the dummies, flashing in the sunlight, revealed the garb of warriors, the townspeople, imagining that their city had been captured from the rear, scattered in flight and left the field to the enemy.[2]

IX. On Attacks from an Unexpected Quarter

Scipio, when fighting before Carthage, approached the walls of the city, just before the turn of the tide, guided, as he said, by some god. Then, when the tide went out in the shallow lagoon, he burst in at that point, the enemy not expecting him there.[3]

Fabius Maximus, son of Fabius Cunctator, finding Arpi occupied by Hannibal's forces, first inspected the site of the town, and then sent six hundred soldiers on a dark night to mount the walls with scaling-ladders at a part of the town which was fortified and therefore less guarded, and to tear down the gates. These men were aided in the execution of their orders by the noise of the falling rain, which deadened the sound of their operations.

bat, iussa peragunt; ipse dato signo ab alia parte adgressus cepit Arpos.

3. C. Marius bello Iugurthino apud flumen Mulucham, cum oppugnaret castellum in monte saxeo situm, quod una et angusta semita adibatur, cetera parte velut consulto praecipiti, nuntiato sibi per Ligurem quendam ex auxiliis gregalem militem, qui forte aquatum progressus, dum per saxa montis cocleas legit, ad summa pervenerat, erepi posse in castellum, paucos centuriones cum velocissimis militibus, quibus perfectissimos aeneatores immiscuerat, misit capite pedibusque nudis, ut prospectus nisusque per saxa facilior foret, scutis gladiisque tergo aptatis. Hi Ligure ducente loris et clavis quibus in ascensu nitebantur adiuti, cum ad posteriora et ob id vacua defensoribus castelli pervenissent, concinere et tumultuari, ut praeceptum erat, coeperunt. Ad quod constitutum Marius constantius adhortatus suos acrius instare castellanis coepit, quos ab imbelli multitudine suorum revocatos, tamquam a tergo capti essent, insecutus castellum cepit.

4. L. Cornelius Rufinus consul complura Sardiniae cepit oppida, dum firmissimas partes copiarum noctu

---

[1] 213 B.C. *Cf.* Livy xxiv. 46-47.
[2] 107 B.C. *Cf.* Sall. *Jug.* 92-94; Flor. III. i. 14.

In another quarter, Fabius himself made an attack at a given signal and captured Arpi.[1]

In the Jugurthine War Gaius Marius was at one time besieging a fortress situated near the Mulucha River. It stood on a rocky eminence, accessible on one side by a single narrow path, while the other side, as though by special design, was precipitous. It happened that a certain Ligurian, a common soldier from among the auxiliaries, had gone out to procure water, and, while gathering snails among the rocks of the mountain, had reached the summit. This man reported to Marius that it was possible to clamber up to the stronghold. Marius accordingly sent a few centurions in company with his fleetest soldiers, including also the most skilful trumpeters. These men went bare-headed and bare-footed, that they might see better and make their way more easily over the rocks; their shields and swords were fastened to their backs. Guided by the Ligurian, and aided by straps and staffs, with which they supported themselves, they made their way up to the rear of the fortress, which, owing to its position, was without defenders, and then began to sound their trumpets and make a great uproar, as they had previously been directed. At this signal, Marius, steadfastly urging on his men, began to advance with renewed fury against the defenders of the fortress. The latter were recalled from the defence by the populace, who had lost heart under the impression that the town had been captured from the rear, so that Marius was enabled to press on and capture the fort.[2]

The consul Lucius Cornelius Rufinus captured numerous towns in Sardinia by landing powerful

exponit, quibus praecipiebat, delitiscerent opperirenturque tempus, quo ipse naves appelleret. Occurrentibus deinde adventanti hostibus et ab ipso per simulationem fugae longius ad persequendum avocatis, illi in relictas ab his urbes impetum fecerunt.

5. Pericles dux Atheniensium, cum oppugnaret quandam civitatem magno consensu defendentium tutam, nocte ab ea parte murorum, quae mari adiacebat, classicum cani clamoremque tolli iussit. Hostes penetratum illic in oppidum rati reliquerunt portas, per quas Pericles destitutas praesidio inrupit.

6. Alcibiades dux Atheniensium Cyzicum oppugnandae eius causa nocte improvisus accessit et ex diversa parte moenium cornicines canere iussit. Sufficere propugnatores murorum poterant. Ad id latus, a quo solo se temptari putabant, cum confluerent, qua non obsistebatur, muros transcendit.

7. Thrasybulus dux Milesiorum, ut portum Sicyoniorum occuparet, a terra subinde oppidanos temptavit et illo, quo lacessebantur, conversis hostibus classe inexspectata portum cepit.

8. Philippus in obsidione cuiusdam maritimae urbis binas naves procul a conspectu contabulavit superstruxitque eis turres; aliis deinde turribus adortus

---

[1] Probably a confusion of names: Lucius Cornelius Scipio invaded Sardinia in 259 B.C. (*cf* III. x 2). Publius Cornelius Rufinus was consul in 277, but waged war only in Italy.

[2] According to Thuc. viii. 107 and Diodor. XIII. xl. 6, Cyzicus was not fortified by walls. This stratagem belongs rather to the taking of Byzantium, 409 B.C. *Cf.* III. xi. 3; Diodor. xiii. 66–67; Plut. *Alcib.* 31.   [3] About 600 B.C.

detachments of troops at night, with instructions to remain in hiding and to wait till he himself drew near land with his ships. Then as the enemy came to meet him at his approach, he led them a long chase by pretending to flee, while his other troops attacked the cities thus abandoned by their inhabitants.[1]

Pericles, the Athenian general, was once besieging a city which was protected by very determined defenders. At night he ordered the trumpet to be sounded and a loud outcry to be raised at a quarter of the walls adjacent to the sea. The enemy, thinking that the town had been entered at that point, abandoned the gates, whereupon, as soon as these were left without defence, Pericles burst into the town.

Alcibiades, the Athenian general, planning to assault Cyzicus, approached the town unexpectedly at night, and commanded his trumpeters to sound their instruments at a different part of the fortifications. The defenders of the walls were ample, but since they all flocked to the side where alone they imagined themselves to be attacked, Alcibiades succeeded in scaling the walls at the point where there was no resistance.[2]

Thrasybulus, general of the Milesians, in his efforts to seize the harbour of the Sicyonians, made repeated attacks upon the inhabitants from the land side. Then, when the enemy directed their attention to the point where they were attacked, he suddenly seized the harbour with his fleet.[3]

Philip, while besieging a certain coast town, secretly lashed ships together in pairs, with a common deck over all, and erected towers on them. Then launching an attack with other towers by land,

SEXTUS JULIUS FRONTINUS

a terra, dum urbis propugnatores distringit, turritas naves a mari applicuit et, qua non resistebatur, subiit muros.

9. Pericles Peloponnesiorum castellum oppugnaturus, in quod duo omnino erant accessus, alterum fossa interclusit, alterum munire instituit. Castellani securiores ab altera parte facti eam solam, quam muniri videbant, custodire coeperunt. Pericles praeparatis pontibus iniectisque super fossam, qua non cavebatur, subiit castellum.

10. Antiochus adversus Ephesios Rhodiis, quos in auxilio habebat, praecepit, ut nocte portum cum magno strepitu invaderent; ad quam partem omni multitudine cum tumultu decurrente, nudatis defensore reliquis munitionibus, ipse a diverso adgressus civitatem cepit.

X. De Disponendis Insidiis, in quas Eliciantur Obsessi

1. Cato in conspectu Lacetanorum, quos obsidebat, reliquis suorum summotis, Suessetanos quosdam ex auxiliaribus maxime imbelles adgredi moenia iussit; hos cum facta eruptione Lacetani facile avertissent et fugientes avide insecuti essent, illis quos occultaverat coortis oppidum cepit.

2. L. Scipio in Sardinia, cuiusdam civitatis propugnatores ut eliceret, cum parte militum, relicta

---

[1] 430 b.c.   *Cf.* i. v. 10 and note.
[2] 195 b.c.   *Cf.* Livy xxxiv. 20.

he distracted the attention of the defenders of the city, till he brought up by sea the ships provided with towers, and advanced against the walls at the point were no resistance was offered.

Pericles, when about to lay siege to a fortress of the Peloponnesians to which there were only two avenues of approach, cut off one of these by a trench and began to fortify the other. The defenders of the fortress, thrown off their guard at one point, began to watch only the other where they saw the building going on. But Pericles, having prepared bridges, laid them across the trench and entered the fortress at the point where no guard was kept.[1]

Antiochus, when fighting against the Ephesians, directed the Rhodians, whom he had as allies, to make an attack on the harbour at night with a great uproar. When the entire population rushed headlong to this quarter, leaving the rest of the fortress without defenders, Antiochus attacked at a different quarter and captured the town.

X. On Setting Traps to Draw out the Besieged

When Cato was besieging the Lacetani, he sent away in full view of the enemy all his other troops, while ordering certain Suessetani, who were the least martial of all his allies, to attack the walls of the town. When the Lacetani, making a sortie, easily repulsed these forces and pursued them eagerly as they fled, the soldiers whom Cato had placed in hiding rose up and by their help he captured the town.[2]

When campaigning in Sardinia, Lucius Scipio, in order to draw out the defenders of a certain city, abandoned the siege which he had begun, and pre-

SEXTUS JULIUS FRONTINUS

oppugnatione quam instruxerat, speciem fugientis praestitit; insecutisque temere oppidanis per eos, quos in proximo occultaverat, oppidum invasit.

3. Hannibal, cum obsideret civitatem Himeram, castra sua capi de industria passus est, iussis recedere Poenis, tamquam praevaleret hostis; quo eventu Himeraeis ita deceptis, ut gaudio impulsi relicta urbe procurrerent ad Punicum vallum, Hannibal vacuam urbem per eos, quos in insidiis ad hanc ipsam occasionem posuerat, cepit.

4. Idem, ut Saguntinos eliceret, rara acie ad muros accedens ad primam eruptionem oppidanorum simulata fuga cessit interpositoque exercitu ab oppido interclusos a suis hostes in medio trucidavit.

5. Himilco Carthaginiensis apud Agrigentum iuxta oppidum partem copiarum in insidiis posuit praecepitque his, ut, cum processissent oppidani, ligna umida incenderent. Deinde cum reliqua parte exercitus luce ad eliciendos hostes progressus simulata fuga persequentis oppidanos longius cedendo protraxit. Insidiatores prope moenia imperatum ignem acervis subiecerunt; unde obortum contemplati fumum Agrigentini incensam civitatem suam existimaverunt, defendendaeque eius gratia dum trepide

---

[1] 259 B.C. *Cf.* III. ix. 4 and note.
[2] 409 B.C. The Hannibal here mentioned is the son of Gisgo. Diodor. xiii. 59–62 represents the Carthaginians as withdrawing in flight rather than executing a stratagem.
[3] 219 B.C. Son of Hamilcar Barca. The identity of names led to the confusion between these two generals.

tended to flee with a detachment of his troops. Then, when the inhabitants followed him pell-mell, he attacked the town with the help of those whom he had placed in hiding near at hand.[1]

When Hannibal was besieging the city of Himera, he purposely allowed his camp to be captured, ordering the Carthaginians to retire, on the ground that the enemy were superior. The inhabitants were so deceived by this turn of affairs that in their joy they came out of the city and advanced against the Carthaginian breast-works, whereupon Hannibal, finding the town vacant, captured it by means of the troops whom he had placed in ambush for this very contingency.[2]

In order to draw out the Saguntines, Hannibal on a certain occasion advanced against their walls with a thin line of troops. Then, at the first sally of the inhabitants, feigning flight, he withdrew, and interposing troops between the pursuing foe and the city, he slaughtered the enemy thus cut off from their fellows between the two forces.[3]

Himilco, the Carthaginian, when campaigning near Agrigentum, placed part of his forces in ambush near the town, and directed them to set fire to some damp wood as soon as the soldiers from the town should come forth. Then, advancing at daybreak with the rest of his army for the purpose of luring forth the enemy, he feigned flight and drew the inhabitants after him for a considerable distance by his retirement. The men in ambush near the walls applied the torch to the wood-piles as directed. The Agrigentines, beholding the smoke ascend, thought their city on fire and ran back in alarm to protect it.

SEXTUS JULIUS FRONTINUS

recurrunt, obviis eis qui insidiati iuxta muros erant et a tergo instantibus quos persecuti fuerant, in medio trucidati sunt.

6. Viriathus disposito per occulta milite paucos misit, qui abigerent pecora Segobrigensium; ad quae illi vindicanda cum frequentes procurrissent simulantesque fugam praedatores persequerentur, deducti in insidias caesique sunt.

7. Scordisci equites, cum Heracleae duarum cohortium[1] praesidio praepositus esset Lucullus, pecora abigere simulantes provocaverunt eruptionem; fugam deinde mentiti sequentem Lucullum in insidias deduxerunt et octingentos cum eo milites occiderunt.

8. Chares dux Atheniensium civitatem adgressurus litori appositam, post quaedam promunturia occulte habita classe, e navibus velocissimam praeter hostilia praesidia ire iussit; qua visa cum omnia navigia, quae pro custodia portus agebant, ad persequendam evolassent, Chares indefensum portum cum reliqua classe invectus etiam civitatem occupavit.

9. Barca dux Poenorum in Sicilia Lilybaeum nostris terra marique obsidentibus partem classis suae procul armatam iussit ostendi; ad eius conspectum cum evolassent nostri, ipse reliquis quas in occulto tenuerat navibus Lilybaei portum occupavit.

[1] cohortium, *Freinsheim;* partium *MSS. and edd.*

---

[1] 406 B.C. *Cf.* Polyaen. v. x. 4. In the Old Testament (Josh. viii.), a similar stratagem is employed by Joshua.
[2] 147-139 B.C.　　　　[3] 366-336 B.C.

Being encountered by those lying in wait for them near the walls, and beset in the rear by those whom they had just been pursuing, they were caught between two forces and so cut to pieces.[1]

Viriathus, on one occasion, having placed men in ambush, sent a few others to drive off the flocks of the Segobrigenses. When the latter rushed out in great numbers to defend their flocks and followed up the marauders, who pretended to flee, they were drawn into an ambush and cut to pieces.[2]

When Lucullus was put in charge of a garrison of two cohorts at Heraclea, the cavalry of the Scordisci, by pretending to drive off the flocks of the inhabitants, provoked a sortie. Then, when Lucullus followed, they drew him into an ambush, feigning flight, and killed him together with eight hundred of his followers.

The Athenian general, Chares, when about to attack a city on the coast, hid his fleet behind certain promontories and then ordered his swiftest ship to sail past the forces of the enemy. At sight of this ship, all the forces guarding the harbour darted out in pursuit, whereat Chares sailed in with the rest of his fleet and took possession of the undefended harbour and likewise of the city itself.[3]

On one occasion when Roman troops were blockading Lilybaeum by land and sea, Barca, general of the Carthaginians in Sicily, ordered a part of his fleet to appear in the offing ready for action. When our men darted out at sight of this, Barca seized the harbour of Lilybaeum with the ships which he had held in hiding.[4]

[4] 249 B.C. *Cf.* Polyb. i. 44, where Hannibal, rather than Barca, is the general.

SEXTUS JULIUS FRONTINUS

XI. De Simulatione Regressus

1. Phormion dux Atheniensium, cum depopulatus esset agros Chalcidensium, legatis eorum de ea re querentibus benigne respondit et nocte, qua dimissurus illos erat, finxit litteras sibi supervenisse civium suorum, propter quas redeundum haberet; ac paulum regressus dimisit legatos. His omnia tuta et abisse Phormionem renuntiantibus, Chalcidenses spe et oblatae humanitatis et abducti exercitus remissa urbis custodia, cum confestim Phormion revertisset, prohibere inexspectatam vim non potuerunt.

2. Agesilaus dux Lacedaemoniorum, cum Phocenses obsideret et intellexisset eos, qui tunc praesidio illis erant, iam gravari belli incommoda, paulum regressus tamquam ad alios actus liberam recedendi occasionem his dedit; non multo post milite reducto destitutos Phocenses superavit.

3. Alcibiades adversus Byzantios, qui se moenibus continebant, insidias disposuit et simulato regressu incautos eos oppressit.

4. Viriathus, cum tridui iter discedens confecisset, idem illud uno die remensus securos Segobrigenses et sacrificio cum maxime occupatos oppressit.

5. Epaminondas Mantineae, cum Lacedaemonios in subsidium hosti venisse animadverteret, ratus posse

---

[1] 432 B.C. *Cf.* Polyaen. III. iv. 1.
[2] 396–394 B.C. *Cf.* Polyaen. II. i. 16.
[3] 409 B.C. *Cf.* Diodor. xiii. 66–67; Plut. *Alcib.* 31.
[4] 147–139 B.C.

XI. On Pretended Retirements

When the Athenian general Phormio had ravaged the lands of the Chalcidians, and their envoys complained of this action, he answered them graciously, and at evening, when he was about to dismiss them, pretended that a letter had come from his fellow-citizens requiring his return. Accordingly he retired a short distance and dismissed the envoys. When these reported that all was safe and that Phormio had withdrawn, the Chalcidians in view of the promised consideration and of the withdrawal of the troops, relaxed the guard of their town. Then Phormio suddenly returned and the Chalcidians were unable to withstand his unexpected attack.[1]

When the Spartan commander, Agesilaus, was blockading the Phocaeans and had learned that those who were then lending them support were weary with the burdens of war, he retired a short distance as though for other objects, thus leaving the allies free opportunity to withdraw. Not long after, bringing back his troops, he defeated the Phocaeans thus left without assistance.[2]

When fighting against the Byzantines, who kept within their walls, Alcibiades laid an ambush and, feigning a retirement, took them off their guard and crushed them.[3]

Viriathus, after retreating for three days, suddenly turned round and traversed the same distance in one day. He thus crushed the Segobrigenses, taking them off their guard at a moment when they were earnestly engaged in sacrifice.[4]

In the operations around Mantinea, Epaminondas, having noticed that the Spartans had come to help

SEXTUS JULIUS FRONTINUS

Lacedaemonem occupari, si clam illo profectus esset, nocte crebros ignes fieri iussit, ut specie remanendi occultaret profectionem. Sed a transfuga proditus, adsecuto exercitu Lacedaemoniorum, itinere quidem quo Spartam petebat destitit, idem tamen consilium convertit ad Mantinienses; aeque enim ignibus factis Lacedaemonios, quasi maneret, frustratus per quadraginta milia passuum revertit Mantineam eamque auxilio destitutam occupavit.

Ex Contrario circa Tutelam Obsessorum

XII. De Excitanda Cura Suorum

1. Alcibiades dux Atheniensium, civitate sua a Lacedaemoniis obsessa, veritus neglegentiam vigilum denuntiavit his qui in stationibus erant, observarent lumen, quod nocte ostenturus esset ex arce, et ad conspectum eius ipsi quoque lumina attollerent; in quo munere qui cessasset, poenam passurum. Dum sollicite exspectatur signum ducis, pervigilatum ab omnibus et suspectae noctis periculum evitatum est.

2. Iphicrates dux Atheniensium, cum praesidio Corinthum teneret et sub adventum hostium ipse vigilias circumiret, vigilem, quem dormientem invenerat, transfixit cuspide; quod factum quibusdam

---

[1] 362 B.C. *Cf.* Polyb. ix. 8; Diodor. xv. 82–84; Plut. *Ages.* 34.
[2] *Cf.* Polyaen. I. xl. 3.

his enemies, conceived the idea that Sparta might be captured, if he should set out against it secretly. Accordingly he ordered numerous watch-fires to be built at night, that, by appearing to remain, he might conceal his departure. But betrayed by a deserter and pursued by the Lacedaemonian troops, he abandoned his march to Sparta, and employed the same scheme against the Mantineans; for by building watch-fires as before, he deceived the Spartans into thinking that he would remain. Meanwhile, returning to Mantinea by a march of forty miles, he found it without defences and captured it.[1]

On the other Hand, Stratagems Connected with the Protection of the Besieged

XII. On Stimulating the Vigilance of One's Own Troops

Alcibiades, the Athenian commander, when his own city was blockaded by the Spartans, fearing negligence on the part of the guards, ordered the men on picket-duty to watch for the light which he should exhibit from the citadel at night, and to raise their own lights at sight of it, threatening that whoever failed in this duty should suffer a penalty. While anxiously awaiting the signal of their general, all maintained constant watch, and so escaped the dangers of the perilous night.[2]

When Iphicrates, the Athenian general, was holding Corinth with a garrison and on one occasion personally made the rounds of the sentries as the enemy were approaching, he found one of the guards asleep at his post and stabbed him with his spear.

SEXTUS JULIUS FRONTINUS

tamquam saevum increpantibus "qualem inveni," inquit, "talem reliqui."

3. Epaminondas Thebanus idem fecisse dicitur.[1]

XIII. De Emittendo et Recipiendo Nuntio

1. Romani, obsessi in Capitolio, ad Camillum auxilio[2] implorandum miserunt Pontium Cominium, qui, ut stationes Gallorum falleret, per saxa Tarpeia demissus tranato Tiberi Veios pervenit et perpetrata legatione similiter ad suos rediit.

2. Campani, diligenter Romanis a quibus obsessi erant custodias agentibus, quendam pro transfuga subornatum miserunt, qui occultatam balteo epistulam inventa effugiendi occasione ad Poenos pertulit.

3. Venationi quoque et pecoribus quidam insuerunt litteras membranis mandatas.[3]

4. Aliqui et iumento in aversam partem infulserunt, dum stationes transeunt.[3]

5. Nonnulli interiora vaginarum inscripserunt.[3]

6. L. Lucullus, Cyzicenos obsessos a Mithridate ut certiores adventus sui faceret, cum praesidiis hostium teneretur introitus urbis, qui unus et angustus ponte modico insulam continenti iungit, militem e suis nandi et nauticae artis peritum iussit insidentem duobus inflatis utribus litteras insutas habentibus,

---

[1] *No. 3 is apparently interpolated.*
[2] *The text is uncertain,* auxilio *Gund*; ab exilio *HP*; ad auxilium *d*; ad Camilli auxilium ab exsilio *Ded.*
[3] *Nos. 3, 4, and 5 are thought to be interpolations.*

---

[1] 393–391 B.C. *Cf.* Nep. *Iphic.* ii. 1–2.
[2] 390 B.C. *Cf.* Livy v. 46; Diodor. xiv. 116; Plut. *Camil.* 25.

STRATAGEMS, III. xii. 2–xiii. 6

When certain ones rebuked this procedure as cruel, he answered: "I left him as I found him."[1]

Epaminondas the Theban is said, on one occasion, to have done the same thing.

XIII. On Sending and Receiving Messages

When the Romans were besieged in the Capitol, they sent Pontius Cominius to implore Camillus to come to their aid. Pontius, to elude the pickets of the Gauls, let himself down over the Tarpeian Rock, swam the Tiber, and reached Veii. Having accomplished his errand, he returned by the same route to his friends.[2]

When the Romans were maintaining careful guard against the inhabitants of Capua, whom they were besieging, the latter sent a certain fellow in the guise of a deserter, and he, finding an opportunity to escape, conveyed to the Carthaginians a letter which he had secreted in his belt.[3]

Some have written messages on skins and then sewed these to the carcasses of game or sheep.

Some have stuffed the message under the tail of a mule while passing the picket-posts.

Some have written on the linings of scabbards.

When the Cyzicenes were besieged by Mithridates, Lucius Lucullus wished to inform them of his approach. There was a single narrow entrance to the city, connecting the island with the mainland by a small bridge. Since this was held by forces of the enemy, he sewed some letters up inside two inflated skins and then ordered one of his soldiers, an adept

---

[3] 211 B.C. Livy xxvi. 7 represents Hannibal as sending the letter to the Capuans.

SEXTUS JULIUS FRONTINUS

quos ab inferiore parte duabus regulis inter se distantibus commiserat, ire septem milia passuum traiectum. Quod ita perite gregalis fecit, ut cruribus velut gubernaculis dimissis cursum dirigeret et procul visentis, qui in statione erant, marinae specie beluae deciperet.

7. Hirtius consul ad Decimum Brutum, qui Mutinae ab Antonio obsidebatur, litteras subinde misit plumbo scriptas, quibus ad brachium religatis milites Scultennam amnem tranabant.

8. Idem columbis, quas inclusas ante tenebris et fame adfecerat, epistulas saeta ad collum religabat easque a propinquo, in quantum poterat, moenibus loco emittebat. Illae lucis cibique avidae altissima aedificiorum petentes excipiebantur a Bruto, qui eo modo de omnibus rebus certior fiebat, utique postquam disposito quibusdam locis cibo columbas illuc devolare instituerat.

XIIII. DE INTRODUCENDIS AUXILIIS ET COMMEATIBUS SUGGERENDIS

1. BELLO civili, cum Ategua urbs in Hispania Pompeianarum partium obsideretur, Maurus inter noctem tamquam Caesarianus tribuni cornicularius vigiles quosdam excitavit; ex quibus cum tesseram accepisset, alios excitans constantia fallaciae suae per medias Caesaris copias praesidium Pompei transduxit.

---

[1] 74 B.C. *Cf.* Flor. III. v. 15–16; Oros. VI. ii. 14.
[2] 43 B.C. *Cf.* Dio xlvi. 36.
[3] 43 B.C. *Cf.* Plin. *N.H.* x. 37.
[4] 45 B.C. *Cf.* Dio xliii. 33–34. According to Dio, this man, Munatius Flaccus, whose real mission is to aid the Ateguans in withstanding the blockade of Caesar's troops, represents to the sentries that he has been sent by Caesar

in swimming and boating, to mount the skins, which he had fastened together at the bottom by two strips some distance apart, and to make the trip of seven miles across. So skilfully did the soldier do this that, by spreading his legs, he steered his course as though by rudder, and deceived those watching from a distance by appearing to be some marine creature.[1]

The consul Hirtius often sent letters inscribed on lead plates to Decimus Brutus, who was besieged by Antonius at Mutina. The letters were fastened to the arms of soldiers, who then swam across the Scultenna River.[2]

Hirtius also shut up pigeons in the dark, starved them, fastened letters to their necks by a hair, and then released them as near to the city walls as he could. The birds, eager for light and food, sought the highest buildings and were received by Brutus, who in that way was informed of everything, especially after he set food in certain spots and taught the pigeons to alight there.[3]

XIV. On Introducing Reinforcements and Supplying Provisions

In the Civil War, when the Spanish city of Ategua, belonging to Pompey's party, was under blockade, one night a Moor, pretending to be a tribune's adjutant belonging to the Caesarian party, roused certain sentries and got from them the password. He then roused others, and by continuing his deception, succeeded in conducting reinforcements for Pompey through the midst of Caesar's troops.[4]

to betray the city. Thus having once learned the password, he secures an easy entrance into the city.

SEXTUS JULIUS FRONTINUS

2. Hannibale obsidente Casilinum, Romani far in doliis secunda aqua Volturni fluminis demittebant, ut ab obsessis exciperetur. Quibus cum obiecta per medium amnem catena Hannibal obstitisset, nuces sparsere. Quae cum aqua ferente ad oppidum defluerent, eo commeatu sociorum necessitatem sustentaverunt.

3. Hirtius Mutinensibus obsessis ab Antonio salem, quo maxime indigebant, cupis conditum per amnem Scultennam intromisit.

4. Idem pecora secunda aqua demisit, quae excepta sustentaverunt necessariorum inopiam.

XV. Quemadmodum Efficiatur ut Abundare Videantur quae Deerunt

1. Romani, cum a Gallis Capitolium obsideretur, in extrema iam fame panem in hostem iactaverunt consecutique, ut abundare commeatibus viderentur, obsidionem, donec Camillus subveniret, toleraverunt.

2. Athenienses adversus Lacedaemonios idem fecisse dicuntur.[1]

3. Hi, qui ab Hannibale Casilini obsidebantur, ad extremam famem pervenisse crediti, cum etiam herbas alimentis eorum Hannibal, arato loco qui erat inter castra ipsius et moenia, praeriperet, semina in praeparatum locum iecerunt, consecuti, ut habere viderentur, quo victum sustentarent usque ad satorum proventum.

---

[1] *No. 2 is thought to be interpolated.*

---

[1] 216 B.C. *Cf.* Livy xxiii. 19.   [2] 43 B.C.
[3] 390 B.C. *Cf.* Livy v. 48; Val. Max. VII. iv. 3; Ovid *Fast.* vi. 350 ff.   [4] 216 B.C. *Cf.* Livy xxiii. 19.

STRATAGEMS, III. xiv. 2–xv. 3

When Hannibal was besieging Casilinum, the Romans sent big jars of wheat down the current of the Volturnus, to be picked up by the besieged. After Hannibal stopped these by throwing a chain across the river, the Romans scattered nuts on the water. These floated down stream to the city and thus sustained the necessities of the allies.[1]

When the inhabitants of Mutina were blockaded by Antonius, and were greatly in need of salt, Hirtius packed some in jars and sent it in to them by way of the Scultenna River.[2]

Hirtius also sent down the river carcasses of sheep, which were received and thus furnished the necessities of life.

XV. How to Produce the Impression of Abundance of what is Lacking

When the Capitol was besieged by the Gauls, the Romans, in the extremity of famine, threw bread among the enemy. They thus produced the impression that they were well supplied with food, and so withstood the siege till Camillus came.[3]

The Athenians are said to have employed the same ruse against the Spartans.

The inhabitants of Casilinum, when blockaded by Hannibal, were thought to have reached the starvation point, since Hannibal had cut off from their food supply even their use of the growing herbs by ploughing the ground between his camp and the city walls. The ground being thus made ready, the beseiged flung seed into it, thus giving the impression that they had enough wherewith to sustain life even till harvest time.[4]

SEXTUS JULIUS FRONTINUS

4. Reliqui ex Variana clade, cum obsiderentur, quia defici frumento videbantur, horrea tota nocte circumduxerunt captivos, deinde praecisis manibus dimiserunt; hi circumsedentibus suis persuaserunt, ne spem maturae expugnationis reponerent in fame Romanorum, quibus ingens alimentorum copia superesset.

5. Thraces in arduo monte obsessi, in quem hostibus accessus non erat, conlato viritim exiguo tritico paucas oves paverunt et egerunt in hostium praesidia; quibus exceptis et occisis, cum frumenti vestigia in visceribus earum apparuissent, opinatus hostis magnam vim tritici superesse eis, qui inde etiam pecora pascerent, recessit ab obsidione.

6. Thrasybulus dux Milesiorum, cum longa obsidione milites sui angerentur ab Alyatte, qui sperabat eos ad deditionem fame posse compelli, sub adventum legatorum Alyattis frumentum omne in forum compelli iussit et conviviis sub id tempus institutis per totam urbem epulas praestitit; atque ita persuasit hosti superesse ipsis copias, quibus diuturnam sustinerent obsidionem.

XVI. Qua Ratione Proditoribus et Transfugis Occurratur

1. Claudius Marcellus, cognito consilio L. Bantii Nolani, qui corrumpere ad defectionem populares

---

[1] *i. e.* the defeat of Varus by Arminius in the Teutoburg Forest in 9 A.D

[2] About 611 B.C. *Cf.* Herod. i. 21–22; Polyaen. vi. 47.

STRATAGEMS, III. xv. 4–xvi. 1

When the survivors of the Varian disaster[1] were under siege and seemed to be running short of food, they spent an entire night in leading prisoners round their store-houses; then, having cut off their hands, they turned them loose. These men persuaded the besieging force to cherish no hope of an early reduction of the Romans by starvation, since they had an abundance of food supplies.

When the Thracians were besieged on a steep mountain inaccessible to the enemy, they got together by individual contributions a small amount of wheat. This they fed to a few sheep which they then drove among the forces of the enemy. When the sheep had been caught and slaughtered, and traces of wheat had been found in their intestines, the enemy raised the siege, imagining that the Thracians had a surplus of wheat, inasmuch as they fed it even to their sheep.

The Milesians were at one time suffering a long siege at the hands of Alyattes, who hoped they could be starved into surrender. But the Milesian commander, Thrasybulus, in anticipation of the arrival of envoys from Alyattes, ordered all the grain to be brought together into the market-place, arranged for banquets to be held on that occasion, and provided sumptuous feasts throughout the city. Thus he convinced the enemy that the Milesians had abundance of provisions with which to sustain a long siege.[2]

XVI. How to Meet the Menace of Treason and Desertion

A certain Lucius Bantius of Nola on one occasion cherished the plan of rousing his fellow-citizens to

SEXTUS JULIUS FRONTINUS

studebat et Hannibali gratificabatur, quod illius beneficio curatus inter Cannenses saucius et ex captivitate remissus ad suos erat, quia interficere eum, ne supplicio eius reliquos concitaret Nolanos, non audebat, arcessitum ad se allocutus est, dicens fortissimum militem eum esse, quod antea ignorasset, hortatusque est, ut secum moraretur, et super verborum honorem equo quoque donavit. Qua benignitate non illius tantum fidem, sed etiam popularium, quae ex illo pendebat, sibi obligavit.

2. Hamilcar dux Poenorum, cum frequenter auxiliares Galli ad Romanos transirent et iam ex consuetudine ut socii exciperentur, fidissimos subornavit ad simulandam transitionem, qui Romanos excipiendorum causa eorum progressos ceciderunt. Quae sollertia Hamilcari non tantum ad praesentem profuit successum, sed in posterum praestitit, ut Romanis veri quoque transfugae forent suspecti.

3. Hanno Carthaginiensium imperator in Sicilia, cum comperisset Gallorum mercennariorum circiter quattuor milia conspirasse ad transfugiendum ad Romanos, quod aliquot mensum mercedes non receperant, animadvertere autem in eos non auderet metu seditionis, promisit prolationis iniuriam liberalitate pensaturum. Quo nomine gratias agentibus Gallis per tempus idoneum praedaturos pollicitus fidelissimum dispensatorem ad Otacilium

---

[1] 216 B.C. *Cf.* Livy xxiii. 15–16; Plut. *Marcel.* 10–11. Variations of this stratagem and its author are found in IV. vii. 36, Plut. *Fab.* 20 and Val. Max. VII. iii. 7.
[2] 260–241 B.C.

revolt, as a favour to Hannibal, by whose kindness he had been tended when wounded among those engaged at Cannae, and by whom he had been sent back from captivity to his own people. Claudius Marcellus, learning of his purpose and not daring to put him to death, for fear that by his punishment he would stir up the rest of the people of Nola, summoned Bantius and talked with him, pronouncing him a very valiant soldier (a fact which Marcellus admitted he had not previously known), and urging him to remain with him. Besides these compliments, he presented him also with a horse. By such kindness he secured the loyalty, not only of Bantius, but also of his townspeople, since their allegiance hinged on his.[1]

When the Gallic auxiliaries of Hamilcar, the Carthaginian general, were in the habit of crossing over to the Romans and were regularly received by them as allies, Hamilcar engaged his most loyal men to pretend desertion, while actually they slew the Romans who came out to welcome them. This device was not merely of present aid to Hamilcar, but caused real deserters to be regarded in future as objects of suspicion in the eyes of the Romans.[2]

Hanno, commander of the Carthaginians in Sicily, learned on one occasion that about four thousand Gallic mercenaries had conspired to desert to the Romans, because for several months they had received no pay. Not daring to punish them, for fear of mutiny, he promised to make good the deferred payment by increasing their wages. When the Gauls rendered thanks for this, Hanno, promising that they should be permitted to go out foraging at a suitable time, sent to the consul Otacilius an

SEXTUS JULIUS FRONTINUS

consulem misit, qui, tamquam rationibus interversis transfugisset, nuntiavit nocte proxima Gallorum quattuor milia, quae praedatum forent missa, posse excipi. Otacilius nec statim credidit transfugae nec tamen rem spernendam ratus disposuit in insidiis lectissimam manum suorum. Ab ea Galli excepti dupliciter Hannonis consilio satisfecerunt; et Romanos ceciderunt et ipsi omnes interfecti sunt.

4. Hannibal simili consilio se a transfugis ultus est. Nam cum aliquos ex militibus suis sciret transisse proxima nocte nec ignoraret exploratores hostium in castris suis esse, palam pronuntiavit non debere transfugas vocari sollertissimos milites, qui ipsius iussu exierint ad excipienda hostium consilia. Auditis quae pronuntiavit, rettulerunt exploratores ad suos. Tum comprehensi a Romanis transfugae et amputatis manibus remissi sunt.

5. Diodotus, cum praesidio Amphipolim tueretur et duo milia Thracum suspecta haberet, quae videbantur urbem direptura, mentitus est paucas hostium naves proximo litori applicuisse easque diripi posse; qua spe stimulatos Thracas emisit ac deinde clausis portis non recepit.

XVII. DE ERUPTIONIBUS

1. ROMANI, qui in praesidio Panormitanorum erant, veniente ad obsidionem Hasdrubale raros ex industria

---

[1] 261 B.C. *Cf.* Diodor. XXIII. viii. 3. Zonar viii. 10 attributes this stratagem to Hamilcar.

[2] 168 B.C. In Livy xliv. 44, the author of the stratagem is called Diodorus.

extremely trustworthy steward, who pretended to have deserted on account of embezzlement, and who reported that on the coming night four thousand Gauls, sent out on a foraging expedition, could be captured. Otacilius, not immediately crediting the deserter, nor yet thinking the matter ought to be treated with disdain, placed the pick of his men in ambush. These met the Gauls, who fulfilled Hanno's purpose in a twofold manner, since they not only slew a number of the Romans, but were themselves slaughtered to the last man.[1]

By a similar plan Hannibal took vengeance on certain deserters; for, being aware that some of his soldiers had deserted on the previous night, and knowing that spies of the enemy were in his camp, he publicly proclaimed that the name of "deserter" ought not to be applied to his cleverest soldiers, who at his order had gone out to learn the designs of the enemy. The spies, as soon as they heard this pronouncement, reported it to their own side. Thereupon the deserters were arrested by the Romans and sent back with their hands cut off.

When Diodotus was holding Amphipolis with a garrison, and entertained suspicions of two thousand Thracians, who seemed likely to pillage the city, he invented the story that a few hostile ships had put in at the shore near by and could be plundered. When he had incited the Thracians at that prospect, he let them out. Then, closing the gates, he refused to admit them again.[2]

XVII. On Sorties

When Hasdrubal came to besiege Panormus, the Romans, who were in possession of the town,

SEXTUS JULIUS FRONTINUS

in muris posuerunt defensores; quorum paucitate contempta cum incautus muris succederet Hasdrubal, eruptione facta ceciderunt eum.

2. Aemilius Paulus, universis Liguribus improviso adortis castra eius, simulato timore militem diu continuit; deinde fatigato iam hoste quattuor portis eruptione facta stravit cepitque Ligures.

3. Livius praefectus Romanorum arcem Tarentinorum tenens misit ad Hasdrubalem legatos, abire uti sibi incolumi liceret; ea simulatione ad securitatem perductum hostem eruptione facta cecidit.

4. Cn. Pompeius circumsessus ad Dyrrachium non tantum obsidione liberavit suos, verum etiam post eruptionem, quam opportuno et loco et tempore fecerat, Caesarem ad castellum, quod duplici munitione instructum erat, avide inrumpentem exterior ipse circumfusus corona obligavit, ut ille inter eos quos obsidebat et eos qui extra circumvenerant medius non leve periculum et detrimentum senserit.

5. Flavius Fimbria in Asia apud Rhyndacum adversum filium Mithridatis, brachiis ab latere ductis, deinde fossa in fronte percussa, quietum in vallo militem tenuit, donec hostilis equitatus intraret angustias munimentorum; tunc eruptione facta sex milia eorum cecidit.

---

[1] 251 B.C. *Cf.* Polyb. i. 40.
[2] 181 B.C. *Cf.* Livy xl. 25, 27–28.
[3] 212–209 B.C.
[4] 48 B.C. *Cf.* Caes. *B.C.* iii. 65–70.
[5] 85 B.C.

purposely placed a scanty number of defenders on the walls. In contempt of their small numbers, Hasdrubal incautiously approached the walls, whereupon they made a sortie and slew him.[1]

When the Ligurians with their entire force made a surprise attack on the camp of Aemilius Paulus, the latter feigned fear and for a long time kept his troops in camp. Then, when the enemy were exhausted, making a sortie by the four gates, he defeated the Ligurians and made them prisoners.[2]

Livius, commander of the Romans, when holding the citadel of the Tarentines, sent envoys to Hasdrubal, requesting the privilege of withdrawing undisturbed. When by this feint he had thrown the enemy off their guard, he made a sortie and cut them to pieces.[3]

Gnaeus Pompey, when besieged near Dyrrhachium, not only released his own men from blockade, but also made a sally at an opportune time and place; for just as Caesar was making a fierce assault on a fortified position surrounded by a double line of works, Pompey, by this sortie, so enveloped him with a cordon of troops that Caesar incurred no slight peril and loss, caught, as he was, between those whom he was besieging and those who had surrounded him from the outside.[4]

Flavius Fimbria, when fighting in Asia near the river Rhyndacus against the son of Mithridates, constructed two lines of works on his flanks and a ditch in front, and kept his soldiers quietly within their entrenchments, until the cavalry of the enemy passed within the confined portions of his fortifications. Then, making a sortie, he slew six thousand of them.[5]

SEXTUS JULIUS FRONTINUS

6. C. Caesar in Gallia, deletis ab Ambiorige Titurii Sabini et Cottae legatorum copiis, cum a Q. Cicerone, qui et ipse oppugnabatur, certior factus cum duabus legionibus adventaret, conversis hostibus metum simulavit militesque in castris, quae artiora solito industria fecerat, tenuit. Galli praesumpta iam victoria velut ad praedam castrorum tendentes fossas implere et vallum detrahere coeperunt; qua re proelio non aptatos Caesar emisso repente undique milite trucidavit.

7. Titurius Sabinus adversus Gallorum amplum exercitum continendo militem intra munimenta praestitit eis suspicionem metuentis. Cuius augendae causa perfugam misit, qui adfirmaret exercitum Romanum in desperatione esse ac de fuga cogitare. Barbari oblata victoriae spe concitati lignis sarmentisque se oneraverunt, quibus fossas complerent, ingentique cursu castra nostra in colle posita petiverunt. Unde in eos Titurius universas immisit copias multisque Gallorum caesis plurimos in deditionem accepit.

8. Asculani, oppugnaturo oppidum Pompeio cum paucos senes et aegros in muris ostendissent, ob id securos Romanos eruptione facta fugaverunt.

9. Numantini obsessi ne pro vallo quidem in-

---

[1] 54 B.C. *Cf.* Caes. *B.G.* v. 37–52; Dio xl. 10; Polyaen. VIII. xxiii. 7.
[2] 56 B.C. *Cf.* Caes. *B.G.* iii. 17–19.
[3] 90 B.C.

STRATAGEMS, III. xvii. 6-9

When the forces of Titurius Sabinus and Cotta, Caesar's lieutenants in Gaul, had been wiped out by Ambiorix, Caesar was urged by Quintus Cicero, who was himself also under siege, to come with two legions to his relief. The enemy then turned upon Caesar, who feigned fear and kept his troops within his camp, which he had purposely constructed on a smaller scale than usual. The Gauls, already counting on victory, and pressing forward as though to plunder the camp, began to fill up the ditches and to tear down the ramparts. Caesar, therefore, as the Gauls were not equipped for battle, suddenly sent forth his own troops from all quarters and cut the enemy to pieces.[1]

When Titurius Sabinus was fighting against a large force of Gauls, he kept his troops within their fortifications, and thus produced upon the Gauls the impression that he was afraid. To further this impression, he sent a deserter to state that the Roman army was in despair and was planning to flee. Spurred on by the hope of victory thus offered, the Gauls loaded themselves with wood and brush with which to fill the trenches, and at top speed started for our camp, which was pitched on the top of an elevation. From there Titurius launched all his forces against them, killing many of the Gauls and receiving large numbers in surrender.[2]

As Pompey was about to assault the town of Asculum, the inhabitants exhibited on the ramparts a few aged and feeble men. Having thus thrown the Romans off their guard, they made a sortie and put them to flight.[3]

When the Numantines were blockaded, they did not even draw up a line of battle in front of the

struxerunt aciem adeoque se continuerunt, ut Popilio Laenati fiducia fieret scalis oppidum adgrediendi; quo deinde suspicante insidias, quia ne tunc quidem obsistebatur, ac suos revocante, eruptione facta aversos et descendentis adorti sunt.

XVIII. De Constantia Obsessorum

1. Romani, adsidente moenibus Hannibale, ostentandae fiduciae gratia supplementum exercitibus, quos in Hispania habebant, diversa porta miserunt.

2. Idem agrum, in quo castra Hannibal habebat, defuncto forte domino venalem ad id pretium licendo perduxerunt, quo is ager ante bellum venierat.

3. Idem, cum ab Hannibale obsiderentur et ipsi obsiderent Capuam, decreverunt, ne nisi capta ea revocaretur inde exercitus.

---

[1] 138 b.c.
[2] 211 b.c. *Cf.* Livy xxvi. 11; Val. Max. iii. vii. 10.
[3] 211 b.c. *Cf.* Livy xxvi. i. 7–8.

entrenchments, but kept so closely within the town that Popilius Laenas was emboldened to attack it with scaling-ladders. But, suspecting a ruse, since not even then was resistance offered, he recalled his men; whereupon the Numantines made a sortie and attacked the Romans in the rear as they were climbing down.[1]

XVIII. Concerning Steadfastness on the Part of the Besieged

The Romans, when Hannibal was encamped near their walls, in order to exhibit their confidence, sent troops out by a different gate to reinforce the armies which they had in Spain.[2]

The land on which Hannibal had his camp having come into the market owing to the death of the owner, the Romans bid the price up to the figure at which the property had sold before the war.[2]

When the Romans were besieged by Hannibal and were themselves besieging Capua, they passed a decree not to recall their army from the latter place until it was captured.[3]

BOOK IV

LIBER QUARTUS[1]

Multa lectione conquisitis strategematibus et non exiguo scrupulo digestis, ut promissum trium librorum implerem, si modo implevi, hoc exhibebo ea, quae parum apte discriptioni priorum ad speciem alligatae subici videbantur et erant exempla potius strategicon quam strategemata; quae idcirco separavi, quia quamvis clara diversae tamen erant substantiae, ne, si qui forte in aliqua ex his incidissent, similitudine inducti praetermissa opinarentur. Et sane velut res residua expedienda fuit, in qua et ipse ordinem per species servare conabor:

I. De disciplina.
II. De effectu disciplinae.
III. De continentia.
IIII. De iustitia.
V. De constantia.
VI. De affectu et moderatione.
VII. De variis consiliis.

[1] *On the authenticity of Book IV, see Introduction, pp. xix. ff.*

BOOK IV

HAVING, by extensive reading, collected examples of stratagems, and having arranged these at no small pains, in order to fulfil the promise of my three books (if only I have fulfilled it), in the present book I shall set forth those instances which seemed to fall less naturally under the former classification (which was limited to special types), and which are illustrations rather of military science in general than of stratagems. Inasmuch as these incidents, though famous, belong to a different subject,[1] I have given them separate treatment, for fear that if any persons should happen in reading to run across some of them, they might be led by the resemblance to imagine that these examples had been overlooked by me. As supplementary material, of course, these topics called for treatment. In presenting them, I shall endeavour to observe the following categories:

  I. On discipline.
 II. On the effect of discipline.
III. On restraint and disinterestedness
 IV. On justice
  V. On determination ("the will to victory").
 VI. On good will and moderation.
VII. On sundry maxims and devices.

[1] That is, different from the class of stratagems proper.

SEXTUS JULIUS FRONTINUS

I. De Disciplina

1. P. Scipio ad Numantiam corruptum superiorum ducum socordia exercitum correxit dimisso ingenti lixarum numero, redactis ad munus cotidiana exercitatione militibus. Quibus cum frequens iniungeret iter, portare complurium dierum cibaria imperabat, ita ut frigora et imbres pati, vada fluminum pedibus traicere adsuesceret miles, exprobrante subinde imperatore timiditatem et ignaviam, frangente delicatioris usus ac parum necessaria expeditioni vasa. Quod maxime notabiliter accidit C. Memmio tribuno, cui dixisse traditur Scipio: "mihi paulisper, tibi et rei publicae semper nequam eris."

2. Q. Metellus bello Iugurthino similiter lapsam militum disciplinam pari severitate restituit, cum insuper prohibuisset alia carne quam assa elixave milites uti.

3. Pyrrhus dilectatori suo fertur dixisse: "tu grandes elige, ego eos fortes reddam."

4. L. Paulo et C. Varrone consulibus milites primo iure iurando adacti sunt; antea enim sacramento tantummodo a tribunis rogabantur, ceterum ipsi inter se coniurabant se fugae atque formidinis causa non

---

[1] 134 B C. *Cf.* Val. Max II. vii. 1; Livy *Per.* 57; Polyaen. VIII. xvi. 2.

[2] 109 B.C. *Cf.* Val. Max. II. vii. 2; Sall. *Jug.* 45. Polyaen. VIII. xvi. 2 and Plut. *Apophth. Scip. Min.* 16 attribute this to Scipio.

[3] The point of the prohibition is not obvious to the modern sense.

STRATAGEMS, IV. i. 1-4

I. On Discipline

When the Roman army before Numantia had become demoralized by the slackness of previous commanders, Publius Scipio reformed it by dismissing an enormous number of camp-followers and by bringing the soldiers to a sense of responsibility through regular daily routine. On the occasion of the frequent marches which he enjoined upon them, he commanded them to carry several days' rations, under such conditions that they became accustomed to enduring cold and rain, and to the fording of streams. Often the general reproached them with timidity and indolence; often he broke utensils which served only the purpose of self-indulgence and were quite unnecessary for campaigning. A notable instance of this severity occurred in the case of the tribune Gaius Memmius, to whom Scipio is said to have exclaimed: "To me you will be worthless merely for a certain period; to yourself and the state for ever!"[1]

Quintus Metellus, in the Jugurthine War,[2] when discipline had similarly lapsed, restored it by a like severity, while in addition he had forbidden the soldiers to use meat, except when baked or boiled.[3]

Pyrrhus is said to have remarked to his recruiting officer: "You pick out the big men! I'll make them brave."

In the consulship of Lucius Paulus and Gaius Varro, soldiers were for the first time compelled to take the *ius iurandum*. Up to that time the *sacramentum* was the oath of allegiance administered to them by the tribunes, but they used to pledge each other not to quit the force by flight, or in conse-

SEXTUS JULIUS FRONTINUS

abituros neque ex ordine recessuros nisi teli petendi feriendive hostis aut civis servandi causa.

5. Scipio Africanus, cum ornatum scutum elegantius cuiusdam vidisset, dixit non mirari se, quod tanta cura ornasset, in quo plus praesidii quam in gladio haberet.

6. Philippus, cum primum exercitum constitueret, vehiculorum usum omnibus interdixit, equitibus non amplius quam singulos calones habere permisit, peditibus autem denis singulos, qui molas et funes ferrent; in aestiva exeuntibus triginta dierum farinam collo portari imperavit.

7. C. Marius recidendorum impedimentorum gratia, quibus maxime exercitus agmen oneratur, vasa et cibaria militis in fasciculos aptata furcis imposuit, sub quibus et habile onus et facilis requies esset; unde et proverbium tractum est "muli Mariani."

8. Theagenes Atheniensis, cum exercitum Megaram duceret, petentibus ordines respondit ibi se daturum. Deinde clam equites praemisit eosque hostium specie impetum in suos retorquere iussit. Quo facto cum quos secum habebat tamquam ad hostium occursum praepararentur, permisit ita ordi-

---

[1] 216 B.C. *Cf.* Livy xxii. 38. Up to the time of the Battle of Cannae, there were two military oaths, the *sacramentum*, which was compulsory and was administered by the consul when the soldier first enlisted, and the *ius iurandum*, a voluntary oath taken before a tribune when the soldiers were assigned to separate divisions. In 216 the two were united, and thereafter the *ius iurandum*, administered by the military tribune, was compulsory. The facts here stated are slightly at variance with the general understanding.

[2] 134 B.C. *Cf.* Livy *Per.* 57; Plut. *Apophth. Scip. Min.* 18; Polyaen. VIII. xvi. 3, 4.

quence of fear, and not to leave the ranks except to seek a weapon, strike a foe, or save a comrade.[1]

Scipio Africanus, noticing the shield of a certain soldier rather elaborately decorated, said he didn't wonder the man had adorned it with such care, seeing that he put more trust in it than in his sword.[2]

When Philip was organizing his first army, he forbade anyone to use a carriage. The cavalrymen he permitted to have but one attendant apiece. In the infantry he allowed for every ten men only one servant, who was detailed to carry the mills and ropes.[3] When the troops marched out to summer quarters, he commanded each man to carry on his shoulders flour for thirty days.

For the purpose of limiting the number of pack animals, by which the march of the army was especially hampered, Gaius Marius had his soldiers fasten their utensils and food up in bundles and hang these on forked poles, to make the burden easy and to facilitate rest; whence the expression "Marius's mules."[4]

When Theagenes, the Athenian, was leading his troops towards Megara and his men inquired as to their place in the ranks, he told them he would assign them their places when they arrived at their destination. Then he secretly sent the cavalry ahead and commanded them, in the guise of enemies, to turn back and attack their comrades. When this plan was carried out and the men whom he had with him made preparations for an encounter with the foe, he permitted the battle-line to be drawn up in

[3] The mills were for grinding corn. The allusion to the ropes is not clear.
[4] *Cf.* Fest. Paul. 24, 2; 148, 6.

nari aciem, ut quo quis voluisset loco consisteret;
cum inertissimus quisque retro se dedisset, strenui
autem in frontem prosiluissent, ut quemque invenerat
stantem, ita ad ordines militiae provexit.

9. Lysander Lacedaemonius egressum via quendam castigabat. Cui dicenti ad nullius rei rapinam se ab agmine recessisse respondit: "ne speciem quidem rapturi praebeas volo."

10. Antigonus, cum filium suum audisset devertisse in mulieris[1] domum, cui tres filiae insignes specie essent, "audio," inquit, "fili, anguste habitare te pluribus dominis domum possidentibus: hospitium laxius accipe." Iussoque commigrare edixit, ne quis minor quinquaginta annos natus hospitio matris familias uteretur.

11. Q. Metellus consul, quamvis nulla lege impediretur, quin filium contubernalem perpetuum haberet, maluit tamen eum in ordine merere.

12. P. Rutilius consul, cum secundum leges in contubernio suo habere posset filium, in legione militem fecit.

13. M. Scaurus filium, quod in saltu Tridentino loco hostibus cesserat, in conspectum suum venire vetuit. Adulescens verecundia ignominiae pressus mortem sibi conscivit.

14. Castra antiquitus Romani ceteraeque gentes passim per corpora cohortium velut mapalia constituere soliti erant, cum solos urbium muros nosset

---

[1] mulieris, *Götz;* eius *MSS.;* anus *Gund.*

---

[1] Polyaen. v. xxviii. 1 attributes this to Theognis; in III. ix. 10, he attributes it to Iphicrates.

[2] 323–321 B C. *Cf.* Plut. *Demetr.* 23, *Apophth. Antig.* 5.

[3] 143 (?) 109 (?) B.C. Sall. *Jug.* lxiv. 4 says that Metellus Numidicus kept his son with him as tent-mate.

such a way that each man took his place where he wished, the most cowardly retiring to the rear, the bravest rushing to the front. He thereupon assigned to each man, for the campaign, the same position in which he had found him.[1]

Lysander, the Spartan, once flogged a soldier who had left the ranks while on the march. When the man said that he had not left the line for the purpose of pillage, Lysander retorted. "I won't have you look as if you were going to pillage."

Antigonus, hearing that his son had taken lodgings at the house of a woman who had three handsome daughters, said: "I hear, son, that your lodgings are cramped, owing to the number of mistresses in charge of your house. Get roomier quarters." Having commanded his son to move, he issued an edict that no one under fifty years of age should take lodgings with the mother of a family.[2]

The consul Quintus Metellus, although not prevented by law from having his son with him as a regular tent-mate, yet preferred to have him serve in the ranks.[3]

The consul Publius Rutilius, though he might by law have kept his son in his own tent, made him a soldier in the legion.[4]

Marcus Scaurus forbade his son to come into his presence, since he had retreated before the enemy in the Tridentine Pass. Overwhelmed by the shame of this disgrace, the young man committed suicide.[5]

In ancient times the Romans and other peoples used to make their camps like groups of Punic huts, distributing the troops here and there by cohorts, since the men of old were not acquainted with walls

[4] 105 B.C. [5] 102 B.C. *Cf.* Val. Max. v. viii. 4.

SEXTUS JULIUS FRONTINUS

antiquitas. Pyrrhus Epirotarum rex primus totum exercitum sub eodem vallo continere instituit. Romani deinde, victo eo in campis Arusinis circa urbem Malventum, castris eius potiti et ordinatione notata paulatim ad hanc usque metationem, quae nunc effecta est, pervenerunt.

15. P. Nasica in hibernis, quamvis classis usus non esset necessarius, ne tamen desidia miles corrumperetur aut per otii licentiam sociis iniuriam inferret, navis aedificare instituit.

16. M. Cato memoriae tradidit in furto comprehensis inter commilitones dextras esse praecisas aut, si lenius animadvertere voluissent, in principiis sanguinem missum.

17. Clearchus dux Lacedaemoniorum exercitui dicebat imperatorem potius quam hostem metui debere, significans eos, qui in proelio dubiam mortem timuissent, certum, si deseruissent, manere supplicium.

18. Appii Claudii sententia senatus eos, qui a Pyrrho rege Epirotarum capti et postea remissi erant, equites ad peditem redegit, pedites ad levem armaturam, omnibus extra vallum iussis tendere, donec bina hostium spolia singuli referrent.

19. Otacilius Crassus consul eos, qui ab Hannibale

---

[1] The modern Benevento. Pyrrhus was defeated here in 275 B.C.

[2] *Cf.* Livy xxxv. 14. Plut. *Pyrrh.* 16 represents Pyrrhus, on the other hand, marvelling at the arrangement of the Roman camp.   [3] 194–193 B.C.

[4] Gell. x. viii. 1 gives an interesting conjecture as to the origin of this second punishment.

[5] 431–401 B.C.   *Cf.* Val. Max. II. vii. extr. 2; Xen. *Anab.* II. vi. 10.

[6] 279 B.C.   *Cf.* Val. Max. II. vii. 15; Eutrop. ii. 13.

except in the case of cities. Pyrrhus, king of the Epirotes, was the first to inaugurate the custom of concentrating an entire army within the precincts of the same entrenchments. Later the Romans, after defeating Pyrrhus on the Arusian Plains near the city of Maleventum,[1] captured his camp, and, noting its plan, gradually came to the arrangement which is in vogue to-day.[2]

At one time, when Publius Nasica was in winter-quarters, although he had no need of ships, yet he determined to construct them, in order that his troops might not become demoralized by idleness, or inflict harm on their allies in consequence of the licence resulting from leisure.[3]

Marcus Cato has handed down the story that, when soldiers were caught in theft, their right hands used to be cut off in the presence of their comrades; or if the authorities wished to impose a lighter sentence, the offender was bled at headquarters.[4]

The Spartan general Clearchus used to tell his troops that their commander ought to be feared more than the enemy, meaning that the death they feared in battle was doubtful, but that execution for desertion was certain.[5]

On motion of Appius Claudius the Senate degraded to the status of foot-soldiers those knights who had been captured and afterwards sent back by Pyrrhus, king of the Epirotes, while the foot-soldiers were degraded to the status of light-armed troops, all being commanded to tent outside the fortifications of the camp until each man should bring in the spoils of two foemen.[6]

The consul Otacilius Crassus ordered those who had been sent under the yoke by Hannibal and had

SEXTUS JULIUS FRONTINUS

sub iugum missi redierant, tendere extra vallum iussit, ut immuniti adsuescerent periculis et adversus hostem audentiores fierent.

20. P. Cornelio Nasica Decimo Iunio consulibus, qui exercitum deseruerant, damnati virgis caesi publice venierunt.

21. Domitius Corbulo in Armenia duas alas et tres cohortes, quae ad castellum Initia hostibus cesserant, extra vallum iussit tendere, donec adsiduo labore et prosperis excursionibus redimerent ignominiam.

22. Aurelius Cotta consul, cum ad opus equites necessitate cogente iussisset accedere eorumque pars detractasset imperium, questus apud censores effecit, ut notarentur; a patribus deinde obtinuit, ne eis praeterita aera procederent; tribuni quoque plebis de eadem re ad populum pertulerunt omniumque consensu stabilita disciplina est.

23. Q. Metellus Macedonicus in Hispania quinque cohortes, quae hostibus cesserant, testamentum facere iussas ad locum reciperandum remisit, minatus non nisi post victoriam receptum iri.

24. P. Valerio consuli senatus praecepit, exercitum ad Sirim victum ducere Saepinum ibique castra munire et hiemem sub tentoriis exigere.

24a. Senatus, cum turpiter fugati eius milites

---

[1] Manius Otacilius Crassus was consul in 263 and 246 B.C. Titus Otacilius Crassus was consul in 261 B.C.
[2] 138 B.C. *Cf.* Livy *Per.* 55.
[3] 58–59 A.D. *Cf.* Tac. *Ann.* xiii. 36.
[4] 252 B.C. Val. Max. II. ix. 7 cites a somewhat similar case of discipline.
[5] 143 B.C. *Cf.* Val. Max. II. vii. 10; Vell. ii. 5.
[6] 280 B.C. Publius Valerius Laevinus.

then returned, to camp outside the entrenchments, in order that they might become used to dangers while without defences, and so grow more daring against the enemy.[1]

In the consulship of Publius Cornelius Nasica and Decimus Junius those who had deserted from the army were condemned to be scourged publicly with rods and then to be sold into slavery.[2]

Domitius Corbulo, when in Armenia, ordered two squadrons and three cohorts, which had given way before the enemy near the fortress of Initia, to camp outside the entrenchments, until by steady work and successful raids they should atone for their disgrace.[3]

When the consul Aurelius Cotta under pressing necessity ordered the knights to participate in a certain work and a part of them renounced his authority, he made complaint before the censors and had the mutineers degraded. Then from the senators he secured an enactment that arrears of their wages should not be paid. The tribunes of the plebs also carried through a bill with the people on the same matter, so that discipline was maintained by the joint action of all.[4]

When Quintus Metellus Macedonicus was campaigning in Spain, and five cohorts on one occasion had given way before the enemy, he commanded the soldiers to make their wills, and then sent them back to recover the lost ground, threatening that they should not be received in camp except after victory.[5]

The Senate ordered the consul Publius Valerius to lead the army, which had been defeated near the river Siris, to Saepinum, to construct a camp there, and to spend the winter under canvas.[6]

When his soldiers had been disgracefully routed

SEXTUS JULIUS FRONTINUS

essent, decrevit, ne auxilia eis summitterentur, nisi captis eius . . .

25. Legionibus, quae Punico bello militiam detractaverant, in Siciliam velut relegatis per septem annos hordeum ex senatus consulto datum est.

26. L. Piso C. Titium praefectum cohortis, quod loco fugitivis cesserat, cinctu togae praeciso, soluta tunica, nudis pedibus in principiis cotidie stare, dum vigiles venirent, iussit, conviviis et balneo abstinere.

27. Sulla cohortem et centuriones, quorum stationem hostis perruperat, galeatos et discinctos perstare in principiis iussit.

28. Domitius Corbulo in Armenia Aemilio Rufo praefecto equitum, quia hostibus cesserat et parum instructam armis alam habebat, vestimenta per lictorem scidit eidemque ut erat foedato habitu perstare in principiis, donec emitteretur,[1] imperavit.

29. Atilius Regulus, cum ex Samnio in Luceriam transgrederetur exercitusque eius obviis hostibus aversus esset, opposita cohorte iussit fugientes pro desertoribus caedi.

30. Cotta consul in Sicilia in Valerium, nobilem tribunum militum ex gente Valeria, virgis animadvertit.

31. Idem P. Aurelium sanguine sibi iunctum, quem

---

[1] emitteretur *Bennett*; mitterentur *MSS. and edd.*

---

[1] In the Second Punic War, after Cannae.
[2] *Cf.* Livy xxiv. 18. The substitution of barley for wheat rations was a common form of punishment; *Cf.* Suet. *Aug.* 24; Plut. *Marc.* 25.
[3] 133 B.C. *Cf.* Val. Max. II. vii. 9.   [4] 58–59 A.D.
[5] 294 B.C. *Cf.* II. viii. 11 and note.   [6] 252 B.C.

the Senate ordered that no reinforcements should be sent them, unless . . .

The legions which had refused to serve in the Punic War[1] were sent into a kind of banishment in Sicily, and by vote of the Senate were put on barley rations for seven years.[2]

Because Gaius Titius, commander of a cohort, had given way before some runaway slaves, Lucius Piso ordered him to stand daily in the headquarters of the camp, barefooted, with the belt of his toga cut and his tunic ungirt, and wait till the nightwatchmen came. He also commanded that the culprit should forgo banquets and baths.[3]

Sulla ordered a cohort and its centurions, through whose defences the enemy had broken, to stand continuously at headquarters, wearing helmets and without uniforms.

When Domitius Corbulo was campaigning in Armenia, a certain Aemilius Rufus, a praefect of cavalry, gave way before the enemy. On discovering that Rufus had kept his squadron inadequately equipped with weapons, Corbulo directed the lictors to strip the clothes from his back, and ordered the culprit to stand at headquarters in this unseemly plight until he should be released.[4]

When Atilius Regulus was crossing from Samnium to Luceria and his troops turned away from the enemy whom they had encountered, Regulus blocked their retreat with a cohort, as they fled, and ordered them to be cut to pieces as deserters.[5]

The consul Cotta, when in Sicily, flogged a certain Valerius, a noble military tribune belonging to the Valerian gens.[6]

The same Cotta, when about to cross over to

SEXTUS JULIUS FRONTINUS

obsidioni Lipararum, ipse ad auspicia repetenda Messanam transiturus, praefecerat, cum agger incensus et capta castra essent, virgis caesum in numerum gregalium peditum referri et muneribus fungi iussit.

32. Fulvius Flaccus censor Fulvium fratrem suum, quia legionem, in qua tribunus militum erat, iniussu consulis dimiserat, senatu movit.

33. M. Cato ab hostili litore, in quo per aliquot dies manserat, cum ter dato profectionis signo classem solvisset et relictus e militibus quidam a terra voce et gestu expostularet, uti tolleretur, circumacta ad litus universa classe, comprehensum supplicio adfici iussit et, quem occisuri per ignominiam hostes fuerant, exemplo potius impendit.

34. Appius Claudius ex his, qui loco cesserant, decimum quemque militem sorte ductum fusti percussit.

35. Fabius Rullus consul ex duabus legionibus, quae loco cesserant, sorte ductos in conspectu militum securi percussit.

36. Aquilius ternos ex centuriis, quarum statio ab hoste perrupta erat, securi percussit.

37. M. Antonius, cum agger ab hostibus incensus esset, ex his, qui in opere fuerant, duarum cohortium

---

[1] *Cf.* Val. Max. II. vii. 4.
[2] 174 B.C. *Cf.* Val. Max. II. vii. 5; Livy xl. 41, xli. 27; Vell. I. x. 6.
[3] 471 B.C. *Cf.* Livy ii. 59; Dionys. ix. 50; Zonar. vii. 17.

Messana to take the auspices afresh, placed in charge of the blockade of the Liparian Islands a certain Publius Aurelius, who was connected with him by ties of blood. But when Aurelius's line of works was burned and his camp captured, Cotta had him scourged with rods and ordered him to be reduced to the ranks and to perform the tasks of a common soldier.[1]

The censor Fulvius Flaccus removed from the Senate his own brother Fulvius, because the latter without the command of the consul had disbanded the legion in which he was tribune of the soldiers.[2]

On one occasion when Marcus Cato, who had lingered for several days on a hostile shore, had at length set sail, after three times giving the signal for departure, and a certain soldier, who had been left behind, with cries and gestures from the land, begged to be picked up, Cato turned his whole fleet back to the shore, arrested the man, and commanded him to be put to death, thus preferring to make an example of the fellow than to have him ignominiously put to death by the enemy.[3]

In the case of those who quitted their places in the line, Appius Claudius picked out every tenth man by lot and had him clubbed to death.

In the case of two legions which had given way before the foe, the consul Fabius Rullus chose men by lot and beheaded them in sight of their comrades.

Aquilius beheaded three men from each of the centuries whose position had been broken through by the enemy.

Marcus Antonius, when fire had been set to his line of works by the enemy, decimated the soldiers of two cohorts of those who were on the works, and

SEXTUS JULIUS FRONTINUS

militem decimavit et in singulos ex his centuriones animadvertit, legatum cum ignominia dimisit, reliquis ex legione hordeum dari iussit.

38. In legionem, quae Regium oppidum iniussu ducis diripuerat, animadversum est ita, ut quattuor milia tradita custodiae necarentur; praeterea senatus consulto cautum est, ne quem ex eis sepelire vel lugere fas esset.

39. L. Papirius Cursor dictator Fabium Rullum magistrum equitum, quod adversum edictum eius quamvis prospere pugnaverat, ad virgas poposcit, caesum securi percussurus; nec contentioni aut precibus militum concessit animadversionem eumque profugientem Romam persecutus est, ne ibi quidem remisso prius supplicii metu, quam ad genua eius et Fabius cum patre provolveretur et pariter senatus ac populus rogarent.

40. Manlius, cui Imperioso postea cognomen fuit, filium, quod is contra edictum patris cum hoste pugnaverat, quamvis victorem in conspectu exercitus virgis caesum securi percussit.

41. Manlius filius, exercitu pro se adversus patrem seditionem parante, negavit tanti esse quemquam, ut propter illum disciplina corrumperetur, et obtinuit, ut ipsum puniri paterentur.

---

[1] 36 B.C. *Cf.* Plut. *Ant.* 39.

[2] When Pyrrhus was in southern Italy, the people of Rhegium applied to Rome for assistance, and the Romans sent them a garrison of four thousand soldiers, levied among the Latin colonies in Campania. In 279 these troops seized the town, killed or expelled the male inhabitants, and took possession of the women and children. *Cf.* Livy xxviii. 28; Val. Max. II. vii. 15; Polyb. I. vii. 6–13; Oros. IV. iii. 3–5.

[3] 325 B.C. *Cf.* Livy viii. 29 ff.; Val. Max. II. vii. 8, III. ii. 9; Eutr. ii. 8.

[4] 340 B.C. *Cf.* Livy viii. 7; Val. Max. II. vii 6; Sall. *Cat.*

STRATAGEMS, IV. i. 37-41

punished the centurions of each cohort. Besides this, he dismissed the commanding officer in disgrace, and ordered the rest of the legion to be put on barley rations.[1]

The legion which had plundered the city of Rhegium without the orders of its commander was punished as follows: four thousand men were put under guard and executed. Moreover the Senate by decree made it a crime to bury any one of these or indulge in mourning for them.[2]

The dictator Lucius Papirius Cursor demanded that Fabius Rullus, his master of the horse, be scourged, and was on the point of beheading him, because he had engaged in battle against orders—successfully withal. Even in the face of the efforts and pleas of the soldiers, Papirius refused to renounce his purpose of punishment, actually following Rullus, when he fled for refuge to Rome, and not even there abandoning his threats of execution until Fabius and his father fell at the knees of Papirius, and the Senate and people alike joined in their petition.[3]

Manlius, to whom the name " The Masterful " was afterwards given, had his own son scourged and beheaded in the sight of the army, because, even though he came out victorious, he had engaged in battle with the enemy contrary to the orders of his father.[4]

The younger Manlius, when the army was preparing to mutiny in his behalf against his father, said that no one was of such importance that discipline should be destroyed on his account, and so induced his comrades to suffer him to be punished.[4]

52 ; Cic. *de Fin.* I. vii. 23, *de Off.* III. xxxi. 112. The father, Titus Manlius Torquatus, was the son of Lucius Manlius, dictator in 363, who had also received the cognomen *Imperiosus* on account of his severity.

SEXTUS JULIUS FRONTINUS

42. Q. Fabius Maximus transfugarum dextras praecidit.

43. C. Curio consul bello Dardanico circa Dyrrachium, cum ex quinque legionibus una seditione facta militiam detractasset secuturamque se temeritatem ducis in expeditionem asperam et insidiosam negasset, quattuor legiones eduxit armatas et consistere ordinibus detectis armis velut in acie iussit. Post hoc seditiosam legionem inermem procedere discinctamque in conspectu armati exercitus stramenta coegit secare, postero autem die similiter fossam discinctos milites facere, nullisque precibus legionis impetrari ab eo potuit, ne signa eius summitteret nomenque aboleret, milites autem in supplementum ceterarum legionum distribueret.

44. Q. Fulvio Appio Claudio consulibus milites ex pugna Cannensi in Siciliam a senatu relegati postulaverunt a consule M. Marcello, ut in proelium ducerentur. Ille senatum consuluit; senatus negavit sibi placere committi his rempublicam, qui eam deseruissent; Marcello tamen permisit facere, quod videretur, dum ne quis eorum munere vacaret neve donaretur neve quod praemium ferret aut in Italiam reportaretur, dum Poeni in ea fuissent.

45. M. Salinator consularis damnatus est a populo, quod praedam non aequaliter diviserat militibus.

---

[1] 142–140 B.C. Quintus Fabius Maximus Servilianus. *Cf.* Val. Max II. vii. 11 ; Oros. v iv. 12.

[2] One of the leading tribes of Illyria, subdued by the Romans. According to Livy *Per.* 92, 95 and Eutrop. vi. 2, Curio was proconsul when he carried on this campaign in 75 B.C.

[3] 212 B.C. *Cf.* Livy xxv. 5–7 ; Val. Max. II. vii. 15 ; Plut. *Marc.* 13. Marcellus was proconsul, not consul. *Cf.* Livy xxvi. 1.

Quintus Fabius Maximus cut off the right hands of deserters.[1]

When the consul Gaius Curio was campaigning near Dyrrhachium in the war against the Dardani,[2] and one of the five legions, having mutinied, had refused service and declared it would not follow his rash leadership on a difficult and dangerous enterprise, he led out four legions in arms and ordered them to take their stand in the ranks with weapons drawn, as if in battle. Then he commanded the mutinous legion to advance without arms, and forced its members to strip for work and cut straw under the eyes of armed guards. The following day, in like manner, he compelled them to strip and dig ditches, and by no entreaties of the legion could he be induced to renounce his purpose of withdrawing its standards, abolishing its name, and distributing its members to fill out other legions.

In the consulship of Quintus Fulvius and Appius Claudius, the soldiers, who after the battle of Cannae had been banished to Sicily by the Senate, petitioned the consul Marcellus to be led to battle. Marcellus consulted the Senate, who declared it was not their pleasure that the public welfare should be trusted to those who had proved disloyal. Yet they empowered Marcellus to do what seemed best to him, provided none of the soldiers should be relieved of duty, honoured with a gift or reward, or conveyed back to Italy, so long as there were any Carthaginians in the country.[3]

Marcus Salinator, when ex-consul, was condemned by the people because he had not divided the booty equally among his soldiers.[4]

[4] 218 B.C. *Cf.* Livy xxvii. 34, xxix. 37.

SEXTUS JULIUS FRONTINUS

46. Cum ab Liguribus in proelio Q. Petilius consul interfectus esset, decrevit senatus, uti ea legio, in cuius acie consul erat occisus, tota infrequens referretur, stipendium ei annuum non daretur, aera reciderentur.

II. DE EFFECTU DISCIPLINAE

1. BRUTI et Cassi exercitus, memoriae proditum est, bello civili cum una per Macedoniam iter facerent priorque Brutus ad fluvium, in quo pontem iungi oportebat, pervenisset, Cassi tamen exercitum et in efficiendo ponte et in maturando transitu praecessisse; qui vigor disciplinae effecit, ne solum in operibus, verum et in summa belli praestarent Cassiani Brutianos.

2. C. Marius, cum facultatem eligendi exercitus haberet ex duobus, qui sub Rutilio et qui sub Metello ac postea sub se ipso meruerant, Rutilianum quamquam minorem, quia certioris disciplinae arbitrabatur, praeoptavit.

3. Domitius Corbulo duabus legionibus et paucissimis auxiliis disciplina correcta Parthos sustinuit.

4. Alexander Macedo XL milibus hominum iam inde a Philippo patre disciplinae adsuefactis orbem terrarum adgressus innumerabiles hostium copias vicit.

---

[1] The term *infrequens* was technically applied to soldiers who were absent from or irregular in attendance on their duties.

[2] 176 B.C. *Cf.* Livy xli. 18; Val. Max. II. vii. 15.
[3] 42 B.C.   [4] 104 B.C.
[5] 55–59 A.D. *Cf.* Tac. *Ann.* xiii. 8, 35.

STRATAGEMS, IV. i. 46–ii. 4

When the consul Quintus Petilius had been killed in battle by the Ligurians, the Senate decreed that that legion in whose ranks the consul had been slain should, as a whole, be reported "deficient";[1] that its year's pay should be withheld, and its wages reduced.[2]

II. On the Effect of Discipline

When, during the Civil War, the armies of Brutus and Cassius were marching together through Macedonia, the story goes that the army of Brutus arrived first at a stream which had to be bridged, but that the troops of Cassius were the first in constructing the bridge and in effecting a passage. This rigorous discipline made Cassius's men superior to those of Brutus not only in constructing military works, but also in the general conduct of the war.[3]

When Gaius Marius had the option of choosing a force from two armies, one of which had served under Rutilius, the other under Metellus and later under himself, he preferred the troops of Rutilius, though fewer in number, because he deemed them of trustier discipline.[4]

By improving discipline, Domitius Corbulo withstood the Parthians with a force of only two legions and a very few auxiliaries.[5]

Alexander of Macedon conquered the world, in the face of innumerable forces of enemies, by means of forty thousand men long accustomed to discipline under his father Philip.[6]

[6] 334 B.C. *Cf.* Livy xxxv. 14; Justin. xi. 6; Plut. *Alex.* 15. The numbers vary in the different authors.

SEXTUS JULIUS FRONTINUS

5. Cyrus bello adversus Persas quattuordecim milibus armatorum immensas difficultates superavit.

6. Epaminondas dux Thebanorum quattuor milibus hominum, ex quibus CCCC tantum equites erant, Lacedaemoniorum exercitum viginti quattuor milium peditum, equitum mille sescentorum vicit.

7. A quattuordecim milibus Graecorum, qui numerus in auxiliis Cyri adversus Artaxerxen fuit, centum milia barbarorum proelio superata sunt.

8. Eadem Graecorum proelio quattuordecim milia amissis ducibus, reditus sui cura uni ex corpore suo Xenophonti Atheniensi demandata, per iniqua et ignota loca incolumia reversa sunt.

9. Xerxes a trecentis Lacedaemoniorum ad Thermopylas vexatus, cum vix eos confecisset, hoc se deceptum aiebat, quod multos quidem homines haberet, viros autem disciplinae tenaces nullos.

III. DE CONTINENTIA

1. M. CATONEM vino eodem quo remiges contentum fuisse traditur.

2. Fabricius, cum Cineas legatus Epirotarum grande pondus auri dono ei daret, non accepto eo dixit malle se habentibus id imperare, quam habere.

3. Atilius Regulus, cum summis rebus praefuisset,

---

[1] 401 B.C. *Cf.* IV. ii. 7; Xen. *Anab.* I. ii. 9; Plut. *Artax.* 6; Diodor. XIV. xix. 6.

[2] 371 B C. Battle of Leuctra. Diodor. XV. 52 ff. and Plut. *Pelop.* 20 give different numbers.

[3] Battle of Cunaxa. *Cf.* IV. ii. 5.

[4] *Cf.* Xen. *Anab.* iii. 1 ff.

[5] 480 B.C. In Herod. vii. 210, the historian himself, not the king, makes this observation.

[6] *Cf.* Val. Max. IV. iii. 11; Plin. *N.H.* 3, 14.

Cyrus in his war against the Persians overcame incalculable difficulties with a force of only fourteen thousand armed men.[1]

With four thousand men, of whom only four hundred were cavalry, Epaminondas, the Theban leader, conquered the Spartan army of twenty-four thousand infantry and sixteen hundred cavalry.[2]

A hundred thousand barbarians were defeated in battle by fourteen thousand Greeks, the number assisting Cyrus against Artaxerxes.[3]

The same fourteen thousand Greeks, having lost their generals in battle, returned unharmed through difficult and unknown places, having committed the management of their retreat to one of their number, Xenophon, the Athenian.[4]

When Xerxes was defied by the three hundred Spartans at Thermopylae and had with difficulty destroyed them, he declared that he had been deceived, because, while he had numbers enough, yet of real men who adhered to discipline he had none.[5]

III. On Restraint and Disinterestedness

The story goes that Marcus Cato was content with the same wine as the men of his crews.[6]

When Cineas, ambassador of the Epirotes, offered Fabricius a large amount of gold, the latter rejected it, declaring that he preferred to rule those who had gold rather than to have it himself.[7]

Atilius Regulus, though he had been in charge

---

[7] 280 B.C. Gell. i. 14 tells this story of Fabricius; usually it is related of Curius. *Cf.* Val. Max. IV. iii. 5; Cic. *de Sen.* xvi. 55; Plut. *Apophth. M'. Curii.*

SEXTUS JULIUS FRONTINUS

adeo pauper fuit, ut se coniugem liberosque toleraret agello, qui colebatur per unum vilicum; cuius audita morte scripsit senatui de successore, destitutis rebus obitu servi necessariam esse praesentiam suam.

4. Cn. Scipio post res prospere in Hispania gestas in summa paupertate decessit, ne ea quidem relicta pecunia, quae sufficeret in dotem filiarum, quas ob inopiam publice dotavit senatus.

5. Idem praestiterunt Athenienses filiis Aristidis post amplissimarum rerum administrationem in maxima paupertate defuncti.

6. Epaminondas dux Thebanorum tantae abstinentiae fuit, ut in suppellectili eius praeter stoream et unicum veru nihil inveniretur.

7. Hannibal surgere de nocte solitus ante noctem non requiescebat; crepusculo demum ad cenam vocabat neque amplius quam duobus lectis discumbebatur apud eum.

8. Idem, cum sub Hasdrubale imperatore militaret, plerumque super nudam humum sagulo tectus somnos capiebat.

9. Aemilianum Scipionem traditur in itinere cum amicis ambulantem accepto pane vesci solitum.

10. Idem et de Alexandro Macedone dicitur.[1]

---

[1] *No. 10 is probably interpolated.*

---

[1] 255 B.C. *Cf.* Val. Max. IV. iv. 6; Livy *Per.* 18; Senec. *ad Helv.* 12; Apul. *Apol.* 18.

[2] Other writers, excepting Seneca, speak of but one daughter, and represent the dowry as given when Scipio was warring in Spain, in 218–211 B.C. *Cf.* Val. Max. IV. iv. 10; Sen. *Qu. Nat.* I. xvii. 9; Apul. *Apol.* 18; Ammian. Marc. XIV. vi. 11.

[3] 468 B.C. *Cf.* Nep. *Aristid.* 3; Plut. *Aristid.* 25.

[4] 362 B.C. *Cf.* Plut. *Fab. Max.* 27; Nep. *Epam.* 3.

of the greatest enterprises, was so poor that he supported himself, his wife, and children on a small farm which was tilled by a single steward. Hearing of the death of this steward, Regulus wrote to the Senate requesting them to appoint someone to succeed him in the command, since his property was left in jeopardy by the death of his slave, and his own presence at home was necessary.[1]

Gnaeus Scipio, after successful exploits in Spain, died in the extremest poverty, not even leaving money enough for a dowry for his daughters. The Senate, therefore, in consequence of their poverty, furnished them dowries at public expense.[2]

The Athenians did the same thing for the daughters of Aristides, who died in the greatest poverty after directing most important enterprises.[3]

Epaminondas, the Theban general, was a man of such simple habits that among his belongings nothing was found beyond a mat and a single spit.[4]

Hannibal was accustomed to rise while it was still dark, but never took any rest before night. At dusk, and not before, he called his friends to dinner; and not more than two couches [5] were ever filled with dinner guests at his headquarters.[6]

The same general, when serving under Hasdrubal as commander, usually slept on the bare ground, wrapped only in a common military cloak.[6]

The story goes that Scipio Aemilianus used to eat bread offered him as he walked along on the march in the company of his friends.

The same story is related of Alexander of Macedon.

---

[5] Frontinus has in mind a Roman *lectus*, or dining-couch, which accommodated three persons.

[6] *Cf.* Livy xxi. 4; Sil. Ital. xii. 559-560.

SEXTUS JULIUS FRONTINUS

11. Masinissam, nonagensimum aetatis annum agentem, meridie ante tabernaculum stantem vel ambulantem capere solitum cibos legimus.

12. M'. Curius, cum victis ab eo Sabinis ex senatus consulto ampliaretur ei modus agri, quem consummati milites accipiebant, gregalium portione contentus fuit, malum civem dicens, cui non esset idem quod ceteris satis.

13. Universi quoque exercitus notabilis saepe fuit continentia, sicut eius qui sub M. Scauro meruit. Namque memoriae tradidit Scaurus pomiferam arborem, quam in pede castrorum fuerat complexa metatio, postero die abeunte exercitu intactis fructibus relictam.

14. Auspiciis Imperatoris Caesaris Domitiani Augusti Germanici[1] bello, quod Iulius Civilis in Gallia moverat, Lingonum opulentissima civitas, quae ad Civilem desciverat, cum adveniente exercitu Caesaris populationem timeret, quod contra exspectationem inviolata nihil ex rebus suis amiserat, ad obsequium redacta septuaginta milia armatorum tradidit mihi.

15. L. Mummius, qui Corintho capta non Italiam solum sed etiam provincias tabulis statuisque exornavit, adeo nihil ex tantis manubiis in suum convertit, ut filiam eius inopem senatus ex publico dotaverit.

---

[1] Germanici eo *d*; *the best MSS., followed by Gund., have* Germanico.

---

[1] 148 B.C. *Cf.* Polyb. xxxvii. 11–12; Cic. *de Sen.* x. 34.

[2] *Cf.* Val. Max. IV. iii. 5; Plin. *N.H.* xviii. 4; Plut. *Apophth. M'. Curii.* Nep. *Thras.* 4 attributes a somewhat similar reply to Pittacus.

[3] For the Memoirs of Scaurus, consul 115 B.C., *cf.* Val. Max. IV. iv. 11; Tac. *Agric.* 1.

[4] 70 A.D. *Cf.* Introduction, p. xx.

We read that Masinissa, when in his ninetieth year, used to eat at noon, standing or walking about in front of his tent.[1]

When, in honour of his defeat of the Sabines, the Senate offered Manius Curius a larger amount of ground than the discharged troops were receiving, he was content with the allotment of ordinary soldiers, declaring that that man was a bad citizen who was not satisfied with what the rest received.[2]

The restraint of an entire army was also often noteworthy, as for example of the troops which served under Marcus Scaurus. For Scaurus has left it on record that a tree laden with fruit, at the far end of the fortified enclosure of the camp, was found, the day after the withdrawal of the army, with the fruit undisturbed.[3]

In the war waged under the auspices of the Emperor Caesar Domitianus Augustus Germanicus and begun by Julius Civilis in Gaul, the very wealthy city of the Lingones, which had revolted to Civilis, feared that it would be plundered by the approaching army of Caesar. But when, contrary to expectation, the inhabitants remained unharmed and lost none of their property, they returned to their loyalty, and handed over to me seventy thousand armed men.[4]

After the capture of Corinth, Lucius Mummius adorned not merely Italy, but also the provinces, with statues and paintings. Yet he refrained so scrupulously from appropriating anything from such vast spoils to his own use that his daughter was in actual need and the Senate furnished her dowry at the public expense.[5]

[5] 146 B.C. *Cf.* Cic. *de Off.* II. xxii. 76; *in Verr.* I. xxi. 55; Plin. *N.H.* xxxiv. vii. 17.

SEXTUS JULIUS FRONTINUS

IIII. De Iustitia

1. Camillo Faliscos obsidenti ludi magister liberos Faliscorum tamquam ambulandi causa extra murum eductos tradidit, dicens retentis eis obsidibus necessario civitatem imperata facturam. Camillus non solum sprevit perfidiam, sed et restrictis post terga manibus magistrum virgis agendum ad parentes tradidit pueris, adeptus beneficio victoriam, quam fraude non concupierat, nam Falisci ob hanc iustitiam sponte ei se dediderunt.

2. Ad Fabricium ducem Romanorum medicus Pyrrhi Epirotarum regis pervenit pollicitusque est daturum se Pyrrho venenum, si merces sibi, in qua operae pretium foret, constitueretur. Quo facinore Fabricius egere victoriam suam non arbitratus, regi medicum detexit atque ea fide meruit, ut ad petendam amicitiam Romanorum compelleret Pyrrhum.

V. De Constantia

1. Cn. Pompeius minantibus direpturos pecuniam militibus, quae in triumphum[1] ferretur, Servilio et Glaucia cohortantibus, ut divideret eam, ne seditio fieret, adfirmavit non triumphaturum se potius et moriturum, quam licentiae militum succumberet, castigatisque oratione gravi laureatos fasces obiecit,

---

[1] triumphum *Bennett;* triumpho *MSS. and edd.*

---

[1] 394 b.c. *Cf.* Val. Max. vi. v. 1; Livy v. 27; Plut. *Camil.* 10; Polyaen. viii. 7.

[2] 279 b.c. *Cf.* Livy *Per.* 13; Cic. *de Off.* iii. xxii. 86; Plut. *Pyrrh.* 21. Val. Max. vi. v. 1 and Gell. iii. 8 represent Fabricius as disclosing the plot, but not the name of the traitor.

IV. On Justice

When Camillus was besieging the Faliscans, a school teacher took the sons of the Faliscans outside the walls, as though for a walk, and then delivered them up, saying that, if they should be retained as hostages, the city would be forced to execute the orders of Camillus. But Camillus not only spurned the teacher's perfidy, but tying his hands behind his back, turned him over to the boys to be driven back to their parents with switches. He thus gained by kindness a victory which he had scorned to secure by fraud; for the Faliscans, in consequence of this act of justice, voluntarily surrendered to him.[1]

The physician of Pyrrhus, king of the Epirotes, came to Fabricius, general of the Romans, and promised to give Pyrrhus poison if an adequate reward should be guaranteed him for the service. Fabricius, not considering that victory called for any such crime, exposed the physician to the king, and by this honourable act succeeded in inducing Pyrrhus to seek the friendship of the Romans.[2]

V. On Determination ("The Will to Victory")

When the soldiers of Gnaeus Pompey threatened to plunder the money which was being carried for the triumph, Servilius and Glaucia urged him to distribute it among the troops, in order to avoid the outbreak of a mutiny. Thereupon Pompey declared he would forgo a triumph, and would die rather than yield to the insubordination of his soldiers; and after upbraiding them in vehement language, he threw in their faces the fasces wreathed with

SEXTUS JULIUS FRONTINUS

ut ab illorum inciperent direptione; eaque invidia redegit eos ad modestiam.

2. C. Caesar, seditione in tumultu civilium armorum facta, maxime animis tumentibus, legionem totam exauctoravit, ducibus seditionis securi percussis; mox eosdem, quos exauctoraverat, ignominiam deprecantis restituit et optimos milites habuit.

3. Postumius consularis cohortatus suos, cum interrogatus esset a militibus, quid imperaret, dixit, ut se imitarentur, et arrepto signo hostis primus invasit; quem secuti victoriam adepti sunt.

4. Claudius Marcellus, cum in manus Gallorum imprudens incidisset, circumspiciendae regionis qua evaderet causa equum in orbem flexit; deinde cum omnia esse infesta vidisset, precatus deos in medios hostis inrupit; quibus inopinata audacia perculsis, ducem quoque eorum trucidavit atque, ubi spes salutis vix superfuerat, inde opima rettulit spolia.

5. L. Paulus, amisso ad Cannas exercitu, offerente equum Lentulo quo fugeret, superesse cladi quamquam non per ipsum contractae noluit, sed in eo saxo, cui se vulneratus acclinaverat, persedit, donec ab hostibus oppressus confoderetur.

6. Varro collega eius vel maiore constantia post

---

[1] 79 B.C. *Cf.* Plut. *Pomp.* 14, *Apophth. Pomp.* 6; Zonar. x. 2. There is no mention of Glaucia in this connection elsewhere.

[2] 49 B.C. *Cf.* Suet. *Caes.* 69; Appian *B.C.* ii. 47; Dio xli. 26 ff. [3] Viridomarus, the Insubrian Gaul.

[4] Spoils taken by a victorious commander from the leader of the enemy.

[5] 222 B.C. *Cf.* Val. Max. III. ii. 5; Livy *Per.* 20; Plut. *Marc.* 6 ff.; Flor. ii. 4.

[6] 216 B.C. *Cf.* Livy xxii. 49.

laurel, that they might start their plundering by seizing these. Through the odium thus aroused he reduced his men to obedience.[1]

When a sedition broke out in the tumult of the Civil War, and feeling ran especially high, Gaius Caesar dismissed from service an entire legion, and beheaded the leaders of the mutiny. Later, when the very men he had dismissed entreated him to remove their disgrace, he restored them and had in them the very best soldiers.[2]

Postumius, when ex-consul, having appealed to the courage of his troops, and having been asked by them what commands he gave, told them to imitate him. Thereupon he seized a standard and led the attack on the enemy. His soldiers followed and won the victory.

Claudius Marcellus, having unexpectedly come upon some Gallic troops, turned his horse about in a circle, looking around for a way of escape. Seeing danger on every hand, with a prayer to the gods, he broke into the midst of the enemy. By his amazing audacity he threw them into consternation, slew their leader,[3] and actually carried away the *spolia opima*[4] in a situation where there had scarcely remained a hope of saving his life.[5]

Lucius Paulus, after the loss of his army at Cannae, being offered a horse by Lentulus with which to effect his escape, refused to survive the disaster, although it had not been occasioned by him, and remained seated on the rock against which he had leaned when wounded, until he was overpowered and stabbed by the enemy.[6]

Paulus's colleague, Varro, showed even greater resolution in continuing alive after the same disaster,

SEXTUS JULIUS FRONTINUS

eandem cladem vixit gratiaeque ei a senatu et populo actae sunt, quod non desperasset rem publicam. Non autem vitae cupiditate, sed rei publicae amore se superfuisse reliquo aetatis suae tempore approbavit; et barbam capillumque summisit et postea numquam recubans cibum cepit. Honoribus quoque, cum ei deferrentur a populo, renuntiavit, dicens felicioribus magistratibus rei publicae opus esse.

7. Sempronius Tuditanus et Cn. Octavius tribuni militum omnibus fusis ad Cannas, cum in minoribus castris circumsederentur, suaserunt commilitonibus stringerent gladios et per hostium praesidia erumperent secum, id sibi animi esse, etiamsi nemini ad erumpendum audacia fuisset, adfirmantes; de cunctantibus XII omnino equitibus, L peditibus qui comitari sustinerent repertis, incolumes Canusium pervenerunt.

8. C. Fonteius Crassus in Hispania cum tribus milibus hominum praedatum profectus locoque iniquo circumventus ab Hasdrubale, ad primos tantum ordines relato consilio, incipiente nocte, quo tempore minime exspectabatur, per stationes hostium erupit.[1]

9. P. Decius tribunus militum bello Samnitico Cornelio consuli iniquis locis deprehenso ab hostibus

---

[1] *Identical with* I. v. 12.

[1] *Cf.* Livy XXII. lxi. 14–15; Val. Max. III. iv. 4, IV. v. 2.

[2] When the Romans reached Cannae, they pitched two camps, the larger on the N.W. bank of the Aufidus, the smaller on the S.E. bank, about a mile and a quarter apart, according to Polybius III. cx. 10. Before the battle, Hannibal transferred his camp from the east side of the river to the west.

[3] 216 B.C. *Cf.* Livy xxii. 50; Appian *Hann.* 16. Livy gives the number of those escaping from the smaller camp

and the Senate and the people thanked him "because," they said, "he did not despair of the commonwealth." But throughout the rest of his life he gave proof that he had remained alive not from desire of life, but because of his love of country. He suffered his beard and hair to remain untrimmed, and never afterwards reclined when he took food at table. Even when honours were decreed him by the people he declined them, saying that the State needed more fortunate magistrates than himself.[1]

After the complete rout of the Romans at Cannae, when Sempronius Tuditanus and Gnaeus Octavius, tribunes of the soldiers, were besieged in the smaller camp,[2] they urged their comrades to draw their swords and accompany them in a dash through the forces of the enemy, declaring that they themselves were resolved on this course, even if no one else possessed the courage to break through. Although among the wavering crowd only twelve knights and fifty foot-soldiers were found who had the courage to accompany them, yet they reached Canusium unscathed.[3]

When Gaius Fonteius Crassus was in Spain, he set out with three thousand men on a foraging expedition and was enveloped in an awkward position by Hasdrubal. In the early part of the night, at a time when such a thing was least expected, having communicated his purpose only to the centurions of the first rank, he broke through the pickets of the enemy.

When the consul Cornelius had been caught in an awkward position by the enemy in the Samnite War,

as six hundred. Appian says ten thousand from the larger camp escaped. One leader only is mentioned by both.

suasit, ut ad occupandum collem qui in propinquo erat modicam manum mitteret, seque ducem his qui mittebantur obtulit. Avocatus in diversum hostis emisit consulem, Decium autem cinxit obseditque. Illas quoque angustias nocte eruptione facta cum eluctatus esset Decius, incolumis cum militibus consuli accessit.[1]

10. Idem fecit sub Atilio Calatino consule, cuius varie traduntur nomina: alii Laberium, nonnulli Q. Caedicium, plurimi Calpurnium Flammam vocitatum scripserunt. Hic cum demissum in eam vallem videret exercitum, cuius latera omniaque superiora hostis insederat, depoposcit et accepit a consule trecentos milites, quos hortatus, ut virtute sua exercitum servarent, in mediam vallem decucurrit; ad opprimendos eos undique descendit hostis longoque et aspero proelio retentus occasionem consuli ad extrahendum exercitum dedit.[2]

11. C. Caesar adversus Germanos et regem Ariovistum pugnaturus, confusis suorum animis pro contione dixit nullius se eo die opera nisi decimae legionis usurum. Quo adsecutus est, ut et decimani tamquam praecipuae fortitudinis testimonio concitarentur et ceteri pudore, ne penes alios gloria virtutis esset.[3]

12. Lacedaemonius quidam nobilis, Philippo de-

---

[1] *Practically identical with* I. v. 14.
[2] *Identical with* I. v 15.
[3] *Identical with* I. xi. 3.

---

[1] *Cf.* I. v. 14 and note.   [2] *Cf.* I. v. 15 and note.
[3] *Cf.* I. xi. 3 and note.

STRATAGEMS, IV. v. 9-12

Publius Decius, tribune of the soldiers, urged him to send a small force to occupy a neighbouring hill, and volunteered to act as leader of those who should be sent. The enemy, thus diverted to a different quarter, allowed the consul to escape, but surrounded Decius and besieged him. Decius, however, extricated himself from this predicament also by making a sortie at night, and escaped unharmed along with his men and rejoined the consul.[1]

Under the consul Atilius Calatinus the same exploit was performed by a man whose name is variously reported. Some say he was called Laberius, some Quintus Caedicius, but most call him Calpurnius Flamma. This man, seeing that the army had entered a valley, the sides and all the commanding parts of which had been occupied by the enemy, asked and received from the consul three hundred soldiers. After exhorting these to save the army by their courage, he hastened to the centre of the valley. To crush him and his followers, the enemy descended from all directions, but being held in check in a long and fierce battle, they thus afforded the consul an opportunity of extricating his army.[2]

Gaius Caesar, when about to fight the Germans and their king Ariovistus, at a time when his own men had been thrown into panic, called his soldiers together and declared to the assembly that on that day he proposed to employ the services of the tenth legion alone. In this way he caused the soldiers of this legion to be stirred by his tribute to their unique heroism, while the rest were overwhelmed with mortification to think that a reputation for courage should be confined to others.[3]

A certain Spartan noble, when Philip declared he

SEXTUS JULIUS FRONTINUS

nuntiante multa se prohibiturum, nisi civitas sibi traderetur, "num," inquit, "et pro patria mori nos prohibebit?"

13. Leonidas Lacedaemonius, cum dicerentur Persae sagittarum multitudine nubes esse facturi, fertur dixisse: "melius in umbra pugnabimus."

14. C. Aelius praetor urbanus, cum ei ius dicenti picus in capite insedisset et haruspices respondissent dimissa ave hostium victoriam fore, necata populum Romanum superiorem, at C. Aelium cum familia periturum, non dubitavit necare picum. Atque nostro exercitu vincente ipse cum quattuordecim Aeliis ex eadem familia in proelio est occisus. Hunc quidam non C. Caelium, sed Laelium fuisse et Laelios, non Caelios perisse credunt.

15. P. Decius, primo pater, postea filius, in magistratu se pro re publica devoverunt admissisque in hostem equis adepti victoriam patriae contulerunt.

16. P. Crassus, cum bellum adversus Aristonicum in Asia gerens inter Elaeam et Myrinam in hostium copias incidisset vivusque abduceretur, exsecratus in consule Romano capitivitatem virga, qua ad equum erat usus, oculum Thracis, a quo tenebatur, eruit

---

[1] *Cf.* Cic. *Tusc.* v. xiv. 42; Val. Max. vi. iv. ext. 4; Plut. *Apophth. Lacon. Ignot.* 50.

[2] *Cf.* Cic. *Tusc.* i. xlii. 101; Val. Max. iii. vii. ext. 8.

[3] *Cf.* Plin. *N.H.* x. xviii. 20. Val. Max. v. vi. 4 says that the Aelian family lost seventeen members at the battle of Cannae. The last sentence of this paragraph is undoubtedly an interpolation.

[4] 340 & 295 B.C. *Cf.* Val. Max. v. vi. 5, 6; Cic. *de Fin.* ii. xix. 61, and frequently; Livy viii. 9, x. 28.

would cut them off from many things, unless the state surrendered to him, asked: "He won't cut us off from dying in defence of our country, will he?"[1]

Leonidas, the Spartan, in reply to the statement that the Persians would create clouds by the multitude of their arrows, is reported to have said: "We shall fight all the better in the shade."[2]

When Gaius Aelius, a city praetor, was holding court on one occasion, a woodpecker lighted upon his head. The soothsayers were consulted and made answer that, if the bird should be allowed to go, the victory would fall to the enemy, but that, if it were killed, the Roman people would prevail, though Gaius and all his house would perish. Aelius, however, did not hesitate to kill the woodpecker. Our army won the day, but Aelius himself, with fourteen others of the same family, was slain in battle. Certain authorities do not believe that the man referred to was Gaius Caelius, but a certain Laelius, and that they were Laelii, not Caelii, who perished.[3]

Two Romans bearing the name Publius Decius, first the father, later the son, sacrificed their lives to save the State during their tenure of office. By spurring their horses against the foe they won victory for their country.[4]

When waging war against Aristonicus in Asia somewhere between Elaea and Myrina, Publius Crassus fell into the hands of the enemy and was being led away alive. Scorning the thought of captivity for a Roman consul, he used the stick, with which he had urged on his horse, to gouge out the eye of the Thracian by whom he was held captive. The Thracian, infuriated with the pain,

atque ab eo per dolorem concitato transverberatus dedecus servitutis, ut voluerat, effugit.

17. M. Cato Censorii filius in acie decidente equo prolapsus, cum se recollegisset animadvertissetque gladium excidisse vaginae, veritus ignominiam rediit in hostem exceptisque aliquot vulneribus, reciperato demum gladio, reversus est ad suos.

18. Petilini a Poenis obsessi parentes et liberos propter inopiam eiecerunt, ipsi coriis madefactis et igne siccatis foliisque arborum et omni genere animalium vitam trahentes undecim menses obsidionem toleraverunt.

19. Hispani Consabrae obsessi eadem omnia passi sunt nec oppidum Hirtuleio tradiderunt.

20. Casilini obsidente Hannibale tantam inopiam perpessi sunt, ut CC denariis murem venisse proditum memoriae sit eiusque venditorem fame perisse, emptorem autem vixisse. Fidem tamen servare Romanis perseveraverunt.

21. Cyzicum cum oppugnaret Mithridates, captivos eius urbis produxit ostenditque obsessis, arbitratus futurum, ut miseratione suorum compelleret ad deditionem oppidanos; at illi cohortati ad patiendam fortiter mortem captivos servare Romanis fidem perseveraverunt.

22. Segovienses, cum a Viriatho his liberi et con-

---

[1] 130 B.C. Crassus was proconsul at the time of his death. *Cf.* Val. Max. III. ii. 12; Flor. II. xx. 4–5; Oros. v. x 1–4.

[2] 168 B.C. Battle of Pydna. *Cf.* Val. Max. III. ii. 16; Justin. xxxiii. 2. Plut. *Aemil.* 21 gives a slightly different story.

[3] 216 B.C. *Cf.* Livy xxiii. 20, 30; Val. Max. VI. vi. ext. 2; Appian *Hann.* 29. [4] 79–75 B.C. [5] About £6 15s.

[6] 216 B.C. *Cf.* Val. Max. VII. vi. 2, 3; Plin. *N.H.* VIII. lvii. 82; Livy xxiii. 19. [7] 74 B.C. *Cf.* Appian *Mith.* 73.

stabbed him to death. Thus, as he desired, Crassus escaped the disgrace of servitude.[1]

Marcus, son of Cato the Censor, in a certain battle fell off his horse, which had stumbled. Cato picked himself up, but noticing that his sword had slipped out of its scabbard and fearing disgrace, went back among the enemy, and though he received a number of wounds, finally recovered his sword and made his way back to his comrades.[2]

The inhabitants of Petelia, when they were blockaded by the Carthaginians, sent away the children and the aged, on account of the shortage of food. They themselves, supporting life on hides, moistened and then dried by the fire, on leaves of trees, and on all sorts of animals, sustained the siege for eleven months.[3]

The Spaniards, when blockaded at Consabra, endured all these same hardships; nor did they surrender the town to Hirtuleius.[4]

The story goes that the inhabitants of Casilinum, when blockaded by Hannibal, suffered such shortage of food that a mouse was sold for two hundred denarii,[5] and that the man who sold it died of starvation, while the purchaser lived. Yet the inhabitants persisted in maintaining their loyalty to the Romans.[6]

When Mithridates was besieging Cyzicus, he paraded the captives from that city and exhibited them to the besieged, thinking thus to force the people of the town to surrender, through compassion for their fellows. But the townspeople urged the prisoners to meet death with heroism, and persisted in maintaining their loyalty to the Romans.[7]

The inhabitants of Segovia, when Viriathus pro-

SEXTUS JULIUS FRONTINUS

iuges redderentur, praeoptaverunt spectare supplicia pignorum suorum, quam a Romanis deficere.

23. Numantini, ne se dederent, fame mori praefixis foribus domuum suarum maluerunt.

VI. DE AFFECTU ET MODERATIONE

1. Q. FABIUS hortante filio, ut locum idoneum paucorum iactura caperet, "visne," inquit, "tu ex illis paucis esse?"

2. Xenophon, cum equo veheretur et pedites iugum quoddam occupare iussisset, unum ex eis obmurmurantem audiens facile tam laboriosa sedentem imperare, desiluit et gregalem equo imposuit, cursu ipse ad destinatum iugum contendens. Cuius facti ruborem cum perpeti miles non posset, inridentibus commilitonibus sponte descendit. Xenophontem vix universi perpulerunt, ut conscenderet equum et laborem suum in necessaria duci munera reservaret.

3. Alexander, cum hieme duceret exercitum, residens ad ignem recognoscere praetereuntis copias coepit; cumque conspexisset quendam prope exanimatum frigore, considere loco suo iussit dixitque ei:

---

[1] 147–139 B.C.
[2] 133 B.C. *Cf.* Livy *Per.* 59; Val. Max. III. ii. ext. 7, VII. vi. ext. 2; Sen. *de Ira* i. 11.
[3] Quintus Fabius Maximus Cunctator. *Cf.* Sil. Ital. vii. 539 ff. Plut. *Apophth. Caec. Metell.* relates a similar reply of Metellus.
[4] 401 B.C. *Cf.* Xen. *Anab.* III. iv. 44–49.

posed to send them back their wives and children, preferred to witness the execution of their loved ones rather than to fail the Romans.[1]

The inhabitants of Numantia preferred to lock the doors of their houses and die of hunger rather than surrender.[2]

VI. On Good Will and Moderation

Quintus Fabius,[3] upon being urged by his son to seize an advantageous position at the expense of losing a few men, asked: "Do you want to be one of those few?"

When Xenophon on one occasion happened to be on horseback and had just ordered the infantry to take possession of a certain eminence, he heard one of the soldiers muttering that it was an easy matter for a mounted man to order such difficult enterprises. At this Xenophon leaped down and set the man from the ranks on his horse, while he himself hurried on foot with all speed to the eminence he had indicated. The soldier, unable to endure the shame of this performance, voluntarily dismounted amid the jeers of his comrades. It was with difficulty, however, that the united efforts of the troops induced Xenophon to mount his horse and to restrict his energies to the duties which devolved upon a commander.[4]

When Alexander was marching at the head of his troops one winter's day, he sat down by a fire and began to review the troops as they passed by. Noticing a certain soldier who was almost dead with the cold, he bade him sit in his place, adding: "If

SEXTUS JULIUS FRONTINUS

" si in Persis natus esses, in regia sella resedisse tibi capital foret, in Macedonia nato conceditur."

4. Divus Augustus Vespasianus, cum quendam adulescentem honeste natum, militiae inhabilem, angustiarum rei familiaris causa deductum ad longiorem ordinem rescisset, censu constituto honesta missione exauctoravit.

VII. De Variis Consiliis

1. C. Caesar dicebat idem sibi esse consilium adversus hostem, quod plerisque medicis contra vitia corporum, fame potius quam ferro superandi.

2. Domitius Corbulo dolabra hostem vincendum esse dicebat.

3. L. Paulus imperatorem senem moribus dicebat esse oportere, significans moderatiora sequenda consilia.

4. Scipio Africanus fertur dixisse, cum eum parum quidam pugnacem dicerent: " imperatorem me mater, non bellatorem peperit."

5. C. Marius Teutono provocanti eum et postulanti, ut prodiret, respondit, si cupidus mortis esset, laqueo posse eum vitam finire; cum deinde instaret, gladiatorem contemptae staturae et prope exactae aetatis obiecit ei dixitque, si eum superasset, cum victore congressurum.

---

[1] *Cf.* Val. Max. v. i. ext. 1; Curt. viii. iv. 15–17.
[2] *Cf.* Veget. iii. 26; Appian *Hisp.* 87.
[3] *Cf.* Livy xliv. 36; Gell. xiii. iii. 6; Plut. *Apophth. Aemil.* 5.

you had been born among the Persians, it would be a capital crime for you to sit on the king's seat; but since you were born in Macedonia, that privilege is yours."[1]

When the Deified Vespasianus Augustus learned that a certain youth, of good birth, but ill adapted to military service, had received a high appointment because of his straitened circumstances, Vespasian settled a sum of money on him, and gave him an honourable discharge.

VII. On Sundry Maxims and Devices.

Gaius Caesar used to say that he followed the same policy towards the enemy as did many doctors when dealing with physical ailments, namely, that of conquering the foe by hunger rather than by steel.[2]

Domitius Corbulo used to say that the pick was the weapon with which to beat the enemy.

Lucius Paulus used to say that a general ought to be an old man in character, meaning thereby that moderate counsels should be followed.[3]

When people said of Scipio Africanus that he lacked aggressiveness, he is reported to have answered: "My mother bore me a general, not a warrior."

When a Teuton challenged Gaius Marius and called upon him to come forth, Marius answered that, if the man was desirous of death, he could end his life with a halter. Then, when the fellow persisted, Marius confronted him with a gladiator of despicable size, whose life was almost spent, and told the Teuton that, if he would first defeat this gladiator, he himself would then fight with him.

SEXTUS JULIUS FRONTINUS

6. Q. Sertorius, quod experimento didicerat imparem se universo Romanorum exercitui, ut barbaros quoque inconsulte pugnam deposcentis doceret, adductis in conspectum duobus equis, eorum praevalido alteri, alteri admodum exili duos admovit iuvenes similiter electos, robustum et gracilem. Ac robustiori imperavit equi exilis universam caudam abrumpere, gracili autem valentioris per singulos pilos vellere; cumque gracili successisset quod imperatum erat, validissimus cum infirmi equi cauda sine effectu luctaretur, "naturam," inquit Sertorius, "Romanarum virium per hoc vobis exemplum ostendi, milites: insuperabiles sunt universas adgredienti; easdem lacerabit et carpet, qui per partes adtemptaverit."[1]

7. Valerius Laevinus consul, cum intra castra sua exploratorem hostium deprehendisset magnamque copiarum suarum fiduciam haberet, circumduci eum iussit: terrendi quidem hostis causa exercitus suos visendos speculatoribus eorum, quotiens voluissent, patere.

8. Caedicius primipilaris, qui in Germania post Varianam cladem obsessis nostris pro duce fuit, veritus, ne barbari ligna quae congesta erant vallo admoverent et castra eius incenderent, simulata lignorum inopia, missis undique qui ea furarentur effecit, ut Germani universos truncos amolirentur.

---

[1] *Practically identical with* i. x. 1.

---

[1] *Cf.* I. x. 1 and note.
[2] 280 B.C. *Cf.* Eutrop. ii. 11; Zonar. viii. 3. Livy xxx. 29, Appian *Pun.* 39 and Polyaen. VIII. xvi. 8 attributes a similar stratagem to Scipio.
[3] The defeat of Varus by Arminius in the Teutoburg Forest in 9 A.D.
[4] *Cf.* Vell. ii. 120.

After Quintus Sertorius had learned by experience that he was by no means a match for the whole Roman army, and wished to prove this to the barbarians also, who were rashly demanding battle, he brought into their presence two horses, one very strong, the other very feeble. Then he brought up two youths of corresponding physique, one robust, the other slight. The stronger youth was commanded to pull out the entire tail of the feeble horse, while the slender youth was ordered to pull out the hairs of the strong horse, one by one. Then, when the slight youth had succeeded in his task, while the strong one was still struggling vainly with the tail of the feeble horse, Sertorius observed: " By this illustration I have exhibited to you, my men, the nature of the Roman cohorts. They are invincible to him who attacks them in a body; but he who assails them by groups, will tear and rend them."[1]

The consul Valerius Laevinus, having caught a spy within his camp, and having entire confidence in his own forces, ordered the man to be led around, observing that, for the sake of terrifying the enemy, his army was open to inspection by the spies of the enemy, as often as they wished.[2]

Caedicius, a centurion of the first rank, who acted as leader in Germany, when, after the Varian disaster,[3] our men were beleaguered, was afraid that the barbarians would bring up to the fortifications the wood which they had gathered, and would set fire to his camp. He therefore pretended to be in need of fuel, and sent out men in every direction to steal it. In this way he caused the Germans to remove the whole supply of felled trees.[4]

SEXTUS JULIUS FRONTINUS

9. Cn. Scipio bello navali amphoras pice et taeda plenas in hostium classem iaculatus est, quarum iactus et pondere foret noxius et diffundendo, quae continuerant, alimentum praestaret incendio.

10. Hannibal regi Antiocho monstravit, ut in hostium classem vascula iacularetur viperis plena, quarum metu perterriti milites a dimicatione et nauticis ministeriis impedirentur.

11. Idem fecit iam cedente classe sua Prusias.[1]

12. M. Porcius Cato, in classem hostium cum transiluisset, deturbatis ex ea Poenis eorumque armis et insignibus inter suos distributis multas naves hostium, quos sociali habitu fefellerat, mersit.

13. Athenienses, cum subinde a Lacedaemoniis infestarentur, diebus festis, quos sacros Minervae extra urbem celebrabant, omnem quidem colentium imitationem expresserunt, armis tamen sub[2] veste celatis. Peracto ritu suo non statim Athenas reversi, sed protinus inde raptim acto Lacedaemonem versus agmine eo tempore, quo minime timebantur, agrum hostium, quibus subinde praedae fuerant, ultro depopulati sunt.

14. Cassius onerarias naves, non magni ad alia usus, accensas opportuno vento in classem hostium misit et incendio eam consumpsit.

15. M. Livius fuso Hasdrubale hortantibus eum

[1] *No. 11 is probably interpolated.*
[2] sub, *inferior MSS.*; et, *better MSS. and edd.*

---

[1] Justin. xxxii. 4 and Nep. *Hann.* 10 represent Hannibal as suggesting this device to Prusias.
[2] 48 B.C. *Cf.* Caes. *B.C.* iii. 101.

STRATAGEMS, IV. vii. 9-15

Gnaeus Scipio, in a naval combat, hurled jars filled with pitch and rosin among the vessels of the enemy, in order that damage might result both from the weight of the missiles and from the scattering of their contents, which would serve as fuel for a conflagration.

Hannibal suggested to King Antiochus that he hurl jars filled with vipers among the ships of the enemy, in order that the crews, through fear of these, might be kept from fighting and from performing their nautical duties.[1]

Prusias did the same, when his fleet was by now giving way [1]

Marcus Porcius Cato, having boarded the ships of the enemy, drove from them the Carthaginians. Then, having distributed their weapons and insignia among his own men, he sank many ships of the enemy, deceiving them by their own equipment.

Inasmuch as the Athenians had been subject to repeated attacks by the Spartans, on one occasion, in the course of a festival which they were celebrating outside the city in honour of Minerva, they studiously affected the rôle of worshippers, yet with weapons concealed beneath their clothing. When the ceremonial was over, they did not immediately return to Athens, but at once marched swiftly upon Sparta at a time when they were least feared, and themselves devastated the lands of an enemy whose victims they had often been.

Cassius set fire to some transports which were of no great use for anything else, and sent them with a fair wind against the fleet of the enemy, thereby destroying it by fire.[2]

When Marcus Livius had routed Hasdrubal, and

SEXTUS JULIUS FRONTINUS

quibusdam, ut hostem ad internecionem persequeretur, respondit: "aliqui et supersint, qui de victoria nostra hostibus nuntient."

16. Scipio Africanus dicere solitus est hosti non solum dandam esse viam ad fugiendum, sed etiam muniendam.

17. Paches Atheniensis adfirmavit incolumes futuros hostes, si deponerent ferrum; eisque obsecutis condicionibus universos, qui in sagulis ferreas fibulas habuissent, interfici iussit.

18. Hasdrubal subigendorum Numidarum causa ingressus fines eorum resistere parantibus adfirmavit ad capiendos se venisse elephantos, quibus ferax est Numidia; ut hoc permitterent, poscentibus pretium cum promisisset, ea persuasione avocatos adortus sub leges redegit.

19. Alcetas Lacedaemonius, ut Thebanorum commeatum facilius ex inopinato adgrederetur, in occulto paratis navibus, tamquam unam omnino haberet triremem, vicibus in ea remigem exercebat; quodam deinde tempore omnis naves in Thebanos transnavigantis immisit et commeatibus eorum potitus est.

20. Ptolomaeus adversus Perdiccam exercitu praevalentem, ipse invalidus, omne pecudum genus, religatis ad tergum quae traherent sarmentis, agendum per paucos curavit equites; ipse praegressus cum copiis quas habebat effecit, ut pulvis, quem pecora

---

[1] 207 B.C. *Cf.* Livy xxvii. 49.
[2] *Cf.* Veget. iii. 21.
[3] Thuc. iii. 34 and Polyaen. iii. 2 cite another instance of the cunning treachery of Paches in 427 B.C.
[4] 377 B.C. *Cf.* Polyaen. ii. 7 Caes. *B.C.* iii. 24 relates a similar stratagem employed by Antony.

certain persons urged him to pursue the enemy to annihilation, he answered: "Let some survive to carry to the enemy the tidings of our victory!"[1]

Scipio Africanus used to say that a road not only ought to be afforded the enemy for flight, but that it ought even to be paved.[2]

Paches, the Athenian, on one occasion declared that the enemy would be spared, if they put aside the steel. When they had all complied with these terms, he ordered the entire number to be executed, since they had steel brooches on their cloaks.[3]

When Hasdrubal had invaded the territory of the Numidians for the purpose of subduing them, and they were preparing to resist, he declared that he had come to capture elephants, an animal in which Numidia abounds. For this privilege they demanded money, and Hasdrubal promised to pay it. Having by these representations thrown them off the scent, he attacked them and brought them under his power.

Alcetas, the Spartan, in order the more easily to make a surprise attack on a supply convoy of the Thebans, got ready his ships in a secret place, and exercised his rowers by turns on a single galley, as though that was all he had. Then at a certain time, as the Theban vessels were sailing past, he sent all his ships against them and captured their supplies.[4]

When Ptolemy with a weak force was contending against Perdiccas's powerful army, he arranged for a few horsemen to drive along animals of all sorts, with brush fastened to their backs for them to trail behind them. He himself went ahead with the forces which he had. As a consequence, the dust

SEXTUS JULIUS FRONTINUS

excitaverant, speciem magni sequentis exercitus moveret, cuius exspectatione territum vicit hostem.

21. Myronides Atheniensis adversus Thebanos equitatu praevalentes pugnaturus in campis suos edocuit manentibus esse spem aliquam salutis, cedentibus autem perniciem certissimam ; qua ratione confirmatis militibus victoriam consecutus est.

22. C. Pinarius in Sicilia praesidio Hennae praepositus, claves portarum quas penes se habebat reposcentibus magistratibus Hennensium, quod suspectos eos tamquam transitionem ad Poenum pararent habebat, petiit unius noctis ad deliberationem spatium indicataque militibus fraude Graecorum, cum praecepisset ut parati postera die signum exspectarent, prima luce adsistentibus militibus redditurum se claves dixit, si idem omnes Hennenses censuissent; ob eam causam universa multitudine convocata in theatrum et idem flagitante, manifesta deficiendi voluntate, signo militibus dato universos Hennenses cecidit.

23. Iphicrates dux Atheniensium classem suam hostili habitu instruxit et ad eos quos suspectos habebat invectus, cum effuso studio exciperetur, deprehensa eorum perfidia oppidum diripuit.

24. Ti. Gracchus, cum edixisset futurum, ut ex

---

[1] 321 B.C. *Cf.* Polyaen. iv. 19. Front. II. iv. 1 relates a somewhat similar device of Papirius Cursor.
[2] 457 B.C. *Cf.* Polyaen. I. xxxv. 2.
[3] 214 B.C. *Cf.* Livy xxiv. 37–39; Polyaen. viii. 21.
[4] 390–389 B.C. *Cf.* Polyaen. III. ix. 58.

raised by the animals produced the appearance of a mighty army following, and the enemy, terrified by this impression, were defeated.[1]

Myronides, the Athenian, when about to fight on an open plain against the Thebans, who were very strong in cavalry, warned his troops that, if they stood their ground, there was some hope of safety, but that, if they gave way, destruction was absolutely certain. In this way he encouraged his men and won the victory.[2]

When Gaius Pinarius was in charge of the garrison of Henna in Sicily, the magistrates of the city demanded the keys of the gates, which he had in his keeping. Suspecting that they were preparing to go over to the Carthaginians, he asked for the space of a single night to consider the matter; and, revealing to his soldiers the treachery of the Greeks, he instructed them to get ready and wait for his signal on the morrow. At daybreak, in the presence of his troops, he announced to the people of Henna that he would surrender the keys, if all the inhabitants of the town should be agreed in their view. When the entire populace assembled in the theatre to settle this matter, and, with the obvious purpose of revolting, made the same demand, Pinarius gave the signal to his soldiers and murdered all the people of Henna.[3]

Iphicrates, the Athenian general, once rigged up his own fleet after the style of the enemy, and sailed away to a certain city whose people he viewed with suspicion. Being welcomed with unrestrained enthusiasm, he thus discovered their treachery and sacked their town.[4]

When Tiberius Gracchus had proclaimed that he

SEXTUS JULIUS FRONTINUS

volonum numero fortibus libertatem daret, ignavos crucibus adfigeret, et quattuor milia ex his, quia segnius pugnaverant, metu poenae in quendam munitum collem coissent, misit qui eis dicerent totum sibi exercitum volonum vicisse videri, quod hostes fudissent; et sic eos et sua fide et ipsorum metu exsolutos recepit.

25. Hannibal post proelium, quo ingentem cladem ad Trasumennum Romani acceperunt, cum sex milia hostium interposita pactione in potestatem suam redegisset, socios Latini nominis benigne in civitates suas dimisit, dictitans se Italiae liberandae causa bellum gerere; eorumque opera aliquot populos in deditionem accepit.

26. Mago, cum Locri obsiderentur a Crispino classis nostrae praefecto, diffudit ad Romana castra rumorem Hannibalem caeso Marcello ad liberandos obsidione Locros venire; clam deinde equites emissos iussit a montibus, qui in conspectu erant, se ostendere; quo facto effecit, ut Crispinus Hannibalem adesse ratus conscenderet naves ac fugeret.

27. Scipio Aemilianus ad Numantiam omnibus non cohortibus tantum, sed centuriis sagittarios et funditores interposuit.

---

[1] 214 B.C. *Cf.* Livy xxiv. 14–16.
[2] *i.e.* the Latin League. The cities of Latium were from very early times united in an alliance with Rome. At first this bond was of a political nature, on a basis of perfect equality; later Rome became the leading power and assumed the supremacy.
[3] 217 B.C. *Cf.* Livy xxii. 6, 7, 13; Polyb. iii. 77, 84–85.
[4] 208 B.C. *Cf.* Livy xxvii. 28.

would confer freedom on such of the volunteer slaves as showed courage, but would crucify the cowards, some four thousand men who had fought rather listlessly, gathered on a fortified hill in fear of punishment. Thereupon Gracchus sent men to tell them that in his opinion the whole force of volunteer slaves had shared in the victory, since they had routed the enemy. By this expression of confidence he freed them from their apprehensions and took them back again.[1]

After the battle of Lake Trasimenus, where the Romans suffered great disaster, Hannibal, having brought six thousand of the enemy under his power by virtue of a covenant he had made, generously allowed the allies of the " Latin Name "[2] to return to their cities, declaring that he was waging war for the purpose of freeing Italy. As a result, by means of their assistance he received in surrender a number of tribes.[3]

When Locri was blockaded by Crispinus, admiral of our fleet, Mago spread the rumour in the Roman camp that Hannibal had slain Marcellus and was coming to relieve Locri from blockade. Then, secretly sending out cavalry, he commanded them to show themselves on the mountains, which were in view. By doing this, he caused Crispinus, in the belief that Hannibal was at hand, to board his vessels and make off.[4]

Scipio Aemilianus, in the operations before Numantia, distributed archers and slingers not only among all his cohorts, but even among all the centuries.[5]

[5] 133 B.C. Veget. i. 15 narrates instances of the important part played in Roman battles by archers and javelin throwers, and emphasizes the necessity of training in archery.

SEXTUS JULIUS FRONTINUS

28. Pelopidas Thebanus, cum a Thessalis in fugam versus flumen, in quo tumultuarium fecerat pontem, liberasset, ne sequentibus hostibus idem transitus maneret, novissimo agmini praecepit, incenderent pontem.

29. Romani, cum Campanis equitibus nullo modo pares essent, Q. Naevius centurio in exercitu Fulvi Flacci proconsulis excogitavit, ut delectos ex toto exercitu, qui velocissimi videbantur et mediocris erant staturae, parmulis non amplis et galeiculis gladiisque ac septenis singulos hastis quaternorum circiter pedum armaret eosque adiunctos equitibus iuberet usque ad moenia provehi, deinde ibi positos, nostris equitibus se recipientibus, inter hostium equitatum proeliari; quo facto vehementer et ipsi Campani adflicti sunt et maxime equi eorum, quibus turbatis prona nostris victoria fuit.

30. P. Scipio in Lydia, cum die ac nocte imbre continuo vexatum exercitum Antiochi videret nec homines tantum aut equos deficere, verum arcus quoque madentibus nervis inhabiles factos, exhortatus est fratrem, ut postero quamvis religioso die committeret proelium; quam sententiam secuta victoria est.

31. Catonem vastantem Hispaniam legati Ilergetum, qui sociorum populus erat, adierunt oraveruntque auxilia. Ille, ne aut negato adiutorio socios

---

[1] 369–364 B.C.
[2] 211 B.C. *Cf.* Livy xxvi. 4; Val. Max. II. iii. 3.
[3] 190 B.C. *Cf.* Livy xxxvii. 37, 39 ff.; Flor. II. viii. 17.

When Pelopidas, the Theban, had been put to flight by the Thessalians and had crossed the river over which he had constructed an emergency bridge, he ordered his rearguard to burn the bridge, in order that it might not serve also as a means of passage to the enemy who were following him.[1]

When the Romans in certain operations were no match for the Campanian cavalry, Quintus Naevius, a centurion in the army of Fulvius Flaccus, the proconsul, conceived the plan of picking from the whole army the men who seemed swiftest of foot and of medium stature, arming them with small shields, helmets, and swords, and giving to each man seven spears, about four feet in length. These men he attached to the cavalry, and commanded them to advance to the very walls, and then, taking their position at that point, to fight amid the cavalry of the enemy, when our cavalry retreated. By this means the Campanians suffered severely, and especially their horses. When these were thrown into confusion, victory became easy for our troops.[2]

When Publius Scipio was in Lydia, and observed that the army of Antiochus was demoralized by the rain, which fell day and night without cessation, and when he further noted that not only were men and horses exhausted, but that even the bows were rendered useless from the effect of the dampness on their strings, he urged his brother to engage in battle on the following day, although it was consecrated to religious observance. The adoption of this plan was followed by victory.[3]

When Cato was ravaging Spain, the envoys of the Ilergetes, a tribe allied with the Romans, came to him and begged for assistance. Cato, unwilling either

SEXTUS JULIUS FRONTINUS

alienaret aut diducto exercitu vires minueret, tertiam partem militum cibaria parare et naves ascendere iussit, dato praecepto, ut causati ventos retro redirent; praecedens interim adventantis auxilii rumor ut Ilergetum excitavit animos, ita hostium consilia discussit.

32. C. Caesar, cum in partibus Pompeianis magna equitum Romanorum esset manus eaque armorum scientia milites conficeret, ora oculosque eorum gladiis peti iussit et sic aversa facie cedere coegit.

33. Voccaei, cum a Sempronio Graccho conlatis signis urgerentur, universas copias cinxere plaustris, quae impleverant fortissimis viris muliebri veste tectis. Sempronium, tamquam adversus feminas audentius ad obsidendos hostis consurgentem, hi qui in plaustris erant adgressi fugaverunt.

34. Eumenes Cardianus, unus[1] ex successoribus Alexandri, in castello quodam clausus, quoniam exercere equos non poterat, certis cotidie horis ita suspendebat, ut posterioribus pedibus innixi, prioribus allevatis, dum naturalem adsistendi appetunt consuetudinem, ad sudorem usque crura iactarent.

35. M. Cato pollicentibus barbaris duces itinerum et insuper praesidium, si magna summa eis promit-

---

[1] *Suggested by Gundermann.*

[1] 195 B.C. *Cf.* Livy xxxiv. 11–13.
[2] 48 B.C. *Cf.* Plut. *Caes.* 44–45, *Pomp.* 69; Polyaen. VIII. xxiii. 25. [3] 179–178 B.C.
[4] 320 B.C. *Cf.* Nep. *Eumen.* 5; Plut. *Eumen.* 11; Diodor. xviii. 42.

to alienate his allies by refusing aid, or to diminish his own strength by dividing his forces, ordered a third part of his soldiers to prepare rations and embark on their ships, directing them to return and to allege head winds as the reason for this action. Meanwhile the report of approaching aid went on before them, raising the hopes of the Ilergetes, and shattering the plans of the enemy.[1]

Since in the army of Pompey there was a large force of Roman cavalry, which by its skill in arms wrought havoc among the soldiers of Gaius Caesar, the latter ordered his troops to aim with their swords at the faces and eyes of the enemy. He thus forced the enemy to avert their faces and retire.[2]

When the Voccaei were hard pressed by Sempronius Gracchus in a pitched battle, they surrounded their entire force with a ring of carts, which they had filled with their bravest warriors dressed in women's clothes. Sempronius rose up with greater daring to assault the enemy, because he imagined himself proceeding against women, whereupon those in the carts attacked him and put him to flight.[3]

When Eumenes of Cardia, one of the successors of Alexander, was besieged in a certain stronghold, and was unable to exercise his horses, he had them suspended during certain hours each day in such a position that, resting on their hind legs and with their fore feet in the air, they moved their legs till the sweat ran, in their efforts to regain their natural posture.[4]

When certain barbarians promised Marcus Cato guides for the march and also reinforcements, provided that a large sum of money should be assured

SEXTUS JULIUS FRONTINUS

teretur, non dubitavit polliceri, quia aut victoribus ex spoliis hostilibus poterat dare aut interfectis exsolvebatur promisso.

36. Q. Maximus transfugere ad hostes volentem Statilium, nobilem clarae operae equitem, vocari ad se iussit eique excusavit, quod invidia commilitonum virtutes illius ad id tempus ignorasset; tum donato ei equo pecuniam insuper largitus obtinuit, ut, quem ex conscientia trepidum arcessierat, laetum dimitteret et ex dubio in reliquum non minus fidelem quam fortem haberet equitem.

37. Philippus, cum audisset Pythian quendam bonum pugnatorem alienatum sibi, quod tres filias inops vix aleret nec a rege adiuvaretur, monentibus quibusdam, uti eum caveret, "quid? si," inquit, "partem aegram corporis haberem, absciderem potius quam curarem?" deinde familiariter secreto elicitum Pythian, accepta difficultate necessitatium domesticarum, pecunia instruxit ac meliorem fidelioremque habuit, quam habuerat antequam offenderet.

38. T. Quintius Crispinus post infaustam adversus Poenos dimicationem, qua collegam Marcellum amiserat, cum comperisset potitum anulo interfecti Hannibalem, litteras circa municipia totius Italiae dimisit, ne crederent epistulis, si quae Marcelli anulo

---

[1] 195 B.C. *Cf.* Plut. *Cat. Maj.* 10.
[2] *Cf.* III. xvi. 1 and note; Plut. *Fab.* 20.
[3] The father of Alexander.

them, he did not hesitate to make the promise, since, if they won, he could reward them from the spoils of the enemy, while, if they were slain, he would be released from his pledge.[1]

When a certain Statilius, a knight of distinguished record, evinced an inclination to desert to the enemy, Quintus Maximus ordered him to be summoned to his presence, and apologized for not having known until then the real merits of Statilius, owing to the jealousy of his fellow-soldiers. Then, giving Statilius a horse and bestowing a large gift of money besides, he succeeded in sending away rejoicing a man who, when summoned, was conscience-stricken; he succeeded also in securing for the future a loyal and brave knight in place of one whose fealty was in doubt.[2]

Philip,[3] having heard that a certain Pythias, an excellent warrior, had become estranged from him because he was too poor to support his three daughters, and was not assisted by the king, and having been warned by certain persons to be on his guard against the man, replied: "What! If part of my body were diseased, should I cut it off, rather than give it treatment?" Then, quietly drawing Pythias aside for a confidential talk, and learning the seriousness of his domestic embarrassments, he supplied him with funds, and found in him a better and more devoted adherent than before the estrangement.

After an unsuccessful battle with the Carthaginians, in which he had lost his colleague Marcellus, Titius Quinctius Crispinus, learning that Hannibal had obtained possession of the ring of the slain hero, sent letters among all the municipal towns of Italy, warning the inhabitants to give credit to no letters which should be brought sealed with the ring of

SEXTUS JULIUS FRONTINUS

signatae perferrentur. Monitione consecutus est, ut Salapia et aliae urbes frustra Hannibalis dolis temptarentur.

39. Post Cannensem cladem perculsis ita Romanorum animis, ut pars magna reliquiarum nobilissimis auctoribus deserendae Italiae iniret consilium, P. Scipio adulescens admodum impetu facto in eo ipso in quo talia agitabantur coetu pronuntiavit manu se sua interfecturum, nisi qui iurasset non esse sibi mentem destituendae rei publicae; cumque ipse se primus religione tali obligasset, stricto gladio mortem uni ex proximis minatus, nisi acciperet sacramentum, illum metu, ceteros etiam exemplo coegit ad iurandum.

40. Volscorum castra cum prope a virgultis silvaque posita essent, Camillus ea omnia, quae conceptum ignem usque in vallum perferre poterant, incendit et sic adversarios exuit castris.[1]

41. P. Crassus bello sociali eodem modo prope cum copiis omnibus interceptus est.[2]

42. Q. Metellus in Hispania castra moturus, cum in agmine milites continere vellet, pronuntiavit comperisse se insidias ab hostibus dispositas; idcirco ne discederent a signis neve agmen laxarent. Quod cum ex disciplina fecisset, exceptus forte veris insidiis, quia praedixerat, interritos milites habuit.

---

[1] *Identical with* II. iv. 15.
[2] *Identical with* II. iv. 16.

---

[1] 208 B.C. *Cf.* Livy xxvii. 28.
[2] 216 B.C. *Cf.* Livy xxii. 53; Val. Max. v. vi. 7; Sil. Ital. x. 426 ff
[3] 389 B.C. *Cf.* Livy VI. ii. 9–11; Plut. *Cam.* 34.
[4] 90 B.C.   [5] 143–142 B.C.

Marcellus. As a result of this advice, Salapia and other cities were assailed in vain by Hannibal's insidious efforts.[1]

After the disaster at Cannae, when the Romans were so terror-stricken that a large part of the survivors thought of abandoning Italy, and that too with the endorsement of nobles of the highest standing, Publius Scipio, then extremely young, in the very assembly where such a course was being discussed, proclaimed with great vehemence that he would slay with his own hand whoever refused to declare on oath that he cherished no purpose of abandoning the State. Having first bound himself with such an oath, he drew his sword and threatened death to one of those standing near unless he too should take the oath. This man was constrained by fear to swear allegiance; the rest were compelled by the example of the first.[2]

When the camp of the Volscians had been pitched near bushes and woods, Camillus set fire to everything which could carry the flames, once started, up to the very fortifications. In this way he deprived the enemy of their camp.[3]

In the Social War Publius Crassus was cut off in almost the same way with all his troops.[4]

When Quintus Metellus was about to break camp in Spain and wished to keep his soldiers in line, he proclaimed that he had discovered that an ambush had been laid by the enemy; therefore the soldiers should not quit the standards nor break ranks. Though he had done this merely for purposes of discipline, yet happening to meet with an actual ambuscade, he found his soldiers unafraid, since he had given them warning.[5]

SEXTUS JULIUS FRONTINUS
THE AQUEDUCTS OF ROME

IVLII FRONTINI
DE AQVIS VRBIS ROMAE
LIBRI II

1. Cum omnis res ab imperatore delegata intentiorem exigat curam, et me seu naturalis sollicitudo seu fides sedula non ad diligentiam modo verum ad amorem quoque commissae rei instigent sitque nunc mihi ab Nerva Augusto, nescio diligentiore an amantiore rei publicae imperatore, aquarum iniunctum officium ad usum, tum ad salubritatem atque etiam securitatem urbis pertinens, administratum per principes semper civitatis nostrae viros, primum ac potissimum existimo, sicut in ceteris negotiis institueram, nosse quod suscepi.

2. Neque enim ullum omnis actus certius fundamentum crediderim, aut aliter quae facienda quaeque vitanda sint posse decerni, aliudve tam indecorum tolerabili viro, quam delegatum officium ex adiutorum

---

[1] Praenomen not in *C* but attested by CIL. vi. 2222, viii. 7066, ix. 6083. 78.

SEXTUS[1] JULIUS FRONTINUS
TWO BOOKS ON
THE AQUEDUCTS OF ROME

Inasmuch as every task assigned by the Emperor demands especial attention; and inasmuch as I am incited, not merely to diligence, but also to devotion, when any matter is entrusted to me, be it as a consequence of my natural sense of responsibility or of my fidelity; and inasmuch as Nerva Augustus (an emperor of whom I am at a loss to say whether he devotes more industry or love to the State) has laid upon me the duties of water commissioner, an office which concerns not merely the convenience but also the health and even the safety of the City, and which has always been administered by the most eminent men of our State; now therefore I deem it of the first and greatest importance to familiarize myself with the business I have undertaken, a policy which I have always made a principle in other affairs.

For I believe that there is no surer foundation for any business than this, and that it would be otherwise impossible to determine what ought to be done, what ought to be avoided; likewise that there is nothing so disgraceful for a decent man as to conduct an office delegated to him, according to the instructions

SEXTUS JULIUS FRONTINUS

agere praeceptis, quod fieri necesse est, quotiens imperitia praepositi ad illorum decurrit usum; quorum etsi necessariae partes sunt ad ministerium, tamen ut manus quaedam et instrumentum agentis. . . .[1] Quapropter ea quae ad universam rem pertinentia contrahere potui, more iam per multa mihi officia servato in ordinem et velut corpus diducta in hunc commentarium contuli, quem pro formula administrationis respicere possem. In aliis autem libris, quos post experimenta et usum composui, succedentium res acta est; huius commentarii pertinebit fortassis et ad successorem utilitas, sed cum inter initia administrationis meae scriptus sit, in primis ad meam[2] institutionem regulamque proficiet.

3. Ac ne quid ad totius rei pertinens notitiam praetermisisse videar, nomina primum aquarum, quae in urbem Romam influunt, ponam; tum per quos quaeque earum et quibus consulibus, quoto post urbem conditam anno perducta sit; dein quibus ex locis et a quoto miliario capta sit[3] ac quantum subterraneo rivo, quantum substructione, quantum opere arcuato; post altitudinem cuiusque modulorumque . . .[4] erogationes ab illis factae sint,

---

[1] *Gap of about eleven letters in C. In this edition, the number of letters omitted in C will be indicated by a corresponding number of dots in the text, except where footnotes are added giving the size of the lacunae. In these cases the lacunae will regularly be indicated by three dots.*
[2] iñā *C*; nostram *B*. [3] *Bennett*; cepisse *C*; coepisset *B*.
[4] *No gap in C, but B suggests supplying* rationes.

---

[1] When it was necessary to carry the water pipes at a high elevation, the arched support was used in order to save

AQUEDUCTS OF ROME

of assistants. Yet precisely this is inevitable whenever a person inexperienced in the matter in hand has to have recourse to the practical knowledge of subordinates For though the latter play a necessary rôle in the way of rendering assistance, yet they are, as it were, but the hands and tools of the directing head. Observing, therefore, the practice which I have followed in many offices, I have gathered in this sketch (into one systematic body, so to speak) such facts, hitherto scattered, as I have been able to get together, which bear on the general subject, and which might serve to guide me in my administration. Now in the case of other books which I have written after practical experience, I consulted the interests of my successors. The present treatise also may be found useful by my successor, but it will serve especially for my own instruction and guidance, being prepared, as it is, at the beginning of my administration.

And lest I seem to have omitted anything requisite to a familiarity with the entire subject, I will first set down the names of the waters which enter the City of Rome; then I will tell by whom, under what consuls, and in what year after the founding of the City each one was brought in; then at what point and at what milestone each water was taken; how far each is carried in a subterranean channel, how far on substructures,[1] how far on arches. Then I will give the elevation[2] of each, [the plan] of the taps, and the distributions that are made from them;

---

masonry; otherwise a low foundation was built, to which the term *substructio* is applied.

[2] *i.e.* at the point of its entrance into the City.

SEXTUS JULIUS FRONTINUS

quantum extra urbem, quantum intra urbem unicuique regioni pro suo modo unaquaeque aquarum serviat; quot castella publica sint, et ex eis quantum publicis operibus, quantum muneribus—ita enim cultiores appellant—quantum lacibus, quantum nomine Caesaris, quantum privatorum usibus beneficio principis detur; quod ius ducendarum tuendarumque sit earum, quae id sanciant poenae lege, senatus consultis, mandatis principum inrogatae.

---

[1] The conventional interpretation of a very uncertain word.

how much each aqueduct brings to points outside the City, what proportion to each quarter within the City; how many public reservoirs there are, and from these how much is delivered to public works, how much to ornamental fountains [1] (*munera*, as the more polite call them), how much to the water-basins; how much is granted in the name of Caesar; how much for private uses by favour of the Emperor; what is the law with regard to the construction and maintenance of the aqueducts, what penalties enforce it, whether established by resolutions of the Senate or by edicts of the Emperors.

BOOK I

LIBER PRIMUS

4. Ab urbe condita per annos quadringentos quadraginta unum contenti fuerunt Romani usu aquarum, quas aut ex Tiberi aut ex puteis aut ex fontibus hauriebant. Fontium memoria cum sanctitate adhuc exstat et colitur; salubritatem aegris corporibus afferre creduntur, sicut Camenarum et Apollinis et Iuturnae. Nunc autem in urbem influunt aqua Appia, Anio Vetus, Marcia, Tepula, Iulia, Virgo, Alsietina quae eadem vocatur Augusta, Claudia, Anio Novus.

5. M. Valerio Maximo P. Decio Mure consulibus, anno post initium Samnitici belli tricesimo aqua Appia in urbem inducta[1] est ab Appio Claudio Crasso censore, cui postea Caeco fuit cognomen, qui et Viam Appiam a Porta Capena usque ad urbem Capuam muniendam curavit. Collegam habuit C. Plautium, cui ob inquisitas eius aquae venas Venocis cognomen datum est. Sed quia is intra annum et sex menses deceptus a collega tamquam idem facturo

[1] *C*; ducta *B*.

---

[1] The location of these is uncertain.
[2] This fountain is close to the Temple of Castor and Pollux, on the south side of the Roman Forum.
[3] 312 B.C.
[4] This gate was on the south side of the City, in the old Servian Wall.

BOOK I

For four hundred and forty-one years from the foundation of the City, the Romans were satisfied with the use of such waters as they drew from the Tiber, from wells, or from springs. Esteem for springs still continues, and is observed with veneration. They are believed to bring healing to the sick, as, for example, the springs of the Camenae,[1] of Apollo,[1] and of Juturna.[2] But there now run into the City: the Appian aqueduct, Old Anio, Marcia, Tepula, Julia, Virgo, Alsietina, which is also called Augusta, Claudia, New Anio.

In the consulship of Marcus Valerius Maximus and Publius Decius Mus,[3] in the thirtieth year after the beginning of the Samnite War, the Appian aqueduct was brought into the City by Appius Claudius Crassus, the Censor, who afterwards received the surname of "the Blind," the same man who had charge of constructing the Appian Way from the Porta Capena[4] as far as the City of Capua. As colleague in the censorship Appius had Gaius Plautius, to whom was given the name of "the Hunter"[5] for having discovered the springs of this water. But since Plautius resigned the censorship within a year and six months,[6] under the mistaken impression that

[5] The English rendering does not reproduce the word play in *venas Venocis*.

[6] Eighteen months was the regular term of office for the censors.

SEXTUS JULIUS FRONTINUS

abdicavit se censura, nomen aquae ad Appii tantum honorem pertinuit, qui multis tergiversationibus extraxisse censuram traditur, donec et viam et huius aquae ductum consummaret. Concipitur Appia in agro Lucullano Via Praenestina[1] inter miliarium septimum et octavum deverticulo sinistrorsus passuum septingentorum octoginta. Ductus eius habet longitudinem a capite usque ad Salinas, qui locus est ad Portam Trigeminam, passuum undecim milium centum nonaginta; ex eo rivus est subterraneus passuum undecim milium centum triginta, supra terram substructio et opus arcuatum proximum Portam Capenam passuum sexaginta. Iungitur ei ad Spem Veterem in confinio hortorum Torquatianorum et Epaphroditianorum[2] ramus Augustae ab Augusto in supplementum eius additus . . . loco nomen . . . denti[3] Gemellorum. Hic Via Praenestina ad miliarium sextum deverticulo sinistrorsus passuum nongentorum octoginta proxime Viam Collatinam[4] accipit fontem. Cuius ductus usque ad Gemellos efficit rivo subterraneo passuum sex milia trecentos octoginta. Incipit distribui Appia imo Publicii Clivo ad Portam Trigeminam, qui locus Salinae appellantur.

[1] Collatina *Lanciani. The emendation is suggested by the location of the supposed sources of the Appia which have been discovered. See map at end of book.*

[2] *Lanciani; gap in C.*

[3] . . . . . toco nomen . . . . denti *C*; imposito cognomine respondenti *Poleni.*

[4] Collatiam *CB.*

---

[1] The conventional rendering of *passus* by "pace" is here followed, although the term applied in strictness to the distance between the outstretched hands, *i.e.* five Roman feet, equivalent to 4 feet 10⅓ inches of our measure.

AQUEDUCTS OF ROME, I. 5

his colleague would do the same, the honour of giving his name to the aqueduct fell to Appius alone, who, by various subterfuges, is reported to have extended the term of his censorship, until he should complete both the Way and this aqueduct. The intake of Appia is on the Lucullan estate, between the seventh and eighth milestones, on the Praenestine Way, on a cross-road, 780 paces [1] to the left.[2] From its intake to the Salinae at the Porta Trigemina,[3] its channel has a length of 11,190 paces, of which 11,130 paces run underground, while above ground sixty paces are carried on substructures and, near the Porta Capena, on arches. Near Spes Vetus,[4] on the edge of the Torquatian and Epaphroditian Gardens, there joins it a branch of Augusta, added by Augustus as a supplementary supply. . . . This branch has its intake at the sixth milestone, on the Praenestine Way, on a cross-road, 980 paces to the left, near the Collatian Way. Its course, by underground channel, extends to 6,380 paces before reaching The Twins.[5] The distribution of Appia begins at the foot of the Publician Ascent, near the Porta Trigemina, at the place designated as the Salinae.[6]

[2] *i.e.* going from Rome.

[3] This was at the northern base of the Aventine Hill, near the Tiber.

[4] The Temple of Spes Vetus was just inside the Aurelian Wall, in the eastern quarter of the City, not far from the Porta Labicana (the modern Porta Maggiore). See p. 485 and plan, p. 490.

[5] The name is evidently derived from the junction of the two aqueducts. "There are considerable remains of two large reservoirs in a garden just outside of the boundary-wall of the Sessorium. These two great reservoirs, so close together in the line of the Aqua Appia, seem to have been the Gemelli mentioned by Frontinus."—*Parker*.

[6] See map, pp. 486–7.

SEXTUS JULIUS FRONTINUS

6. Post annos quadraginta quam Appia perducta est, anno ab urbe condita quadringentesimo octogesimo uno M'. Curius Dentatus, qui censuram cum Lucio Papirio Cursore gessit, Anionis qui nunc Vetus dicitur aquam perducendam in urbem ex manubiis de Pyrro captis locavit, Spurio Carvilio Lucio Papirio consulibus iterum. Post biennium deinde actum est in senatu de consummando eius aquae opere referente . . . norumi . . . praetore. Tum[1] ex senatus consulto duumviri aquae perducendae creati sunt Curius, qui eam locaverat et Fulvius Flaccus. Curius intra quintum diem quam erat duumvirum creatus decessit; gloria perductae pertinuit ad Fulvium. Concipitur Anio Vetus supra Tibur vicesimo miliario extra Portam . . . RRa . . . nam,[2] ubi partem dat in Tiburtium usum. Ductus eius habet longitudinem, ita exigente libramento, passuum quadraginta trium milium: ex eo rivus est subterraneus passuum quadraginta duum milium septingentorum septuaginta novem, substructio supra terram passuum ducentorum viginti unius.

7. Post annos centum viginti septem, id est anno ab urbe condita sexcentesimo octavo, Ser. Sulpicio

---

[1] *B*; irefent . . . . norumi . . . . praetorium *C*.

[2] *Various conjectures have been made in the attempt to identify this gate. Fea, with reference to Hor. Epist. i. xiv. 3, suggested* Varianam; *other readings that have been proposed are* Baranam, Raranam, Romanam, Arretinam, Reatinam. *A passage of Nicodemus,* Storia di Tivoli, *in which he recalls a village named Barana, lying to the east of Tivoli, seems to Lanciani to give some support to* Baranam. *Burn thinks this gate was probably on the site of the modern Porta S. Giovanni.*

---

[1] 273 B.C.   [2] *Cf.* 13.

AQUEDUCTS OF ROME, I. 6-7

Forty years after Appia was brought in, in the four hundred and eighty-first year [1] from the founding of the City, Manius Curius Dentatus, who held the censorship with Lucius Papirius Cursor, contracted to have the waters of what is now called Old [2] Anio brought into the City, with the proceeds of the booty captured from Pyrrhus. This was in the second consulship of Spurius Carvilius and Lucius Papirius. Then two years later the question of completing the aqueduct was discussed in the Senate on the motion . . . of the praetor. At the close of the discussion, Curius, who had let the original contract, and Fulvius Flaccus were appointed by decree of the Senate as a board of two to bring in the water Within five days of the time he had been appointed, one of the two commissioners, Curius, died; thus the credit of achieving the work rested with Flaccus. The intake of Old Anio is above Tibur [3] at the twentieth milestone outside the . . . Gate, where it gives a part of its water to supply the Tiburtines. Owing to the exigence of elevation,[4] its conduit has a length of 43,000 paces. Of this, the channel runs underground for 42,779 paces, while there are above ground substructures for 221 paces.

One hundred and twenty-seven years later, that is in the six hundred and eighth year from the founding of the City,[5] in the consulship of Servius Sulpicius

---

[3] The modern Tivoli, about eighteen miles to the east of Rome. See map pp. 492–3.

[4] All ancient aqueducts are constructed on the principle of flow, not of pressure. The fall was necessarily very gradual. Consequently, when the intake was at a considerable elevation, long detours became necessary in bringing the water to the City.

[5] 146 B.C., but Galba and Cotta were consuls in 144 B.C.

SEXTUS JULIUS FRONTINUS

Galba Lucio Aurelio Cotta consulibus cum Appiae Anionisque ductus vetustate quassati privatorum etiam fraudibus interciperentur, datum est a senatu negotium Marcio, qui tum praetor inter cives ius dicebat, eorum ductuum reficiendorum ac vindicandorum. Et quoniam incrementum urbis exigere videbatur ampliorem modum aquae, eidem mandatum a senatu est, ut curaret, quatenus alias aquas posset in urbem perducere.[1] . . . Priores[2] ductus restituit et tertiam illis salubriorem[3] . . . duxit, cui ab auctore Marciae nomen est. Legimus apud Fenestellam, in haec opera Marcio decretum sestertium milies octingenties, et quoniam ad consummandum negotium non sufficiebat spatium praeturae, in annum alterum est[4] prorogatum. Eo tempore decemviri, dum aliis ex causis libros Sibyllinos inspiciunt, invenisse dicuntur, non esse fas[5] aquam Marciam seu potius Anionem—de hoc enim constantius traditur—in Capitolium perduci, deque ea re in senatu M. Lepido pro collegio[6] verba faciente actum Appio Claudio Q. Caecilio consulibus, eandemque post annum tertium a Lucio Lentulo retractatam C. Laelio Q. Servilio consulibus, sed utroque

---

[1] quatenus alias aquas posset in urbem perducere *Schultze*; quatenus alias aquas quas posset in urbem perduceret *CB*.
[2] . . . . ores *C*; priores *B*.
[3] *Bennett*; tertiam illiobriorum . . . . *CB*.
[4] *C*; omitted by *B*.
[5] *Schöne* (*Hermes vi. 248*); omitted in *C* and *B*.
[6] *Schöne*; collega *CB*.

---

[1] *Praetor urbanus.*
[2] A Roman historian; he died in 21 A.D.

AQUEDUCTS OF ROME, I. 7

Galba and Lucius Aurelius Cotta, when the conduits of Appia and Old Anio had become leaky by reason of age, and water was also being diverted from them unlawfully by individuals, the Senate commissioned Marcius, who at that time administered the law as praetor between citizens,[1] to reclaim and repair these conduits; and since the growth of the City was seen to demand a more bountiful supply of water, the same man was charged by the Senate to bring into the City other waters so far as he could. . . . He restored the old channels and brought in a third supply, more wholesome than these, . . . which is called Marcia after the man who introduced it. We read in Fenestella,[2] that 180,000,000 sesterces[3] were granted to Marcius for these works, and since the term of his praetorship was not sufficient for the completion of the enterprise, it was extended for a second year. At that time the Decemvirs,[4] on consulting the Sibylline Books for another purpose, are said to have discovered that it was not right for the Marcian water, or rather the Anio (for tradition more regularly mentions this) to be brought to the Capitol. The matter is said to have been debated in the Senate, in the consulship of Appius Claudius and Quintus Caecilius,[5] Marcius Lepidus acting as spokesman for the Board of Decemvirs; and three years later the matter is said to have been brought up again by Lucius Lentulus, in the consulship of Gaius Laelius and Quintus

---

[3] About £1,500,000. The sesterce at this period was worth about two pence.

[4] A board of ten men who had charge of the Sibylline Books.

[5] 143 B.C.

SEXTUS JULIUS FRONTINUS

tempore vicisse gratiam Marci Regis; atque ita in Capitolium esse aquam perductam. Concipitur Marcia Via Valeria ad miliarium tricesimum sextum deverticulo euntibus ab urbe Roma dextrorsus milium passuum trium. Sublacensi autem, quae sub Nerone principe primum strata est, ad miliarium tricesimum octavum sinistrorsus intra passus ducentos fontium . . . sub . . . bus petrei . . . stat immobilis stagni modo colore[1] praeviridi. Ductus eius habet longitudinem a capite ad urbem passuum sexaginta milium et mille septingentorum decem et semis; rivo subterraneo passuum quinquaginta quattuor milium ducentorum quadraginta septem semis, opere supra terram passuum septem milium quadringentorum sexaginta trium: ex eo longius ab urbe pluribus locis per vallis opere arcuato passuum quadringentorum sexaginta trium, propius urbem a septimo miliario substructione passuum quingentorum viginti octo, reliquo opere arcuato passuum sex milium quadringentorum septuaginta duum.

8. Cn. Servilius Caepio et L. Cassius Longinus, qui Ravilla appellatus est, censores anno post urbem conditam sexcentesimo vicesimo septimo, M. Plautio Hypsaeo M. Fulvio Flacco cos., aquam quae vocatur Tepula ex agro Lucullano, quem quidam Tusculanum credunt, Romam et in Capitolium adducendam curaverunt. Tepula concipitur Via Latina ad decimum miliarium deverticulo euntibus ab Roma dextrorsus milium passuum duum . . . . inde suo rivo in urbem perducebatur.

[1] fontin . . . . sub . . . . . bus petrei . . . . . .
statim . . . . stagnimo . . . . . . colore *C*.

[1] 140 B.C.   [2] 127 B.C.   [3] 125 B.C.

AQUEDUCTS OF ROME, I. 7-8

Servilius,[1] but on both occasions the influence of Marcius Rex carried the day; and thus the water was brought to the Capitol. The intake of Marcia is at the thirty-sixth milestone on the Valerian Way, on a cross-road, three miles to the right as you come from Rome. But on the Sublacensian Way, which was first paved under the Emperor Nero, at the thirty-eighth milestone, within 200 paces to the left [a view of its source may be seen]. Its waters stand like a tranquil pool, of deep green hue. Its conduit has a length, from the intake to the City, of 61,710½ paces; 54,247½ paces of underground conduit; 7,463 paces on structures above ground, of which, at some distance from the City, in several places where it crosses valleys, there are 463 paces on arches; nearer the City, beginning at the seventh milestone, 528 paces on substructures, and the remaining 6,472 paces on arches.

The Censors, Gnaeus Servilius Caepio and Lucius Cassius Longinus, called Ravilla, in the year 627 [2] after the founding of the City, in the consulate of Marcius Plautius Hypsaeus and Marcus Fulvius Flaccus,[3] had the water called Tepula brought to Rome and to the Capitol, from the estate of Lucullus, which some persons hold to belong to Tusculan [4] territory. The intake of Tepula is at the tenth milestone on the Latin Way, near a cross-road, two miles to the right as you proceed from Rome. . . . From that point it was conducted in its own [5] channel to the City.

---

[4] The country around Tusculum (the modern Frascati), a town in Latium about twenty miles south-east of Rome.

[5] Later it flowed in the same channel with Julia. *Cf.* note on **9.**

SEXTUS JULIUS FRONTINUS

9. Post . . . . . Agrippa aedilis post primum consulatum imperatore Caesare Augusto II L. Volcatio cos., anno post urbem conditam septingentesimo nono decimo ad miliarium ab urbe duodecimo Via Latina deverticulo euntibus ab Roma dextrorsus milium passuum duum alterius aquae proprias vires collegit et Tepulae rivum intercepit. Adquisitae aquae ab inventore nomen Iuliae datum est, ita tamen divisa erogatione, ut maneret Tepulae appellatio. Ductus Iuliae efficit longitudinem passuum quindecim milium quadringentorum viginti sex S.: opere supra terram passuum septem milium: ex eo in proximis urbem locis a septimo miliario substructione passuum quingentorum viginti octo, reliquo opere arcuato passuum sex milium quadringentorum septuaginta duum. Praeter caput Iuliae transfluit aqua quae vocatur Crabra. Hanc Agrippa omisit, seu quia improbaverat, sive quia Tusculanis possessoribus relinquendam credebat; ea namque est quam omnes villae tractus eius per vicem in dies modulosque certos dispensatam accipiunt. Sed non eadem moderatione aquarii nostri partem eius semper in supplementum Iuliae vindicaverunt, nec ut Iuliam augerent, quam hauriebant largiendo compendi sui gratia. Exclusi ergo Crabram et totam iussu imperatoris reddidi Tusculanis, qui nunc, forsitan non sine admiratione, eam sumunt ignari cui causae insolitam

---

[1] 33 B.C.   [2] 35 B.C.   [3] *Cf.* note 4 on ch. 98.
[4] Apparently the name *Julia et Tepula* was applied to it. "The Julia was admitted into the channel of the Tepula at the tenth milestone. At the sixth milestone the compound water was again divided into two conduits, proportioned to the volume of the springs."—*Lanciani.*

AQUEDUCTS OF ROME, I. 9

Later . . . in the second consulate[1] of the Emperor Caesar Augustus, when Lucius Volcatius was his colleague, in the year 719[2] after the foundation of the City, [Marcus] Agrippa, when aedile, after his first consulship,[3] took another independent source of supply, at the twelfth milestone from the City on the Latin Way, on a cross-road two miles to the right as you proceed from Rome, and also tapped Tepula. The name Julia was given to the new aqueduct by its builder, but since the waters were again divided for distribution, the name Tepula remained.[4] The conduit of Julia has a length of 15,426½ paces; 7,000 paces on masonry above ground, of which 528 paces next the City, beginning at the seventh milestone, are on substructures, the other 6,472 paces being on arches. Past the intake of Julia flows a brook, which is called Crabra. Agrippa refrained from taking in this brook either because he had condemned it, or because he thought it ought to be left to the proprietors at Tusculum, for this is the water which all the estates of that district receive in turn, dealt out to them on regular days and in regular quantities. But our water-men,[5] failing to practise the same restraint, have always claimed a part of it to supplement Julia, not, however, thus increasing the actual flow of Julia, since they habitually exhausted it by diverting its waters for their own profit. I therefore shut off the Crabra brook and at the Emperor's command restored it entirely to the Tusculan proprietors, who now, possibly not without surprise, take its waters, without knowing to what cause to ascribe the unusual

[5] The water-men are the men who receive the water from the State and in turn furnish it to the consumers.

SEXTUS JULIUS FRONTINUS

abundantiam debeant. Iulia autem revocatis derivationibus, per quas surripiebatur, modum suum quamvis notabili siccitate servavit. Eodem anno Agrippa ductus Appiae, Anionis, Marciae paene dilapsos restituit et singulari cura compluribus salientibus instruxit urbem.

10. Idem cum iam tertio consul fuisset, C. Sentio Q. Lucretio consulibus, post annum tertium decimum quam Iuliam deduxerat, Virginem quoque in agro Lucullano collectam Romam perduxit. Dies quo primum in urbe[1] responderit, quintus idus Iunias invenitur. Virgo appellata est, quod quaerentibus aquam militibus puella[2] virguncula venas quasdam monstravit, quas secuti qui foderant, ingentem aquae modum invenerunt. Aedicula fonti apposita hanc originem pictura ostendit. Concipitur Virgo Via Collatina[3] ad miliarium octavum palustribus locis, signino circumiecto continendarum scaturiginum causa. Adiuvatur et compluribus aliis adquisitionibus. Venit per longitudinem passuum decem quattuor milium centum quinque: ex eo rivo subterraneo passuum decem duum milium octingentorum sexaginta quinque, supra terram per passus mille ducentos quadraginta: ex eo substructione rivorum locis compluribus passuum quingentorum quadraginta, opere arcuato passuum septingentorum. Adquisitionum ductus rivi subterranei efficiunt passus mille quadringentos quinque.

[1] *Poleni, Jordan*; urbem *CB*.
[2] *C*; omitted by *B*.
[3] Collatia *CB*.

---

[1] Agrippa was consul for the third time in 27 B.C. Gaius

AQUEDUCTS OF ROME, I. 9-10

abundance. The Julian aqueduct, on the other hand, by reason of the destruction of the branch pipes through which it was secretly plundered, has maintained its normal quantity even in times of most extraordinary drought. In the same year, Agrippa repaired the conduits of Appia, Old Anio, and Marcia, which had almost worn out, and with unique forethought provided the City with a large number of fountains.

The same man, after his own third consulship, in the consulship of Gaius Sentius and Quintus Lucretius,[1] twelve years after he had constructed the Julian aqueduct, also brought Virgo to Rome, taking it from the estate of Lucullus. We learn that June 9 was the day that it first began to flow in the City. It was called Virgo, because a young girl pointed out certain springs to some soldiers hunting for water, and when they followed these up and dug, they found a copious supply. A small temple, situated near the spring, contains a painting which illustrates this origin of the aqueduct. The intake of Virgo is on the Collatian Way at the eighth milestone, in a marshy spot, surrounded by a concrete enclosure for the purpose of confining the gushing waters. Its volume is augmented by several tributaries. Its length is 14,105 paces. For 12,865 paces of this distance it is carried in an underground channel, for 1,240 paces above ground. Of these 1,240 paces, it is carried for 540 paces on substructures at various points, and for 700 paces on arches. The underground conduits of the tributaries measure 1,405 paces.

Sentius Saturninus and Quintus Lucretius were consuls in 19 B.C.

SEXTUS JULIUS FRONTINUS

11. Quae ratio moverit Augustum, providentissimum principem, perducendi Alsietinam aquam, quae vocatur Augusta, non satis perspicio, nullius gratiae, immo etiam parum salubrem ideoque nusquam in usus populi fluentem; nisi forte cum opus Naumachiae adgrederetur, ne quid salubrioribus aquis detraheret, hanc proprio opere perduxit et quod Naumachiae coeperat superesse, hortis adiacentibus et privatorum usibus ad inrigandum concessit. Solet tamen ex ea in Transtiberina regione, quotiens pontes reficiuntur et a citeriore ripa aquae cessant, ex necessitate in subsidium publicorum salientium dari. Concipitur ex lacu Alsietino Via Claudia miliario quarto decimo deverticulo dextrorsus passuum sex milium quingentorum. Ductus eius efficit longitudinem passuum viginti duum milium centum septuaginta duorum, opere arcuato passuum trecentorum quinquaginta octo.

12. Idem Augustus in supplementum Marciae, quotiens siccitates egerent auxilio, aliam aquam eiusdem bonitatis opere subterraneo perduxit usque ad Marciae rivum, quae ab inventore appellatur Augusta. Nascitur ultra fontem Marciae. Cuius ductus donec Marciae accedat, efficit passus octingentos.

13. Post hos C. Caesar, qui Tiberio successit, cum parum et publicis usibus et privatis voluptatibus

---

[1] Naumachia was the name given to the artificial lakes prepared for exhibitions of sham naval battles; the same name was applied to the contests themselves. See map on pp. 486–7.

[2] Bridges sometimes served as carriers for the water pipes. Among the bridges crossing the Anio, Ponte Lupo near Gallicano served as transit for four waters, Marcia, Anio

AQUEDUCTS OF ROME, I. 11-13

I fail to see what motive induced Augustus, a most sagacious sovereign, to bring in the Alsietinian water, also called Augusta. For this has nothing to commend it,—is in fact positively unwholesome, and for that reason is nowhere delivered for consumption by the people. It may have been that when Augustus began the construction of his Naumachia,[1] he brought this water in a special conduit, in order not to encroach on the existing supply of wholesome water, and then granted the surplus of the Naumachia to the adjacent gardens and to private users for irrigation. It is customary, however, in the district across the Tiber, in an emergency, whenever the bridges[2] are undergoing repairs and the water supply is cut off from this side the river, to draw from Alsietina to maintain the flow of the public fountains. Its source is the Alsietinian Lake, at the fourteenth milestone, on the Claudian Way, on a cross-road, six miles and a half to the right. Its conduit has a length of 22,172 paces, with 358 paces on arches.

To supplement Marcia, whenever dry seasons required an additional supply, Augustus also, by an underground channel, brought to the conduit of Marcia another water of the same excellent quality, called Augusta from the name of its donor. Its source is beyond the springs of Marcia; its conduit, up to its junction with Marcia, measures 800 paces.

After these aqueducts, Gaius Caesar,[3] the successor of Tiberius, in the second year of his reign, in the

Vetus, Anio Novus and Claudia, besides a carriage-way and a bridle-path. At Lyons there are the ruins of a Roman bridge, which still contains lead pipes.

[3] Caligula, who reigned from 37 to 41 A.D.

353

SEXTUS JULIUS FRONTINUS

septem ductus aquarum sufficere viderentur, altero imperii sui anno, M. Aquila Iuliano P. Nonio Asprenate cos., anno urbis conditae septingentesimo nonagesimo uno [1] duos ductus incohavit. Quod opus Claudius magnificentissime consummavit dedicavitque Sulla et Titiano consulibus, anno post urbem conditam octingentesimo tertio Kalendis Augustis. Alteri nomen, quae ex fontibus Caerulo et Curtio perducebatur, Claudiae datum. Haec bonitatis proximae est Marciae. Altera, quoniam duae Anionis in urbem aquae fluere coeperant, ut facilius appellationibus dinoscerentur, Anio Novus vocitari coepit; priori Anioni cognomen Veteris adiectum.

14. Claudia concipitur Via Sublacensi ad miliarium tricesimum octavum deverticulo sinistrorsus intra passus trecentos ex fontibus duobus amplissimis et speciosis, Caeruleo qui a similitudine appellatus est, et Curtio. Accipit et eum fontem qui vocatur Albudinus, tantae bonitatis, ut Marciae quoque adiutorio quotiens opus est ita sufficiat, ut adiectione sui nihil ex qualitate eius mutet. Augustae fons, quia Marciam sibi sufficere apparebat, in Claudiam derivatus est, manente nihilo minus praesidiario in Marciam, ut ita demum Claudiam aquam adiuvaret Augusta, si eam ductus Marciae non caperet. Claudiae ductus habet longitudinem passuum quadraginta sex milium quadringentorum sex:

---

[1] nonagesimo uno *Bennett*; nonagesimo *C*; undenonagesimo *Poleni, B*. The year 38 A.D., the 2nd year of Caligula's reign, is the 791st year A.U.C.

---

[1] 38 A.D.   [2] *Cf.* Suet. *Claud.* 20.
[3] 52 A.D.   [4] 50 A.D.   [5] "The Blue."

AQUEDUCTS OF ROME, I. 13-14

consulate of Marcus Aquila Julianus and Publius Nonius Asprenas, in the year 791[1] after the founding of the City, began two others, inasmuch as the seven then existing seemed insufficient to meet both the public needs and the luxurious private demands of the day. These works Claudius completed on the most magnificent scale,[2] and dedicated in the consulship of Sulla and Titianus,[3] on the 1st of August in the year 803[4] after the founding of the City. To the one water, which had its sources in the Caerulean and Curtain springs, was given the name Claudia. This is next to Marcia in excellence. The second began to be designated as New Anio, in order the more readily to distinguish by title the two Anios that had now begun to flow to the City. To the former Anio the name of "Old" was added.

The intake of Claudia is at the thirty-eighth milestone on the Sublacensian Way, on a cross-road, less than three hundred paces to the left. The water comes from two very large and beautiful springs, the Caerulean,[5] so designated from its appearance, and the Curtian. Claudia also receives the spring which is called Albudinus, which is of such excellence that, when Marcia, too, needs supplementing, this water answers the purpose so admirably that by its addition there is no change in Marcia's quality. The spring of Augusta was turned into Claudia, because it was plainly evident that Marcia was of sufficient volume by itself. But Augusta remained, nevertheless, a reserve supply to Marcia, the understanding being that Augusta should run into Claudia only when the conduit of Marcia would not carry it. Claudia's conduit has a length of 46,406 paces, of

ex eo rivo subterraneo passuum triginta sex milium ducentorum triginta, opere supra terram passuum decem milium centum septuaginta sex: ex eo opere arcuato in superiori parte pluribus locis passuum trium milium septuaginta sex, et prope urbem a septimo miliario substructione rivorum per passus sexcentos novem, opere arcuato passuum sex milium quadringentorum nonaginta et unius.

15. Anio Novus Via Sublacensi ad miliarium quadragesimum secundum in Simbruino excipitur ex flumine, quod cum terras cultas circa se habeat soli pinguis et inde ripas solutiores, etiam sine pluviarum iniuria limosum et turbulentum fluit. Ideoque a faucibus ductus interposita est piscina limaria, ubi inter amnem et specum consisteret et liquaretur aqua. Sic quoque quotiens imbres superveniunt, turbida pervenit in urbem. Iungitur ei rivus Herculaneus oriens eadem via ad miliarium tricesimum octavum e regione fontium Claudiae trans flumen viamque. Natura est purissimus, sed mixtus gratiam splendoris sui amittit. Ductus Anionis Novi efficit passuum quinquaginta octo milia septingentos: ex eo rivo subterraneo passuum quadraginta novem milia trecentos, opere supra terram passuum novem milia quadringentos: ex eo substructionibus aut opere arcuato superiore parte pluribus locis passuum duo milia trecentos, et propius urbem a septimo miliario substructione rivorum passus sexcentos novem, opere arcuato passuum sex milia quadringentos nonaginta unum. Hi sunt arcus altissimi, sublevati in quibusdam locis pedes centum novem.

16. Tot aquarum tam multis necessariis molibus

---

[1] The Simbruvian Hills were about thirty miles to the north-east of Rome.

AQUEDUCTS OF ROME, I. 14-16

which 36,230 are in a subterranean channel, 10,176 on structures above ground; of these last there are at various points in the upper reaches 3,076 paces on arches; and near the City, beginning at the seventh milestone, 609 paces on substructures and 6,491 on arches.

The intake of New Anio is at the forty-second milestone on the Sublacensian Way, in the district of Simbruvium.[1] The water is taken from the river, which, even without the effect of rainstorms, is muddy and discoloured, because it has rich and cultivated fields adjoining it, and in consequence loose banks. For this reason, a settling reservoir was put in beyond the inlet of the aqueduct, in order that the water might settle there and clarify itself, between the river and the conduit. But even despite this precaution, the water reaches the City in a discoloured condition whenever there are rains. It is joined by the Herculanean Brook, which has its source on the same Way, at the thirty-eighth milestone, opposite the springs of Claudia, beyond the river and the highway. This is naturally very clear, but loses the charm of its purity by admixture with New Anio. The conduit of New Anio measures 58,700 paces, of which 49,300 are in an underground channel, 9,400 paces above ground on masonry; of these, at various points in the upper reaches are 2,300 paces on substructures or arches; while nearer the city, beginning at the seventh milestone, are 609 paces on substructures, 6,491 paces on arches. These are the highest arches, rising at certain points to 109 feet.

With such an array of indispensable structures carrying so many waters, compare, if you will, the

357

SEXTUS JULIUS FRONTINUS

pyramidas videlicet otiosas compares aut cetera
inertia sed fama celebrata opera Graecorum.

17. Non alienum mihi visum est, longitudines
quoque rivorum cuiusque ductus etiam per species
operum complecti. Nam cum maxima huius officii
pars in tutela eorum sit, scire praepositum oportet,
quae maiora impendia exigant. Nostrae quidem
sollicitudini non suffecit, singula oculis subiecisse;
formas quoque ductuum facere curavimus, ex quibus
apparet ubi valles quantaeque, ubi flumina traice-
rentur, ubi montium lateribus specus applicitae
maiorem adsiduamque tuendi ac muniendi rivi exigant curam. Hinc illa contingit utilitas, ut rem
statim veluti in conspectu habere possimus et
deliberare tamquam adsistentes.

18. Omnes aquae diversa in urbem libra perve-
niunt. Inde fluunt quaedam altioribus locis et
quaedam erigi in eminentiora non possunt; nam
et colles sensim propter frequentiam incendiorum
excreverunt rudere. Quinque sunt quarum altitudo
in omnem partem urbis adtollitur, sed ex his aliae
maiore, aliae leviore pressura coguntur. Altissimus
est Anio Novus, proxima Claudia, tertium locum
tenet Iulia, quartum Tepula, dehinc Marcia, quae
capite etiam Claudiae libram aequat. Sed veteres
humiliore directura perduxerunt, sive nondum ad
subtile explorata arte librandi, seu quia ex industria

---

[1] *i.e.*, how much under ground; how much on arches, etc.

idle Pyramids or the useless, though famous, works of the Greeks!

It has seemed to me not inappropriate to include also a statement of the lengths of the channels of the several aqueducts, according to the kinds of construction.[1] For since the chief function of this office of water-commissioner lies in their upkeep, the man in charge of them ought to know which of them demand the heavier outlay. My zeal was not satisfied with submitting details to examination; I also had plans made of the aqueducts, on which it is shown where there are valleys and how great these are; where rivers are crossed; and where conduits laid on hillsides demand more particular constant care for their maintenance and repair. By this provision, one reaps the advantage of being able to have the works before one's eyes, so to speak, at a moment's notice, and to consider them as though standing by their side.

The several aqueducts reach the City at different elevations. In consequence certain ones deliver water on higher ground, while others cannot rise to the loftier points; for the hills have gradually grown higher with rubbish in consequence of frequent conflagrations. There are five whose head rises to every point in the City, but of these some are forced up with greater, others with lesser pressure. The highest is New Anio; next comes Claudia; the third place is taken by Julia; the fourth by Tepula; the last by Marcia, although at its intake this mounts even to the level of Claudia. But the ancients laid the lines of their aqueducts at a lower elevation, either because they had not yet nicely worked out the art of levelling, or because they purposely sunk

SEXTUS JULIUS FRONTINUS

infra terram aquas mergebant, ne facile ab hostibus interciperentur, cum frequentia adhuc contra Italicos bella gererentur. Iam tamen quibusdam locis, sicubi ductus vetustate dilapsus est, omisso circuito subterraneo vallium brevitatis causa substructionibus arcuationibusque traiciuntur. Sextum tenet librae locum Anio Vetus, similiter suffecturus etiam altioribus locis urbis, si, ubi vallium summissarumque regionum condicio exigit, substructionibus arcuationibusve erigeretur. Sequitur huius libram Virgo, deinde Appia; quae cum ex urbano agro perducerentur, non in tantum altitudinis erigi potuerunt. Omnibus humilior Alsietina est, quae Transtiberinae regioni et maxime iacentibus locis servit.

19. Ex eis sex Via Latina intra septimum miliarium contectis piscinis excipiuntur, ubi quasi respirante rivorum cursu limum deponunt. Modus quoque earum mensuris ibidem positis initur. Tres autem earum, Iulia, Marcia, Tepula quae intercepta, sicut supra demonstravimus, rivo Iuliae accesserat, nunc a piscina eiusdem Iuliae modum accipit ac proprio canali et nomine venit—hae tres a piscinis in eosdem arcus recipiuntur. Summus in his est Iuliae, inferior Tepulae, dein Marcia. Quae ad libram Collis Viminalis sco . . . ntea fluentes[1] ad Viminalem usque Portam deveniunt. Ibi rursus emergunt. Prius tamen pars Iuliae ad Spem

[1] *B.*; libram . . . . minalis sco . . . . ntea . . . . ntes. *C.*

---

[1] *i.e.*, when old aqueducts were rebuilt, they were carried *across* valleys on arches or low foundations, instead of going *around* the valleys, as the original underground structures had done.   [2] Namely, in the basins.
[3] *Cf.* 9 and footnote.   [4] See illustration on p. 489.

their aqueducts in the ground, in order that they might not easily be cut by the enemy, since frequent wars were still waged with the Italians. But now, wherever a conduit has succumbed to old age, it is the practice to carry it in certain parts on substructures or on arches, in order to save length, abandoning the subterranean loops in the valleys.[1] The sixth rank in height is held by Old Anio, which would likewise be capable of supplying even the higher portions of the City, if it were raised up on substructures or arches, wherever the nature of the valleys and low places demands. Its elevation is followed by that of Virgo, then by that of Appia. These, since they were brought from points near the City, could not rise to such high elevations. Lowest of all is Alsietina, which supplies the ward across the Tiber and the very lowest districts.

Of these waters, six are received in covered catch-basins, this side the seventh milestone on the Latin Way. Here, taking fresh breath, so to speak, after the run, they deposit their sediment. Their volume also is determined by gauges set up at the same point.[2] Three of these, Julia, Marcia, and Tepula, are carried by the same arches from the catch-basins onward. Tepula, which, as we have above explained,[3] was tapped and added to the conduit of Julia, now leaves the basin of this same Julia, receives its own quota of water, and runs in its own conduit, under its own name. The topmost of these three is Julia; next below is Tepula; then Marcia.[4] These flowing [under ground] reach the level of the Viminal Hill, and in fact even of the Viminal Gate. There they again emerge. Yet a part of Julia is

SEXTUS JULIUS FRONTINUS

Veterem excepta castellis Caelii montis diffunditur. Marcia autem partem sui post hortos Pallantianos in rivum qui vocatur Herculaneus deicit. Is per Caelium ductus, ipsius montis usibus nihil ut inferior subministrans, finitur supra Portam Capenam.

20 Anio Novus et Claudia a piscinis in altiores arcus recipiuntur ita ut superior sit Anio. Finiuntur arcus earum post hortos Pallantianos et inde in usum urbis fistulis diducuntur. Partem tamen sui Claudia prius in arcus qui vocantur Neroniani ad Spem Veterem transfert. Hi directi per Caelium montem iuxta templum Divi Claudii terminantur. Modum quem acceperunt aut circa ipsum montem aut in Palatium Aventinumque et regionem Transtiberinam dimittunt.

21. Anio Vetus citra quartum miliarium infra Novum, qui a Via Latina in Labicanam inter arcus traicit, et ipse piscinam habet. Inde intra secundum miliarium partem dat in specum qui vocatur Octavianus et pervenit in regionem Viae Novae ad hortos Asinianos, unde per illum tractum distribuitur. Rectus vero ductus secundum Spem veniens intra Portam Esquilinam in altos rivos per urbem diducitur.

22. Nec Virgo nec Appia nec Alsietina conceptacula, id est piscinas, habent. Arcus Virginis initium habent sub hortis Lucullanis, finiuntur in Campo Martio secundum frontem Saeptorum. Rivus

---

[1] On the Esquiline.
[2] See illustration on p. 491.
[3] *Cf.* 76, 87.
[4] South of the Caelian, near the Baths of Caracalla.
[5] On the Pincian.

AQUEDUCTS OF ROME, I. 19-22

first diverted at Spes Vetus, and distributed to the reservoirs of Mount Caelius. But Marcia delivers a part of its waters into the so-called Herculanean Conduit, behind the Gardens of Pallas.[1] This conduit, carried along the Caelian, affords no service to the occupants of the hill, on account of its low level; it ends beyond the Porta Capena.

New Anio and Claudia are carried together from their catch-basins on lofty arches, Anio being above.[2] Their arches end behind the Gardens of Pallas, and from that point their waters are distributed in pipes to serve the City. Yet Claudia first transfers a part of its waters near Spes Vetus to the so-called Neronian Arches. These arches pass along the Caelian Hill and end near the Temple of the Deified Claudius.[3] Both aqueducts deliver the volume which they receive, partly about the Caelian, partly on the Palatine and Aventine, and to the ward beyond the Tiber.

Old Anio, this side the fourth milestone, passes under New Anio, which here shifts from the Latin to the Labican Way; it has its own catch-basin. Then, this side the second milestone, it gives a part of its waters to the so-called Octavian Conduit and reaches the Asinian Gardens[4] in the neighbourhood of the New Way, whence it is distributed throughout that district. But the main conduit, which passes Spes Vetus, comes inside the Esquiline Gate and is distributed to high-lying mains throughout the city.

Neither Virgo, nor Appia, nor Alsietina has a receiving reservoir or catch-basin. The arches of Virgo begin under the Lucullan Gardens,[5] and end on the Campus Martius in front of the Voting

363

SEXTUS JULIUS FRONTINUS

Appiae sub Caelio monte et Aventino actus emergit, ut diximus, infra Clivum Publicii. Alsietinae ductus post Naumachiam, cuius causa videtur esse factus, finitur.

23. Quoniam auctores cuiusque aquae et aetates, praeterea origines et longitudines rivorum et ordinem librae persecutus sum, non alienum mihi videtur, etiam singula subicere et ostendere quanta sit copia quae publicis privatisque non solum usibus et auxiliis verum etiam voluptatibus sufficit, et per quot castella quibusque regionibus diducatur, quantum extra urbem, quantum in urbe, et ex eo quantum lacibus, quantum muneribus, quantum operibus publicis, quantum nomine Caesaris, quantum privatis usibus erogetur. Sed rationis existimo, priusquam nomina quinariarum centenariarumque et ceterorum modulorum, per quos mensura constituta est, proferamus, et indicare quae sit eorum origo, quae vires et quid quaeque appellatio significet, propositaque regula, ad quam ratio eorum et initium computatur, ostendere qua ratione discrepantia invenerim et quam emendandi viam sim secutus.

24. Aquarum moduli aut ad digitorum aut ad unciarum mensuram instituti sunt. Digiti in Cam-

---

[1] *Cf.* 5.

[2] The ajutage was the nozzle, fitted to the water-pipe. The size and character of the ajutage, therefore, were important factors in the measurement of the water discharged. The ajutage was gauged according to various principles. *Cf.* 26.

[3] One of the most serious abuses practised by the watermen at Rome was connected with the size of the pipes used in the receiving and the distribution of water. *Cf.* 112, 113, 114. Since the size and position (*cf.* 36) of the ajutage controlled the amount discharged, it was necessary to know

Porticoes. The conduit of Appia, running along the base of the Caelian and Aventine, emerges, as we have said above,[1] at the foot of the Publician Ascent. The conduit of Alsietina terminates behind the Naumachia, for which it seems to have been constructed.

Since I have given in detail the builders of the several aqueducts, their dates, and, in addition, their sources, the lengths of their channels, and their elevations in sequence, it seems to me not out of keeping to add also some separate details, and to show how great is the supply which suffices not only for public and private uses and purposes, but also for the satisfaction of luxury; by how many reservoirs it is distributed and in what wards; how much water is delivered outside the City; how much in the City itself; how much of this latter amount is used for water-basins, how much for fountains, how much for public buildings, how much in the name of Caesar, how much for private consumption. But before I mention the names *quinaria, centenaria,* and those of the other ajutages[2] by which water is gauged, I deem it appropriate to state what is their origin, what their capacities, and what each name means; and, after setting forth the rule according to which their proportions and capacities are computed, to show in what way I discovered their discrepancies, and what course I pursued in correcting them.

The ajutages to measure water are arranged according to the standard either of digits or of inches.[3] Digits are the standard in Campania and

the exact capacity of each type, and Frontinus, therefore, enumerates these first of all.

SEXTUS JULIUS FRONTINUS

pania et in plerisque Italiae locis, unciae in pApula [1] cita huc observatur. Est autem digitus, ut convenit, sextadecima pars pedis, uncia duodecima. Quemadmodum autem inter unciam et digitum diversitas, ita et ipsius digiti non simplex observatio est. Alius vocatur quadratus, alius rotundus. Quadratus tribus quartisdecimis suis rotundo maior, rotundus tribus undecimis suis quadrato minor est, scilicet quia anguli deteruntur.

25. Postea modulus nec ab uncia nec ab alterutro digitorum originem accipiens inductus, ut quidam putant, ab Agrippa, ut alii, a plumbariis per Vitruvium architectum in usum urbis exclusis prioribus venit, appellatus quinariae nomine. Qui autem Agrippam auctorem faciunt, dicunt, quod quinque antiqui moduli exiles et velut puncta, quibus olim aqua cum exigua esset dividebatur, in unam fistulam coacti sint; qui Vitruvium et plumbarios, ab eo quod plumbea lammina plana quinque digitorum latitudinem habens circumacta in rotundum hunc fistulae modulum efficiat. Sed hoc incertum est, quoniam cum circumagitur, sicut interiore parte adtrahitur, ita per illam, quae foras spectat, extenditur. Maxime probabile est, quinariam dictam a diametro quinque quadrantum, quae ratio in sequentibus quoque

[1] *BC.*; Apulia *Scaliger.*

---

[1] The Roman foot measured 11·6 English inches, 0·296 m.

[2] The difference between the areas of a square digit and a round digit whose diameter is equal to the side of the square digit is easily seen: |◯|.

[3] The *quinaria* was a measure not of volume but of capacity, *i.e.* as much water as would flow through a pipe one and a quarter digits in diameter, constantly discharging under pressure. "A *quinaria* was about 5,000 or 6,000

AQUEDUCTS OF ROME, I. 24-25

in most parts of Italy; inches are the standard in ... Now the digit, by common understanding, is $\frac{1}{16}$ part of a foot;[1] the inch $\frac{1}{12}$ part. But precisely as there is a difference between the inch and the digit, just so the standard of the digit itself is not uniform. One is called square; another, round. The square digit is larger than the round digit by $\frac{3}{14}$ of its own size, while the round is smaller than the square by $\frac{3}{11}$ of its size, obviously because the corners are cut off.[2]

Later on, an ajutage called a *quinaria*[3] came into use in the City, to the exclusion of the former measures. This was based neither on the inch, nor on either of the digits, but was introduced, as some think, by Agrippa, or, as others believe, by plumbers at the instance of Vitruvius, the architect. Those who represent Agrippa as its inventor, declare it was so designated because five small ajutages or punctures, so to speak, of the old sort, through which water used to be distributed when the supply was scanty, were now united in one pipe. Those who refer it to Vitruvius and the plumbers, declare that it was so named from the fact that a flat sheet of lead 5 digits wide, made up into a round pipe, forms this ajutage. But this is indefinite, because the plate, when made up into a round pipe, will be extended on the exterior surface and contracted on the interior surface. The most probable explanation is that the *quinaria*[4] received its name from having a diameter of $\frac{5}{4}$ of a digit, a standard which holds in

United States gallons per twenty-four hours, plus or minus 2,000 or 3,000 gallons, according to circumstances, favourable or unfavourable" (Herschel).

[4] *i. e.* "a fiver."

367

SEXTUS JULIUS FRONTINUS

modulis usque ad vicenariam durat, diametro per singulos adiectione singulorum quadrantum crescente: ut in senaria, quae sex quadrantes in diametro habet, et septenaria, quae septem, et deinceps simili incremento usque ad vicenariam.

26. Omnis autem modulus colligitur aut diametro aut perimetro aut areae mensura, ex quibus et capacitas apparet. Differentiam unciae, digiti quadrati et digiti rotundi, et ipsius quinariae ut facilius dinoscamus, utendum est substantia quinariae, qui modulus et certissimus et maxime receptus est. Unciae ergo modulus habet diametri digitum unum et trientem digiti; capit plus, quam quinaria, quinariae octava, hoc est sescuncia quinariae et scripulis tribus et besse scripuli. Digitus quadratus in rotundum redactus habet diametri digitum unum et digiti sescunciam sextulam; capit quinariae dextantem. Digitus rotundus habet diametri digitum unum; capit quinariae septuncem semunciam sextulam.

27. Ceterum moduli, qui a quinaria oriuntur, duobus generibus incrementum accipiunt. Est unum, cum ipsa multiplicatur, id est eodem lumine plures quinariae includuntur, in quibus secundum adiectionem quinariarum amplitudo luminis crescit. Est

---

[1] *Cf.* 38 ff.

[2] Frontinus's fractions and the symbols which represent them are as follows. The total value in each case is the *sum* of the various members.

Uncia,	$\frac{1}{12}$, — or ·		Bes,	$\frac{2}{3}$, S =
Sextans,	$\frac{1}{6}$, = or Z		Dodrans,	$\frac{3}{4}$, S = —
Quadrans,	$\frac{1}{4}$, = — or : ·		Dextans,	$\frac{10}{12}$, S = =
Triens,	$\frac{1}{3}$, = = or : :		Deunx,	$\frac{11}{12}$, S = = —
Quincunx,	$\frac{5}{12}$, = = —		Semuncia,	$\frac{1}{2} \cdot \frac{1}{12}$ ($\frac{1}{24}$), $\underline{S}$ or £
Semissis,	$\frac{1}{2}$, S		Scripulus,	$\frac{1}{288}$, Ɔ
Septunx,	$\frac{7}{12}$, S —			

AQUEDUCTS OF ROME, I. 25–27

the following ajutages also up to the 20-pipe, the diameter of each pipe increasing by the addition of ¼ of a digit. For example the 6-pipe is six quarters in diameter, a 7-pipe seven quarters, and so on by a uniform increase up to a 20-pipe.

Every ajutage, now, is gauged either by its diameter or circumference, or by its area of clear cross-section, from any of which factors its capacity becomes evident. That we may distinguish the more readily between the inch ajutage, the square digit, the circular digit, and the *quinaria* itself, use must be made of the value of the *quinaria*, the ajutage which is most accurately determined and best known. Now the inch ajutage, has a diameter of $1\frac{1}{3}$ digits.[1] Its capacity is [slightly] more than $1\frac{1}{8}$ *quinariae*, *i.e.* $1\frac{1}{2}$ twelfths of a *quinaria* plus $\frac{3}{288}$ plus $\frac{2}{3}$ of $\frac{1}{288}$ more. The square digit, reduced to the circle is 1 digit plus $1\frac{1}{2}$ twelfths of a digit plus $\frac{1}{72}$ in diameter; its capacity is $\frac{10}{12}$ of a *quinaria*. The circular digit is 1 digit in diameter; its capacity is $\frac{7}{12}$ plus $\frac{1}{2}$ twelfth plus $\frac{1}{72}$ of a *quinaria*.[2]

Now the ajutages which are derived from the *quinaria* increase on two principles. One principle is that the *quinaria* itself is taken a given number of times, *i.e.* in one orifice the equivalent of several *quinariae* is included, in which case the size of the orifice increases according to the increase in the number of *quinariae*. This principle is regularly

He also uses Sescuncia, $1\frac{1}{2} \times \frac{1}{12}$, (or $\frac{1}{8}$); Duella, $\frac{1}{36}$; Sicilicus, $\frac{1}{48}$; and Sextula, $\frac{1}{72}$; and depends on combinations of these to express exact terms. Owing to corruptions in text, Frontinus's figures are often at variance with obvious facts.

SEXTUS JULIUS FRONTINUS

autem fere tum in usu, cum plures quinariae impetratae, ne rivus saepius convulneretur, una fistula excipiuntur in castellum, ex quo singuli suum modum recipiunt.

28. Alterum genus est, quotiens non ad quinariarum necessitatem fistula incrementum capit, sed ad diametri sui mensuram, secundum quod et nomen accipit et capacitatem ampliat: ut puta quinaria, cum adiectus est ei ad diametrum quadrans, senariam facit. Nec iam in solidum capacitatem ampliat; capit enim quinariam unam et quincuncem sicilicum. Et deinceps eadem ratione quadrantibus diametro adiectis, ut supra dictum est, crescunt septenaria, octonaria usque ad vicenariam.

29. Subsequitur illa ratio, quae constat ex numero digitorum quadratorum, qui area, id est lumine, cuiusque moduli continentur, a quibus et nomen fistulae accipiunt. Nam quae habet areae, id est luminis in rotundum coacti, digitos quadratos viginti quinque, vicenum quinum appellatur: similiter tricenaria et deinceps pari incremento quinorum digitorum quadratorum usque ad centenum vicenum.

30. In vicenaria fistula, quae in confinio utriusque rationis posita est, utraque ratio paene congruit.

---

[1] Frontinus merely means that, instead of tapping the main conduit for individual consumers, the Romans delivered a given number of *quinariae* to a reservoir, from which the water was delivered to the consumer.

[2] It seems advisable to restate, for clearness' sake, the two principles of increase referred to by Frontinus. In the first class we have pipes, whose capacity is some multiple of the *quinaria*. In the second class, we have an ascending series of pipes each of which increases beyond the next smaller by a diameter of ¼ inch. In other words the first class is

employed, whenever several *quinariae* are delivered by one pipe and received in a reservoir, from which consumers receive their individual supply,—this being done in order that the conduit may not be tapped too often.[1]

The second principle is followed, whenever the pipe does not increase according to some necessary multiple of *quinariae*, but according to the size of diameters, in conformity with which principle they enlarge their capacity and receive their names; as for example, when a quarter [of a digit] is added to the diameter of a *quinaria*, we get as a result the *senaria*,[2] but its capacity is not increased by a whole *quinaria*, for it contains a *quinaria* plus $\frac{5}{12}$ plus $\frac{1}{48}$. So on, by adding successive quarters of a digit to the diameter, as was said above, we get by gradual increases, a 7-pipe (*septenaria*), an 8-pipe (*octonaria*), and up to the 20-pipe (*vicenaria*).

After that [3] we have the method of gauging which is based on the number of square digits contained in the cross-section, that is, the orifice of each ajutage, from which number of square digits the pipes also get their names. Thus those which in cross-section, that is, in circular orifice, have 25 square digits, are called 25-pipes. Similarly we have the 30-pipe (*tricenaria*), and so on, by a regular increase of 5 square digits, up to the 120-pipe.

In the case of the 20-pipe, which is on the border line between the two methods of gauging,[4] the two methods almost coincide. For according to the

based on *multiples* of volume; the second on *slight increases* in diameters.

[3] *i.e.* for the pipes above the 20-pipe.
[4] *i.e.* those mentioned in 28 and 29.

SEXTUS JULIUS FRONTINUS

Nam habet secundum eam computationem, quae in antecedentibus modulis servanda est, in diametro quadrantes viginti, cum diametri eiusdem digiti quinque sint; et secundum eorum modulorum rationem, qui sequuntur, aream habet digitorum quadratorum exiguo minus viginti.

31. Ratio fistularum quinariarum usque ad centenum vicenum per omnes modulos ita se habet, ut ostendimus, et omni genere inita constat sibi. Convenit et cum is modulis, qui in commentariis invictissimi et piissimi principis positi et confirmati sunt. Sive itaque ratio sive auctoritas sequenda est, utroque commentariorum moduli praevalent. Sed aquarii cum manifestae rationi in pluribus consentiant, in quattuor modulis novaverunt, duodenaria et vicenaria et centenaria et centenum vicenum.

32. Et duodenariae quidem nec magnus error nec usus frequens est. Cuius diametro adiecerunt digiti semunciam sicilicum, capacitati quinariae quadrantem. In reliquis autem tribus modulis plus deprenditur. Vicenariam exiguiorem faciunt diametro digiti semisse et semuncia, capacitate quinariis tribus et quadrante et semuncia, quo modulo plerumque erogatur. Centenaria autem et centenum vicenum, quibus adsidue accipiunt, non minuuntur sed augentur. Diametro enim centenariae adiciunt digiti bessem et semunciam, capacitati quinarias decem semissem semunciam. Centenum vicenum diametro adiciunt digitos tres septuncem semunciam sicilicum, capacitati quinarias sexaginta sex sextantem.

---

[1] By successive additions of a quarter of a digit to the diameter of a *quinaria*, the diameter of the *vicenaria* becomes five digits or *twenty* quarter digits. The number of square digits in the cross section, therefore, would be $\pi(\frac{5}{2})^2 = 19 \cdot 6$ (almost *twenty*) square digits.

AQUEDUCTS OF ROME, I. 30–32

reckoning to be used in the first-named set of ajutages, it is twenty quarter digits in diameter, inasmuch as its diameter is 5 digits; while according to the computation to be applied to the higher ajutages, it has an area of 20 square digits, less a fraction.[1]

The gauging of the entire series of ajutages from the 5-pipe (*quinaria*) up to the 120-pipe, is determined in the way I have explained, and in each class the principle adopted is adhered to for that class. It conforms also to the ajutages set down and verified in the records of our most puissant and patriotic emperor.[2] Whether, therefore, computation or authority is to be followed, on either ground the ajutages of the records are of greater weight. But the water-men, while they conform to the obvious reckoning in most ajutages, have made deviation in the case of four of them, namely: the 12-, 20-, 100-, and 120-pipe.

In case of the 12-pipe, the error is not great, nor is its use frequent. They have added $\frac{1}{24}$ plus $\frac{1}{48}$ to its diameter, and to its capacity $\frac{1}{4}$ of a *quinaria*. A greater discrepancy is detected in case of the three remaining ajutages. These water-men diminish the 20-pipe in its diameter by $\frac{1}{2}$ plus $\frac{1}{24}$ of a digit, its capacity by 3 *quinariae* plus $\frac{1}{4}$ plus $\frac{1}{24}$; and common use is made of this ajutage for delivery. But in case of the 100-pipe and 120-pipe, through which they [3] regularly receive water, the pipes are not diminished but *enlarged!* For to the diameter of the 100-pipe they add $\frac{2}{3}$ plus $\frac{1}{24}$ of a digit, and to the capacity, 10 *quinariae* plus $\frac{1}{2}$ plus $\frac{1}{24}$. To the diameter of the 120-pipe they add 3 digits plus $\frac{7}{12}$ plus $\frac{1}{24}$ plus $\frac{1}{48}$; to its capacity, 66 *quinariae* plus $\frac{1}{8}$.

[2] Trajan is meant. [3] The water-men.

SEXTUS JULIUS FRONTINUS

33. Ita dum aut vicenariae, qua subinde erogant, detrahunt aut centenariae et centenum vicenum adiciunt, quibus semper accipiunt, intercipiuntur in centenaria quinariae viginti septem, in centenum vicenum quinariae octoginta sex. Quod cum ratione approbetur, re quoque ipsa manifestum est. Nam et vicenaria, quam Caesar pro quinariis sedecim assignat, non plus erogant quam tredecim, et ex centenaria, quam ampliaverunt eque centenum vicenum certum est illos non erogare nisi ad artiorem numerum, quia Caesar secundum suos commentarios, cum ex quaque centenaria explevit quinarias octoginta unam semissem, item ex centenum vicenum quinarias nonaginta octo, tamquam exhausto modulo desinit distribuere.

34. In summa moduli sunt XX quinque. Omnes consentiunt et rationi et commentariis, exceptis his quattuor, quos aquarii novaverunt. Omnia autem quae mensura continentur, certa et immobilia congruere sibi debent; ita enim universitati ratio constabit. Et quemadmodum verbi gratia sextarii ratio ad cyathos, modii vero et ad sextarios et ad

---

[1] Frontinus's reckoning is as follows: The capacity of a 20-pipe is $16\frac{7}{24}$ *quinariae* (*cf.* 46); the capacity of the 100-pipe is $81\frac{6 \cdot 5}{144}$ *quinariae* (*cf.* 62); the capacity of five 20-pipes, therefore, practically equals that of one 100-pipe. Now, if the gain resulting from *selling* by short measure was $3\frac{7}{24}$ *quinariae* in one 20-pipe, it will have been $16\frac{11}{24}$ *quinariae* in five 20-pipes (or one 100-pipe). In the same way, since the capacity of the 120-pipe is $97\frac{3}{4}$ *quinariae* (*cf.* 63), it is equal to six 20-pipes, and the gain in this case will have been $19\frac{3}{4}$ *quinariae*. But by the increase of the pipes through which they *receive* water (*cf.* 32), the gain was $10\frac{13}{24}$ *quinariae* in case of the 100-pipe, and $66\frac{1}{3}$ *quinariae* in case of the 120-pipe; so that by adding the gains made at both ends of the bargain, we

Thus by diminishing the size of the 20-pipe by which they constantly deliver, and enlarging the 100- and 120-pipes, by which they always receive, they steal in case of the 100-pipe 27 *quinariae*, and in case of the 120-pipe 86 *quinariae*.[1] While this is proved by computation, it is also obvious from the facts. For from the 20-pipe, which Caesar rates[2] at 16 *quinariae*, they do not deliver more than 13; and it is equally certain that from the 100-pipe and the 120-pipe, which they have expanded, they deliver only up to a limited amount, since Caesar, as his records show, has made delivery according to his grant,[3] when out of each 100-pipe he furnishes $81\frac{1}{2}$ *quinariae*, and similarly out of a 120-pipe, 98.

In all there are 25 ajutages. They all conform to their computed and recorded capacities, barring these four which the water-men have altered. But everything embraced under the head of mensuration ought to be fixed, unchanged, and constant. For only so will any special computation accord with general principles. Just as a *sextarius*,[4] for example, has a regular ratio to a *cyathus*,[5] and

---

arrive at an aggregate gain of 27 *quinariae* in case of the 100-pipe, and of $85\frac{11}{12}$, practically 86, *quinariae* in case of the 120-pipe.

[2] *Cf.* 31.

[3] Literally: "stops distributing, as though the ajutage had run dry." Whoever wished to draw water for private uses had to seek for a grant and bring to the water-commissioner a writing from the sovereign (*cf.* 103, 105). Now, the records show that Caesar's grants from a 100-pipe amounted only to 81 *quinariae*, and from a 120 pipe to only 98 *quinariae*, leaving a surplus to be accounted for.

[4] The Roman pint.   [5] About a gill.

SEXTUS JULIUS FRONTINUS

cyathos respondet; ita et quinariarum multiplicatio in amplioribus modulis servare consequentiae suae regulam debet. Alioqui cum in erogatorio modulo minus invenitur, in acceptorio plus, apparet non errorem esse sed fraudem.

35. Meminerimus omnem aquam, quotiens ex altiore loco venit et intra breve spatium in castellum cadit, non tantum respondere modulo suo sed etiam exuberare; quotiens vero ex humiliore, id est minore pressura, longius ducitur, segnitia ductus modum quoque deperdere; et ideo secundum hanc rationem aut onerandam esse erogatione aut relevandam.

36. Sed et calicis positio habet momentum. In rectum et ad libram conlocatus modum servat, ad cursum aquae oppositus et devexus amplius rapit, ad latus praetereuntis aquae conversus et supinus, id est ad haustum pronior, segniter et exiguum sumit. Est autem calix modulus aeneus, qui rivo vel castello induitur; huic fistulae applicantur. Longitudo eius habere debet digitos non minus duodecim, lumen capacitatem quanta imperata fuerit. Excogitatus videtur, quoniam rigor aeris difficilior ad flexum non temere potest laxari vel coartari.

37. Formulas modulorum qui sunt omnes viginti et quinque subieci, quamvis in usu quindecim tantum frequentes sint, derectas ad rationem de qua locuti

---

[1] The Roman peck.

[2] *i.e.* to make the pipe discharge the normal quantity allotted to a pipe of that size.

[3] *i.e.* in any particular instance.

similarly a *modius*[1] to both a *cyathus* and *sextarius*, so also the multiplication of the *quinariae* in case of the larger ajutages must follow a regular progression. However, when less is found in the delivery ajutages and more in the receiving ajutages, it is obvious that there is not error, but fraud.

Let us remember that every stream of water, whenever it comes from a higher point and flows into a reservoir after a short run, not only comes up to its measure, but actually yields a surplus; but whenever it comes from a lower point, that is, under less pressure, and is conducted a longer distance, it shrinks in volume, owing to the resistance of its conduit; and that, therefore, on this principle it needs either a check or a help in its discharge.[2]

But the position of the *calix* is also a factor. Placed at right angles and level, it maintains the normal quantity. Set against the current of the water, and sloping downward, it will take in more. If it slopes to one side, so that the water flows by, and if it is inclined with the current, that is, is less favourably placed for taking in water, it will receive the water slowly and in scant quantity. The *calix*, now, is a bronze ajutage, inserted into a conduit or reservoir, and to it the service pipes are attached. Its length ought not to be less than 12 digits, while its orifice ought to have such capacity as is specified.[3] Bronze seems to have been selected, since, being hard, it is more difficult to bend, and is not easily expanded or contracted.

I have described below all the 25 ajutages that there are (although only 15 of them are in use), gauging them according to the method of com-

SEXTUS JULIUS FRONTINUS

sumus, emendatis quattuor, quos aquarii novaverant. Secundum quod et fistulae omnes, quae opus facient, derigi debent aut, si haec fistulae manebunt, ad quinarias quot capient computari. Qui non sint in usu moduli, in ipsis est adnotatum.

38. Uncia habet diametri digitum unum et trientem digiti; capit plus quam quinaria, quinariae sescuncia et scripulis tribus et besse scripuli. Digitus quadratus in latitudine et longitudine aequalis est. Digitus quadratus in rotundum redactus habet diametri digitum unum et digiti sescunciam sextulam, capit quinariae dextantem. Digitus rotundus habet diametri digitum unum, capit quinariae septuncem et semiunciam sextulam.[1]

39. Fistula quinaria: diametri digitum unum $=-$, perimetri digitos tres S $==-\ni$ III, capit quinariam unam.

40. Fistula senaria: diametri digitum unum semis, perimetri digitos IIII S $=\mathcal{L}\ni$ II, capit quinariam I $==-\ni$ VII.

41. Fistula septenaria: diametri digitum I S∴, perimetri digitos V S, capit quinariam I S $==-\mathcal{L}$; in usu non est.

42. Fistula octonaria: diametri digitos duos, perimetri digitos sex $=-\ni$ X, caput quinarias II S $\mathcal{L}\ni$ quinque.

43. Fistula denaria: diametri digitos duos et semis, perimetri digitos septem S :: $\ni$ VII, capit quinarias IIII.

44. Fistula duodenaria: diametri digitos III, perimetri digitos VIIII $==-\ni$ III, capit quinarias quinque S $=-\ni$ III; in usu non est. Apud aquarios

---

[1] *Bennett*; sextam *CB*.

AQUEDUCTS OF ROME, I. 37-44

putation spoken of,[1] and correcting the four which the water-men have altered. To these specifications all ajutages in use ought to conform, or if those four remain in use, they ought to be gauged by the number of *quinariae* which they contain. The ajutages that are not in use are so referred to.

The inch ajutage[2] is 1 digit plus $\frac{1}{3}$ of a digit in diameter; it contains more than a *quinaria* by $1\frac{1}{2}$ twelfths of a *quinaria* plus $\frac{3}{288}$ plus $\frac{2}{3}$ of $\frac{1}{288}$. The square digit has the same height as breadth. The square digit converted into its equivalent circle is 1 digit plus $1\frac{1}{2}$ twelfths of a digit plus $\frac{1}{72}$ in diameter; it measures $\frac{10}{12}$ of a *quinaria*. The circular digit is 1 digit in diameter; and measures $\frac{7}{12}$ plus a $\frac{1}{2}$ twelfth plus $\frac{1}{72}$ of a *quinaria* in area.

The *quinaria*: 1 digit plus $\frac{3}{12}$ in diameter; 3 digits plus $\frac{1}{2}$ plus $\frac{5}{12}$ plus $\frac{3}{288}$ in circumference; it has a capacity of 1 *quinaria*.

The 6-pipe: $1\frac{1}{2}$ digits in diameter; 4 digits plus $\frac{1}{2}$ plus $\frac{2}{12}$ plus $\frac{1}{24}$ plus $\frac{2}{288}$ in circumference; it has a capacity of 1 *quinaria* plus $\frac{5}{12}$ plus $\frac{7}{288}$.

The 7-pipe: 1 digit plus $\frac{1}{2}$ plus $\frac{3}{12}$ in diameter; 5 digits plus $\frac{1}{2}$ in circumference; it has a capacity of 1 *quinaria*, plus $\frac{1}{2}$ plus $\frac{5}{12}$ plus $\frac{1}{24}$; is not in use.

The 8-pipe: 2 digits in diameter; 6 digits plus $\frac{3}{12}$ plus $\frac{10}{288}$ in circumference; it has a capacity of 2 *quinariae* plus $\frac{1}{2}$ plus $\frac{1}{24}$ plus $\frac{5}{288}$.

The 10-pipe: $2\frac{1}{2}$ digits in diameter; 7 digits plus $\frac{1}{2}$ plus $\frac{4}{12}$ plus $\frac{7}{288}$ in circumference; it has a capacity of 4 *quinariae*.

The 12-pipe: 3 digits in diameter; 9 digits plus $\frac{5}{12}$ plus $\frac{3}{288}$ in circumference; it has a capacity of 5 *quinariae* plus $\frac{1}{2}$ plus $\frac{3}{12}$ plus $\frac{3}{288}$; is not in use.

[1] *Cf.* 26 ff.   [2] *Cf.* 26.

SEXTUS JULIUS FRONTINUS

habebat diametri digitos III ℒ Ƹ VI, capacitatis quinarias sex.

45. Fistula quinum denum: diametri digitos III S ⎓ —, perimetri digitos XI S — ⎓ Ƹ X, capit quinarias novem.

46. Fistula vicenaria: diametri digitos quinque ℒ Ƹ, perimetri digitos XV S :: Ƹ VI, capit quinarias sedecim ⎓ — ℒ. Apud aquarios habebat diametri digitos IIII S, capacitatis quinarias XIII.

47. Fistula vicenum quinum: diametri digitos quinque S · ℒ Ƹ V, perimetri digitos decem et septem S ⎓ ℒ Ƹ VII, capit quinarias XX ⎓ ⎓ Ƹ VIIII; in usu non est.

48. Fistula tricenaria: diametri digitos sex ⎓ Ƹ III, perimetri digitos decem et novem ⎓ ⎓ —, capit quinarias viginti quattuor ⎓ ⎓ — Ƹ quinque.

49. Fistula tricenum quinum: diametri digitos sex S ⎓ Ƹ II, perimetri digitos XX S ⎓ ⎓ — ℒ Ƹ IIII, capit quinarias XXVIII S Ƹ III; in usu non est

50. Fistula quadragenaria: diametri digitos septem — ℒ Ƹ III, perimetri digitos XXII ⎓ ⎓ —, capit quinarias XXXII S —.

51. Fistula quadragenum quinum: diametri digitos septem S ℒ Ƹ octo, perimetri digitos XXIII S ⎓ — ℒ, capit quinarias XXXVI S — ℒ Ƹ octo; in usu non est.

52. Fistula quinquagenaria: diametri digitos septem S ⎓ ⎓ — ℒ Ƹ quinque, perimetri digitos XXV ℒ Ƹ VII, capit quinarias XL S ⎓ ℒ Ƹ V.

53. Fistula quinquagenum quinum: diametri digitos octo ⎓ ⎓ Ƹ decem, perimetri digitos XXVI ⎓ — ℒ,

AQUEDUCTS OF ROME, I. 44-53

But with the water-men it measured 3 digits plus $\frac{1}{24}$ plus $\frac{6}{288}$ in diameter, containing 6 *quinariae*.

The 15-pipe: 3 digits plus $\frac{1}{2}$ plus $\frac{3}{12}$ in diameter; 11 digits plus $\frac{1}{2}$ plus $\frac{3}{12}$ plus $\frac{10}{288}$ in circumference; it has a capacity of 9 *quinariae*.

The 20-pipe: 5 digits plus $\frac{1}{24}$ plus $\frac{1}{288}$ in diameter; 15 digits plus $\frac{1}{2}$ plus $\frac{4}{12}$ plus $\frac{6}{288}$ in circumference; it has a capacity of 16 *quinariae* plus $\frac{3}{12}$ plus $\frac{1}{24}$. With the water-men it measured 4 digits plus $\frac{1}{2}$ in diameter, holding 13 *quinariae*.

The 25-pipe: 5 digits plus $\frac{1}{2}$ plus $\frac{1}{12}$ plus $\frac{1}{24}$ plus $\frac{5}{288}$ in diameter; 17 digits plus $\frac{1}{2}$ plus $\frac{2}{12}$ plus $\frac{1}{24}$ plus $\frac{7}{288}$ in circumference; it has a capacity of 20 *quinariae* plus $\frac{4}{12}$ plus $\frac{9}{288}$; is not in use.

The 30-pipe: 6 digits plus $\frac{2}{12}$ plus $\frac{3}{288}$ in diameter; 19 digits plus $\frac{1}{12}$ in circumference; it has a capacity of 24 *quinariae* plus $\frac{5}{12}$ plus $\frac{5}{288}$.

The 35-pipe: 6 digits plus $\frac{1}{2}$ plus $\frac{2}{12}$ plus $\frac{2}{288}$ in diameter; 20 digits plus $\frac{1}{2}$ plus $\frac{5}{12}$ plus $\frac{1}{24}$ plus $\frac{4}{288}$ in circumference; it has a capacity of 28 *quinariae* plus $\frac{1}{2}$ plus $\frac{3}{288}$; is not in use.

The 40-pipe: 7 digits plus $\frac{1}{12}$ plus $\frac{1}{24}$ plus $\frac{3}{288}$ in diameter; 22 digits plus $\frac{5}{12}$ in circumference; it has a capacity of 32 *quinariae* plus $\frac{1}{2}$ plus $\frac{1}{12}$.

The 45-pipe: 7 digits plus $\frac{1}{2}$ plus $\frac{1}{24}$ plus $\frac{8}{288}$ in diameter; 23 digits plus $\frac{1}{2}$ plus $\frac{3}{12}$ plus $\frac{1}{24}$ in circumference; it has a capacity of 36 *quinariae* plus $\frac{1}{2}$ plus $\frac{1}{12}$ plus $\frac{1}{24}$ plus $\frac{8}{288}$; is not in use.

The 50-pipe: 7 digits plus $\frac{1}{2}$ plus $\frac{5}{12}$ plus $\frac{1}{24}$ plus $\frac{5}{288}$ in diameter; 25 digits plus $\frac{7}{24}$ plus $\frac{7}{288}$ in circumference; it has a capacity of 40 *quinariae* plus $\frac{1}{2}$ plus $\frac{2}{12}$ plus $\frac{1}{24}$ plus $\frac{5}{288}$.

The 55-pipe: 8 digits plus $\frac{4}{12}$ plus $\frac{10}{288}$ in diameter; 26 digits plus $\frac{3}{12}$ plus $\frac{1}{24}$ in circumference;

SEXTUS JULIUS FRONTINUS

capit quinarias XLIIII S ⏁ — ℒ ꝯ II; in usu non est.

54. Fistula sexagenaria: diametri digitos VIII S ⏁ ℒ ꝯ octo, perimetri digitos XXVII ⏁ ⏁ — ℒ, capit quinarias XL octo S ⏁ ⏁ ꝯ XI.

55. Fistula sexagenum quinum: diametri digitos novem — ꝯ III, perimetri digitos XX octo S —, capit quinarias quinquaginta duas S ⏁ — ℒ ꝯ octo; in usu non est.

56. Fistula septuagenaria: diametri digitos novem ⏁ ⏁ — ꝯ sex, perimetri digitos XXIX S ⏁, capit quinarias LVII ꝯ V.

57. Fistula septuagenum quinum: diametri digitos novem S ⏁ — ꝯ sex, perimetri digitos XXX S ⏁ ꝯ VIII, capit quinarias LXI — ꝯ II; in usu non est.

58. Fistula octogenaria: diametri digitos decem — ꝯ II, perimetri digitos XXXI S ⏁ ℒ, capit quinarias LXV ⏁.

59. Fistula octogenum quinum: diametri digitos decem ⏁ ⏁ ⏁ ℒ ꝯ septem, perimetri digitos XXXII S ⏁ ꝯ IIII, capit quinarias LXVIIII ⏁ —; in usu non est.

60. Fistula nonagenaria: diametri digitos decem S ⏁ ꝯ X, perimetri digitos triginta tres S — ℒ ꝯ II, capit quinarias septuaginta tres ⏁ — ℒ ꝯ V.

61. Fistula nonagenum quinum: diametri digitos X S ⏁ ⏁ ⏁ — ℒ ꝯ VIIII, perimetri digitos XXXIIII S ℒ, capit quinarias LXXVII ⏁ ⏁ ℒ ꝯ II; in usu non est.

62. Fistula centenaria: diametri digitos XI ⏁ — ꝯ VIIII, perimetri digitos XXXV ⏁ ⏁ — ℒ, capit

382

AQUEDUCTS OF ROME, I. 53-62

it has a capacity of 44 *quinariae* plus ½ plus $\frac{3}{12}$ plus $\frac{1}{24}$ plus $\frac{2}{288}$; is not in use.

The 60-pipe: 8 digits plus ½ plus $\frac{1}{12}$ plus $\frac{1}{24}$ plus $\frac{8}{288}$ in diameter; 27 digits plus $\frac{5}{12}$ plus $\frac{1}{24}$ in circumference; it has a capacity of 48 *quinariae* plus ½ plus $\frac{4}{12}$ plus $\frac{11}{288}$.

The 65-pipe: 9 digits plus $\frac{1}{12}$ plus $\frac{3}{288}$ in diameter; 28 digits plus ½ plus $\frac{1}{12}$ in circumference; it has a capacity of 52 *quinariae* plus ½ plus $\frac{3}{12}$ plus $\frac{1}{24}$ plus $\frac{8}{288}$; is not in use.

The 70-pipe: 9 digits plus $\frac{5}{12}$ plus $\frac{6}{288}$ in diameter; 29 digits plus ½ plus $\frac{2}{12}$ in circumference; it has a capacity of 57 *quinariae* plus $\frac{5}{288}$.

The 75-pipe: 9 digits plus ½ plus $\frac{3}{12}$ plus $\frac{6}{288}$ in diameter; 30 digits plus ½ plus $\frac{2}{12}$ plus $\frac{8}{288}$ in circumference; it has a capacity of 61 *quinariae* plus $\frac{1}{12}$ plus $\frac{2}{288}$; is not in use.

The 80-pipe: 10 digits plus $\frac{1}{12}$ plus $\frac{2}{288}$ in diameter; 31 digits plus ½ plus $\frac{1}{12}$ plus $\frac{1}{24}$ in circumference; it has a capacity of 65 *quinariae* plus $\frac{5}{12}$.

The 85-pipe: 10 digits plus $\frac{4}{12}$ plus $\frac{1}{24}$ plus $\frac{7}{288}$ in diameter; 32 digits plus ½ plus $\frac{2}{12}$ plus $\frac{4}{288}$ in circumference; it has a capacity of 69 *quinariae* plus $\frac{3}{12}$; is not in use.

The 90-pipe: 10 digits plus ½ plus $\frac{2}{12}$ plus $\frac{10}{288}$ in diameter; 33 digits plus ½ plus $\frac{1}{12}$ plus $\frac{1}{24}$ plus $\frac{2}{288}$ in circumference; it has a capacity of 73 *quinariae* plus $\frac{3}{12}$ plus $\frac{1}{24}$ plus $\frac{5}{288}$.

The 95-pipe: 10 digits plus ½ plus $\frac{5}{12}$ plus $\frac{1}{24}$ plus $\frac{9}{288}$ in diameter; 34 digits plus ½ plus $\frac{1}{24}$ in circumference; it has a capacity of 77 *quinariae* plus $\frac{1}{12}$ plus $\frac{1}{24}$ plus $\frac{2}{288}$; is not in use.

The 100-pipe: 11 digits plus $\frac{3}{12}$ plus $\frac{9}{288}$ in diameter; 35 digits plus $\frac{5}{12}$ plus $\frac{1}{24}$ in circumference;

383

SEXTUS JULIUS FRONTINUS

quinarias octoginta unam $==-\ni$ X. Apud aquarios habebat diametri digitos XII, capacitatis quinarias nonaginta II.

63. Fistula centenum vicenum: diametri digitos duodecim $== \ni$ VI, perimetri digitos XXXVIII S$==$, capit quinarias LXXXXVII S$==-$. Apud aquarios habebat diametri digitos XVI, capacitatis quinarias centum sexaginta tres S$==-$, qui modus duarum centenariarum est.

AQUEDUCTS OF ROME, I. 62-63

it has a capacity of 81 *quinariae* plus $\frac{5}{12}$ plus $\frac{10}{288}$. With the water-men it had a diameter of 12 digits; having a capacity of 92 *quinariae*.

The 120-pipe: 12 digits plus $\frac{4}{12}$ plus $\frac{6}{288}$ in diameter; 38 digits plus $\frac{1}{2}$ plus $\frac{4}{12}$ in circumference; it has a capacity of 97 *quinariae* plus $\frac{1}{2}$ plus $\frac{3}{12}$. With the water-men it had a diameter of 16 digits, having a capacity of 163 *quinariae* plus $\frac{1}{2}$ plus $\frac{5}{12}$, which is the measure of two 100-pipes.

BOOK II

LIBER ALTER

64. Persecutus ea quae de modulis dici fuit necessarium, nunc ponam, quem modum quaeque aqua, ut principum commentariis comprehensum est, usque ad nostram curam habere visa sit quantumque erogaverit; deinde quem ipsi scrupulosa inquisitione praeeunte providentia optimi diligentissimique Nervae principis invenerimus. Fuerunt ergo in commentariis in universo quinariarum decem duo milia septingentae quinquaginta quinque, in erogatione decem quattuor milia decem et octo: plus in distributione quam in accepto computabatur quinariis mille ducentis sexaginta tribus. Huius rei admiratio, cum praecipuum officii opus in exploranda fide aquarum atque copia crederem, non mediocriter me convertit ad scrutandum, quemadmodum amplius erogaretur, quam in patrimonio, ut ita dicam, esset. Ante omnia itaque capita ductuum metiri adgressus sum, sed longe, id est circiter quinariarum decem milibus, ampliorem quam in commentariis modum inveni, ut per singulas demonstrabo.

65. Appiae in commentariis adscriptus est modus

---

[1] Trajan, not Nerva, is meant. From the allusions in chapter 93 to Caesar Nerva Trajan Augustus, and in 102 to Divus Nerva, it is concluded that this work which was begun under Nerva was finished under Trajan.

[2] *Cf.* 31. [3] *Cf.* footnote to 74.

[4] In this and several following paragraphs, Frontinus points out various discrepancies which existed in the accounting for the different water supplies. These are (1) the difference between the amount of water *found by actual measurement* to exist, and the amount *attributed* to each

BOOK II

HAVING detailed those facts which it was necessary to state with reference to the ajutages, I will now set down what discharge each aqueduct, according to the imperial records, was thought to have up to the time of my administration, and also how much it actually did deliver; then the true measure, which I reached by careful investigation, acting on the suggestion of that best and most industrious emperor, Nerva.[1] Now there were, in the aggregate, 12,755 *quinariae* set down in the records,[2] but 14,018 *quinariae* actually delivered; that is, 1,263 more *quinariae* were reported as delivered than were reckoned as received.[3] Since I considered it the most important function of my office to determine the facts concerning the water-supply, my astonishment at this state of affairs stirred me profoundly and led me to investigate how it happened that more was being delivered than belonged to the property, so to speak. Accordingly, I first of all undertook measurements of the intakes of the conduits and discovered a total supply far greater—that is, by about 10,000 *quinariae*—than I found in the records, as I shall explain in connection with each aqueduct.

In the records Appia is credited with 841 *quinariae*.[4]

water supply *in the imperial records;* (2) the difference between the amount of water *recorded as received* and that *delivered;* (3) the difference between the amount *proved by measurement* and that *delivered.* In some cases, where the waters were received in catch-basins, or reservoirs (*cf.* 19), he computes separately the amounts lost before the water reached the basin, and that lost afterwards.

SEXTUS JULIUS FRONTINUS

quinariarum octingentarum quadraginta unius. Cuius aquae ad caput inveniri mensura non potuit, quoniam ex duobus rivis constat. Ad Gemellos tamen, qui locus est infra Spem Veterem, ubi iungitur cum ramo Augustae, inveni altitudinem aquae pedum quinque, latitudinem pedis unius dodrantis; fiunt areae pedes octo dodrans, centenariae viginti duae et quadragenaria, quae efficiunt quinarias mille octingentas viginti quinque; amplius quam commentarii habent quinariis nongentis octoginta quattuor. Erogabat quinarias septingentas quattuor; minus quam in commentariis adscribitur quinariis centum triginta septem, et adhuc minus quam ad Gemellos mensura respondet quinariis mille centum viginti una. Intercidit tamen aliquantum e ductus vitio, qui cum sit depressior, non facile manationes ostendit, quas esse ex eo apparet quod in plerisque urbis partibus probata aqua observatur, quae ex ea manat. Sed et quasdam fistulas intra urbem inlicitas deprehendimus. Extra urbem autem propter pressuram librae, cum sit infra terram ad caput pedibus quinquaginta, nullam accipit iniuriam.

66. Anioni Veteri adscriptus est in commentariis modus quinariarum mille quingentarum [1] quadraginta unius. Ad caput inveni quattuor milia trecentas nonaginta octo praeter eum modum qui in proprium

---

[1] B; quadringentarum C.

---

[1] *Cf.* 5.

[2] The area of a cross section of the Appia would be 2,240 square digits (*cf.* 24). The area of twenty-two 100-pipes would be 2,199 square digits (*cf.* 62), and that of one 40-pipe would be 39·98 square digits (*cf.* 50), so that the area of a cross section of the Appia would practically equal

AQUEDUCTS OF ROME, II. 65–66

A gauging of this aqueduct could not be taken at the intake, since there it consists of two channels. But at The Twins, which is below Spes Vetus,[1] where it joins with a branch of Augusta, I found a depth of water of 5 feet, and a width of 1¾ feet, making an area of 8¾ square feet,[2] twenty-two 100-pipes plus a 40-pipe, which makes 1,825 *quinariae*,— more than the records have it by 984 *quinariae*. It was delivering 704 *quinariae*,—137 *quinariae* less than credited in the records; and, furthermore, 1,121 *quinariae* less than given by the gauging at The Twins.[3] A considerable amount of this, however, is lost by leaks in the conduit, which, being deeply buried, does not clearly exhibit them. And yet their presence is plainly indicated by the fact that in very many parts of the City excellent water is met with, which leaks from that aqueduct. But we also detected some illicit pipes within the City. Outside the City, however, on account of the depth of the level, which at the intake is 50 feet underground, the conduit suffers no depredations.

Old Anio is credited in the records with the amount of 1,541 *quinariae*. At the intake I found 4,398 *quinariae*, exclusive of the quantity which is

---

the sum of the areas of the cross sections of these pipes. Now the capacity of twenty-two 100-pipes, 1,791·93 *quinariae* (*cf.* 62), added to that of one 40-pipe, 32·58 *quinariae* (*cf.* 50), gives approximately 1,825 *quinariae* as the capacity of the Appia at that point.

[3] In the Appia (1) actual measurements showed 1,825 *quinariae*, the records called for 841 *quinariae*—a discrepancy of 984 *quinariae*; (2) while the records credited 841 *quinariae*, only 704 *quinariae* were delivered, a discrepancy of 137 *quinariae*; (3) 1,825 *quinariae* were found by measurement, but only 704 were delivered, a discrepancy of 1,121 *quinariae*.

SEXTUS JULIUS FRONTINUS

ductum Tiburtium derivatur, amplius quam in commentariis est quinariis duobus milibus octingentis quinquaginta septem. Erogabantur antequam ad piscinam veniret quinariae ducentae sexaginta duae. Modus in piscina, qui per mensuras positas initur, efficit quinariarum duo milia trecentas sexaginta duas. Intercidebant ergo inter caput et piscinam quinariae mille septingentae septuaginta quattuor. Erogabat post piscinam quinarias mille trecentas quadraginta octo; amplius quam in commentariis conceptionis modum significari diximus quinariis sexaginta novem; minus quam recipi in ductum post piscinam posuimus quinariis mille decem quattuor. Summa quae inter caput et piscinam et post piscinam intercidebat quinariae duo milia septingentae octoginta octo, quod errore mensurae fieri suspicarer, nisi invenissem ubi averterentur.

67. Marciae in commentariis adscriptus est modus quinariarum duum milium centum sexaginta duarum. Ad caput mensus inveni quinarias quattuor milia sexcentas nonaginta, amplius quam in commentariis est quinariis duobus milibus quingentis viginti octo. Erogabantur antequam ad piscinam perveniret quinariae nonaginta quinque, et dabantur in adiutorium Tepulae quinariae nonaginta duae, item in Anionem quinariae centum sexaginta quattuor. Summa quae erogabatur ante piscinam quinariae trecentae quin-

---

[1] *Cf.* 6.
[2] Old Anio was received into a catch-basin. (1) The measurements at the intake showed 4.398 *quinariae*, the records credited 1,541 *quinariae*, a discrepancy of 2,857

AQUEDUCTS OF ROME, II. 66–67

diverted into the special conduit of the Tiburtines,[1]— 2,857 *quinariae* more than is recorded. There were distributed 262 *quinariae*, before the aqueduct reaches its settling-reservoir. The quantity at the reservoir, determined from the gauges placed there, was 2,362 *quinariae*, so that 1,774 *quinariae* were lost between the intake and the reservoir. Down-stream from the settling-reservoir, 1,348 *quinariae* were delivered, —more than we have stated to be the capacity according to the records by 69 *quinariae*, but less than we have shown was received into the conduit from the settling-reservoir by 1,014 *quinariae*. The total which was lost between the intake and the settling-reservoir, and below the settling-reservoir, amounted to 2,788 *quinariae*, which I should have suspected resulted from an error of measurement, had I not discovered where it was diverted.[2]

In the records Marcia is credited with the quantity of 2,162 *quinariae*. Gauging it at the intake, I found 4,690 *quinariae*,—2,528 *quinariae* more than appear in the records. There were delivered, before it reaches the settling-reservoir, 95 *quinariae*; and 92 *quinariae* were given to supplement Tepula; likewise 164 to Anio. The total delivered before the settling-reservoir is reached, was 351 *quinariae*. The

---

quinariae; (2) between the intake and the reservoir 262 *quinariae* were distributed, and measurements at the reservoir showed 2,362 *quinariae*, revealing a loss of 1,774 *quinariae* above the reservoir; (3) above and below the reservoir, there were delivered in all 1,610 *quinariae*, 69 *quinariae* more than the records showed; but since measurements at the reservoir found 2,362 *quinariae*, and only 1,348 were delivered below that, 1,014 *quinariae* remained unaccounted for and the loss above and below the reservoir totalled 2,788 *quinariae*.

SEXTUS JULIUS FRONTINUS

quaginta una. Modus qui in piscina mensuris positis initur cum eo quod[1] circa piscinam ductum[2] eodem canali in arcus excipitur, efficit quinarias duo milia nongentas quadraginta quattuor. Summa quae aut erogatur ante piscinam aut in arcus recipitur quinariarum tria milia ducentae nonaginta quinque; amplius quam in conceptis commentariorum positum est quinariis mille centum triginta tribus, minus, quam mensurae ad caput actae efficiunt quinariis mille trecentis nonaginta quinque. Erogabat post piscinam quinarias mille octingentas quadraginta; minus quam in commentariis conceptionis modum significari diximus quinariis ducentis viginti septem, minus quam ex piscina in arcus recipiuntur quinariis mille centum quattuor. Summa utraque quae intercidebat aut inter caput et piscinam aut post piscinam quinariarum duo milia ID, quas sicut in ceteris pluribus locis intercipi deprehendimus. Non enim eas cessare manifestum est et ex hoc quod ad caput praeter eam mensuram, quam comprehendisse nos capacitate ductus posuimus, effunduntur amplius trecentae quinariae.

68. Tepulae in commentariis adscriptus est modus quinariarum quadringentarum. Huius aquae fontes nulli sunt; venis quibusdam constabat, quae interceptae sunt in Iulia. Caput ergo eius observandum

---

[1] *C*; qui *B*.   [2] *C*; ductus *B*.

---

[1] Marcia was received into a catch-basin. In the following computations, Frontinus sometimes includes and sometimes disregards the amounts given to Tepula and Anio. (1) The measurements at the intake showed 4,690 *quinariae*, the records credited 2,162 *quinariae*, a discrepancy of 2,528 *quinariae*; (2) between the intake and the reservoir 351 *quinariae* were delivered (including those given to Tepula and Anio), and measurements at the reservoir showed 2,944

quantity which is computed at the reservoir from the gauges set up there, along with what is carried around the reservoir and received in the same channel on arches, is 2,944 *quinariae*. The aggregate of what is delivered above the reservoir or is received on arches is 3,295 *quinariae*,—more than is set down in the scheduled capacity by 1,133 *quinariae*, and less than given by the gaugings made at the intake by 1,395 *quinariae*. After passing the reservoir, it delivered 1,840 *quinariae*,—227 *quinariae* less than we said was set down in the scheduled capacity, and 1,104 *quinariae* less than is taken from the reservoir upon the arches. The aggregate of what was lost either between the intake and the reservoir or downstream from the reservoir, was 2,499 *quinariae*, the diversion of which, as in case of the other aqueducts, we discovered at several places. For that there is no lack of water is manifest also from the fact that at the intake, besides the volume which we noted that we found from the capacity of the conduit, over 300 *quinariae* are wasted.[1]

Tepula is credited in the records with 400 *quinariae*. This aqueduct has no springs; it consists only of some veins of water taken from Julia.[2] Its intake is therefore to be set down as beginning

---

quinariae, thus revealing a loss of 1,395 *quinariae* above the reservoir; (3) above and below the reservoir, there were delivered in all 95 plus 1,840 *quinariae* (this does *not* take account of the contributions to Tepula and Anio), 227 fewer *quinariae* than the amount credited in the records; and the amount delivered below the reservoir was 1,104 *quinariae* less than the amount measured at the reservoir. Since the amount at the intake was 4,690 *quinariae*, and the total distribution (including the amounts given to Tepula and Anio) amounted to 2,191 *quinariae*, there remained 2,499 *quinariae* still unaccounted for.   [2] *Cf.* 9, 19.

SEXTUS JULIUS FRONTINUS

est a piscina Iuliae. Ex ea enim primum accipit quinarias centum nonaginta, deinde statim ex Marcia quinarias nonaginta duas, praeterea ex Anione Novo ad hortos Epaphroditianos quinarias centum sexaginta tres. Fiunt omnes quinariae quadringentae quadraginta quinque, amplius quam in commentariis quinariis quadraginta quinque, quae in erogatione comparent.

69. Iuliae in commentariis adscriptus est modus quinariarum sexcentarum quadraginta novem. Ad caput mensura iniri non potuit, quoniam ex pluribus adquisitionibus constat, sed ad sextum ab urbe miliarium universa in piscinam recipitur, ubi modus eius manifestis mensuris efficit quinarias mille ducentas sex, amplius quam in commentariis quinariis quingentis quinquaginta septem. Praeterea accipit prope urbem post hortos Pallantianos ex Claudia quinarias centum sexaginta duas. Est omne Iuliae in acceptis quinariae mille trecentae sexaginta octo. Ex eo dat in Tepulam quinarias centum nonaginta, erogat suo nomine octingentas tres. Fiunt quas erogat quinariae nongentae nonaginta tres; amplius quam in commentariis habet quinariis trecentis quadraginta quattuor; minus quam in piscina habere posuimus ducentis decem tribus, quas ipsas apud eos, qui sine beneficiis principis usurpabant, deprehendimus.

70. Virgini in commentariis adscriptus est modus quinariarum sexcentarum quinquaginta duarum.

---

[1] That is, Tepula received and delivered 445 *quinariae*, while the records credited only 400.

[2] Julia was received into a catch-basin. (1) The measurements there taken showed its volume to be 1,206 *quinariae*,

AQUEDUCTS OF ROME, II. 68–70

with the Julian reservoir, for from this it first receives 190 *quinariae*; then immediately thereafter 92 *quinariae* from Marcia, and further from New Anio at the Epaphroditian Gardens 163 *quinariae*. This makes in all 445 *quinariae*,—more than the records show by 45 *quinariae*,—which appear in the delivery.[1]

Julia is credited in the records with a measure of 649 *quinariae*. At the intake the gaugings could not be made, because the intake is composed of several tributaries. But at the sixth mile-stone from the City, Julia is wholly taken into the settling reservoir, at which place its volume, according to the plainly visible gauges, amounts to 1,206 *quinariae*,—more than set down in the records by 557 *quinariae*. Besides this, near the City, behind the Gardens of Pallas, it receives from Claudia 162 *quinariae*, making the whole number of *quinariae* received by Julia 1,368. Of this amount, it discharges 190 into Tepula, and delivers on its own account 803 *quinariae*; from this we get a total of 993 *quinariae* which it delivers,—more than the records credit by 344 *quinariae*; less than we set it down as having at the reservoir by 213, which is precisely the amount we found diverted by those who were taking water without grant from the sovereign.[2]

Virgo is credited in the records with a measure of 652 *quinariae*. I could not take a gauging of this

---

while the records credited only 649 *quinariae*, a discrepancy of 557 *quinariae*; (2) there were delivered in all 993 *quinariae*, 344 more than were credited in the records, but 213 less than the gauges showed at the reservoir. (The 162 *quinariae* received from Claudia are not reckoned in computing this loss.)

397

SEXTUS JULIUS FRONTINUS

Huius mensuram ad caput invenire non potui, quoniam ex pluribus adquisitionibus constat et lenior rivum intrat. Prope urbem tamen ad miliarium septimum in agro qui nunc est Ceionii Commodi, ubi velociorem cursum habet, mensuram egi quae efficit quinariarum duo milia quingentas quattuor, amplius quam in commentariis quinariis mille octingentis quinquaginta duabus. Approbatio nostra expeditissima est; erogat enim omnes quas mensura deprendimus, id est duo milia quingentas quattuor.

71. Alsietinae conceptionis modus nec in commentariis adscriptus est nec in re praesenti certus inveniri potuit, cum ex lacu Alsietino et deinde circa Careias ex Sabatino[1] quantum aquarii temperaverunt. Alsietina erogat quinarias trecentas nonaginta duas.

72. Claudia abundantior aliis maxime iniuriae exposita est. In commentariis habet non plus quinariis duobus milibus octingentis quinquaginta quinque, cum ad caput invenerim quinariarum quattuor milia sexcentas septem; amplius quam in commentariis mille septingentis quinquaginta duabus. Adeo autem nostra certior est mensura, ut ad septimum ab urbe miliarium in piscina, ubi indubitatae mensurae sunt, inveniamus quinarias tria milia trecentas decem duas; plus quam in commentariis quadringentis quinquaginta septem, quamvis et ex beneficiis ante piscinam eroget et plurimum subtrahi deprehenderimus ideoque minus inveniatur, quam re vera esse debeat, quinariis mille ducentis nonaginta quinque. Et circa erogationem autem fraus apparet, quae neque ad commentariorum fidem neque ad eas quas ad caput egimus mensuras,

[1] *addendum praeeunte Schultzio* tantum accipiat *B.*

AQUEDUCTS OF ROME, II. 70–72

at the intake, because Virgo is made up of several tributaries, and enters its channels with too slow a current. Near the City, however, at the seventh mile-stone, on the land which now belongs to Cejonius Commodus, where Virgo has a greater velocity, I made a gauging, and it amounted to 2,504 *quinariae*,—1,852 *quinariae* more than was set down in the records. The correctness of our gauging is very easily proved; for Virgo discharges all the *quinariae* which we found by gauging, that is, 2,504.

The measure of the capacity of Alsietina is not set down in the records, nor could it be accurately arrived at under present conditions, because [it receives] from Lake Alsietinus, and afterwards in the vicinity of Careiae from Sabatinus as much water as the water-men arrange for. Alsietina delivers 392 *quinariae*.

Claudia, flowing more abundantly than the others, is especially exposed to depredation. In the records it is credited with only 2,855 *quinariae*, although I found at the intake 4,607 *quinariae*,—1,752 *quinariae* more than are recorded. Our gauging, however, is confirmed by the fact that at the seventh mile-stone from the City, at the settling reservoir, where the gauging is without question, we find 3,312 *quinariae*, —457 more than are recorded, although, before reaching the reservoir, not only are deliveries made, to satisfy private grants, but also, as we detected, a great deal is taken secretly, and therefore 1,295 *quinariae* less are found than there really ought to be. Moreover, in the delivery of the water also it is manifest that there is fraud, since the amount actually delivered does not agree either with the statements of the records or with the gaugings

SEXTUS JULIUS FRONTINUS

neque ad illas saltem ad piscinam post tot iniurias sunt, convenit. Solae enim quinariae mille septingentae quinquaginta erogantur; minus quam commentariorum ratio dat quinariis mille centum quinque; minus autem quam mensurae ad caput factae demonstraverunt quinariis duobus milibus octingentis quinquaginta septem; minus etiam quam in piscina invenitur quinariis mille quingentis sexaginta duabus. Ideoque cum sincera in urbem proprio rivo perveniret, in urbe miscebatur cum Anione Novo, ut confusione facta et conceptio earum et erogatio esset obscurior. Quod si qui forte me adquisitionum mensuris blandiri putant, admonendi sunt adeo Curtium et Caeruleum fontes aquae Claudiae sufficere ad praestandas ductui suo quinarias quas significavi quattuor milia sexcentas septem, ut praeterea mille sexcentae effundantur. Nec eo infitias quin ea quae superfluunt non sint proprie horum fontium; capiuntur enim ex Augusta, quae inventa [1] in Marciae supplementum, dum illa non indiget, adicitur [2] fontibus Claudiae, quamvis ne huius quidem ductus omnem aquam recipiat.

73. Anio Novus in commentariis habere ponebatur quinarias tria milia ducentas sexaginta tres. Mensus ad caput repperi quinarias quattuor milia septingentas triginta octo, amplius quam in conceptis commentariorum est, quinariis mille quadringentis septuaginta quinque. Quarum adquisitionem non avide me amplecti quo alio modo manifestius

---

[1] *Poleni*; quem inventum *C*; quam inventam *B*.
[2] *Poleni*; adiecimus *CB*.

---

[1] Claudia was received into a catch-basin. (1) The measurements at the intake showed 4,607 *quinariae*, the records credited 2,855 *quinariae*, a discrepancy of 1,752 *quinariae*; (2) measurements at the reservoir showed 3,312 *quinariae*, a loss of 1,295 *quinariae* between the intake and

AQUEDUCTS OF ROME, II. 72-73

made by us at the intake, or even with those made at the settling-basins, after so many depredations. For there are only 1,750 *quinariae* delivered,—less than the computation given in the records by 1,105 *quinariae;* also less than is shown by the gauging made at the intake by 2,857 *quinariae,* and less also than is found at the reservoir by 1,562 *quinariae*.[1] For this reason, although it arrived in the City perfectly clear in its own conduit, it was mixed within the City with the New Anio, so that by creating confusion, the quantity as well as the distribution of the two might be obscured. But should any one think that I exaggerate the measure of the water received, such a person must be reminded that the Caerulean and Curtian sources[2] of the Claudian aqueduct are so ample for supplying to their conduit the 4,607 *quinariae* which I have indicated, that 1,600 besides go to waste.[3] But at the same time I do not deny that this superabundance does not really belong to these springs, for it comes from Augusta. This was devised to supplement Marcia, but is turned into the sources of Claudia, when Marcia does not need it, though not even the conduit of Claudia itself can carry all this water.[4]

New Anio was put down in the records as having 3,263 *quinariae*. Gauging at the intake I found 4,738 *quinariae*,—more than the scheduled capacity by 1,475 *quinariae*. In what other way could I more clearly show that I do not exaggerate the number of *quinariae* at the intake than by the fact that in

---

the reservoir; (3) 1,750 *quinariae* were delivered, a smaller amount than that indicated by the records or by either set of measurements. (In the 1,750 *quinariae* were included the 162 given to Julia. *Cf.* 69.)

[2] *Cf.* 13, 14.

[3] *i.e.* at the sources of the aqueduct. [4] *Cf.* 12, 14.

401

SEXTUS JULIUS FRONTINUS

probem, quam quod in erogatione ipsorum commentariorum maior pars earum continetur? Erogantur enim quinariarum quattuor milia ducentae, cum alioquin in eisdem commentariis inveniatur conceptio non amplius quam trium milium ducentarum sexaginta trium. Praeterea intercipi non tantum quingentas XXXVIII, quae inter mensuras nostras et erogationem intersunt, sed longe ampliorem modum deprendi. Ex quo apparet etiam exuberare comprehensam a nobis mensuram. Cuius rei ratio est, quod vis aquae rapacior, ut ex largo et celeri flumine excepta, velocitate ipsa ampliat modum.

74. Non dubito aliquos adnotaturos, quod longe maior copia actis mensuris inventa sit, quam erat in commentariis principum. Cuius rei causa est error

---

[1] The 4,200 *quinariae* included the 163 given to Tepula. *Cf.* 68.

[2] Literally: "the amount estimated by me even exceeds (the figures which I gave)."

[3] "The most troublesome point of ignorance which Frontinus had to contend with was a total inability to measure the velocity of water, or even rightly and fully to grasp the idea of such velocity, whether as flowing in an open channel or in closed pipes. He accordingly compares streams of water merely by the area of their cross sections, and then worries himself into all sorts of explanations as to why his gaugings by areas, made irrespective of heads and velocities, do not balance. The frauds of the water-men, of the plumbers and of others who draw water unlawfully, always furnish a handy explanation, however."—*Herschel.*

[4] The following table will show how Frontinus arrived at the figures which he gives:—

AQUEDUCTS OF ROME, II. 73-74

the records of delivery most of this water is actually accounted for? For it is stated that 4,200 *quinariae* are delivered,[1] although elsewhere in the same records the amount taken in is put down as only 3,263. Besides this, I have discovered that not only 538 *quinariae* (the difference between our gauging and the recorded delivery) are stolen, but a far greater quantity. Whence it appears that the total found by me is none too large.[2] The explanation of this is, that the swifter current of water, coming as it does from a large and rapidly flowing river, increases the volume by its very velocity.[3]

I do not doubt that many will be surprised that according to our gaugings, the quantity of water was found to be much greater than that given in the imperial records.[4] The reason for this is to be found

Quinariae assigned to the various aqueducts in chapters 65-73:

| | In records | By measurement ||| Delivery |
		At intake	At reservoir	Elsewhere	
Appia	841			1,825	704
Old Anio	1,541	4,398	(2,362)		262 1,348 }
Marcia	2,162	4,690	(2,944)		95 1,840 }[1]
Tepula	400			{ 190 92 163 }	445
Julia	649		1,206	(162)	803 [2]
Virgo	652		2,504		2,504
Alsietina	392 [5]			392(?)	392
Claudia	2,855	4,607	(3,312)		1,588 [3]
New Anio	3,263	4,738			4,037 [4]
	12,755	18,433	3,710	2,662	14,018

[1] Besides 92 to Tepula. and 164 to Anio (*cf.* 67).
[2] Besides 190 to Tepula (*cf.* 68).
[3] Besides 162 to Julia (*cf.* 69).
[4] Besides 163 to Tepula (*cf.* 68).
[5] Assumed from number delivered.

771

403

eorum, qui ab initio parum diligenter uniuscuiusque aquae fecerunt aestimationem. Ac ne metu aestatis aut siccitatum in tantum a veritate eos recessisse credam, obstat quod ipse actis mensuris Iulio mense hanc uniuscuiusque copiam, quae supra scripta est, tota deinceps aestate durantem exploravi. Quaecumque tamen est causa quae praecedit, illud utique detegitur, decem milia quinariarum intercidisse, dum beneficia sua principes secundum modum in commentariis adscriptum temperant.

75. Sequens diversitas est quod alius modus concipitur ad capita, alius nec exiguo minor in piscinis, minimus deinde distributione continetur. Cuius rei causa est fraus aquariorum, quos aquas ex ductibus publicis in privatorum usus derivare deprehendimus. Sed et plerique possessorum, e quorum agris aqua circumducitur, formas rivorum perforant, unde fit ut ductus publici hominibus privatis vel ad hortorum usus itinera suspendant.

76. Ac de vitiis eiusmodi nec plura nec melius dici possunt, quam a Caelio Rufo dicta sunt in ea contione, cui titulus est "De Aquis," quae nunc nos omnia simili licentia usurpata utinam non per offensas probaremus; inriguos agros, tabernas, cenacula etiam, corruptelas denique omnes perpetuis salientibus instructas invenimus. Nam quod falsis titulis aliae

---

[1] 85–48 B.C. See Cicero, *Epistulae ad Familiares*, viii, 6, 4 (c. 50 B.C.).

in the blunders of those who carelessly computed each of these waters at the outset. Moreover, I am prevented from believing that it was from fear of droughts in the summer that they deviated so far from the truth, for the reason that I myself made my gaugings in the month of July, and found the above-recorded supply of each one remaining constant throughout the entire remainder of the summer. But whatever the reason may be, it has at any rate been discovered that 10,000 *quinariae* were intercepted, while the amounts granted by the sovereign are limited to the quantities set down in the records.

Another variance consists in this, that one measure is used at the intake, another, considerably smaller, at the settling-reservoir, and the smallest at the point of distribution. The cause of this is the dishonesty of the water-men, whom we have detected diverting water from the public conduits for private use. But a large number of landed proprietors also, past whose fields the aqueducts run, tap the conduits; whence it comes that the public water-courses are actually brought to a standstill by private citizens, just to water their gardens.

Concerning misdemeanours of this sort, nothing more nor better needs to be said than was said by Caelius Rufus,[1] in his speech, which is entitled " Concerning Waters." And would that we were not having daily experience by actual infringement of the law that all these misdemeanours are committed just as flagrantly now as then. We have found irrigated fields, shops, garrets even, and lastly all disorderly houses fitted up with fixtures through which a constant supply of flowing water might be assured. For that some waters should be delivered under a

SEXTUS JULIUS FRONTINUS

pro aliis aquae erogabantur, etiam sunt leviora ceteris vitia. Inter ea tamen quae emendationem videbantur exigere, numerandum est, quod fere circa montem Caelium et Aventinum accidit. Qui colles, priusquam Claudia perduceretur, utebantur Marcia et Iulia. Sed postquam Nero imperator Claudiam opere arcuato ad Spem exceptam usque ad templum Divi Claudii perduxit, ut inde distribueretur, priores non ampliatae sed omissae sunt; nulla enim castella adiecit, sed isdem usus est, quorum quamvis mutata aqua vetus appellatio mansit.

77. Satis iam de modo cuiusque et velut nova quadam adquisitione aquarum et fraudibus et vitiis quae circa ea erant dictum est. Superest ut erogationem, quam confertam et, ut sic dicam, in massa invenimus, immo etiam falsis nominibus positam, per nomina aquarum, uti quaeque se habet, et per regiones urbis digeramus. Cuius comprensionem scio non ieiunam tantum sed etiam perplexam videri posse, ponemus tamen quam brevissime, ne quid velut formulae officii desit. Eis quibus sufficiet cognovisse summa, licebit transire leviora.

78. Fit ergo distributio quinariarum quattuordecim milium decem et octo, ita ut quinariae DCCLXXI,

---

[1] *Cf.* 20.

[2] Since Frontinus in 87 speaks of the rebuilding of Marcia under Trajan it is probable that after Claudia was brought to this quarter, the aqueducts previously supplying it were allowed to fall into disuse.

[3] *i.e.* by discovering and correcting frauds.

[4] *i.e.* the supply of one aqueduct had been accredited to another.

forged name in place of other waters belongs to the lesser misdemeanours. But among the frauds that seemed to demand correction, is to be mentioned what took place in the vicinity of the Caelian and Aventine Hills. These hills, before the construction of Claudia, utilized the waters of Marcia and Julia; but after the Emperor Nero led Claudia over the arches at Spes Vetus up to the Temple of the Deified Claudius,[1] in order to distribute it from there, the first named waters, instead of being augmented by this new supply, were themselves allowed to go unused[2]; for he did not build new reservoirs for Claudia, but used those that already existed; and the old name of these remained, although the water had become a new one.

With this, enough has been said about the volume of each aqueduct, and, if I may so express it, about a new way of acquiring water[3]; about frauds and about offences committed in connection with all this. It remains to account in detail for the supply delivered (which we found given collectively and in a lump sum, so to speak,—and even set down under false entries),[4] and to do this according to the several aqueducts and to the several wards of the City. I know very well that such an enumeration will appear not only dry but also complicated; nevertheless, I will make it—but as short as possible—that nothing may be lacking to the data of this office. Those who are satisfied with knowing the totals, may skip the details.

Now the distribution of the 14,018 *quinariae* is so recorded that the 771 *quinariae*[5] which are transferred

[5] These 771 *quinariae* are made up as follows:

92	from Marcia	to	Tepula,
164	,, ,,	,,	Anio (Vetus),
190	,, Julia	,,	Tepula,
163	,, Anio Novus	to	Tepula,
162	,, Claudia	,,	Julia.

771 *quinariae*.

SEXTUS JULIUS FRONTINUS

quae ex quibusdam aquis in adiutorium aliarum dantur et bis in speciem erogationis cadunt, semel in computationem veniant. Ex his dividuntur extra urbem quinariae quattuor milia sexaginta tres: ex quibus nomine Caesaris quinariae mille septingentae decem et octo, privatis quinariae ∞ ∞ CCCXXXXV. Reliquae intra urbem VIIII milia nongentae quinquaginta quinque distribuebantur in castella ducenta quadraginta septem: ex quibus erogabantur sub nomine Caesaris quinariae mille septingentae septem semis, privatis quinariae tria milia octingentae quadraginta septem, usibus publicis quinariae quattuor milia quadringentae una: ex eo castris . . .[1] quinariae ducentae septuaginta novem, operibus publicis septuaginta quinque quinariae ∞ ∞ CCCI, muneribus triginta novem quinariae CCCLXXXVI, lacibus quingentis nonaginta uni quinariae ∞ trecentae triginta quinque. Sed et haec ipsa dispensatio per nomina aquarum et regiones urbis partienda est.

79. Ex quinariis ergo quattuordecim milibus decem et octo, quam summam erogationibus omnium aquarum seposuimus, dantur nomine Appiae extra urbem quinariae tantummodo quinque, quoniam humilior turetia metitoribus.[2] Reliquae quinariae sescentae nonaginta novem intra urbem dividebantur per regiones secundam IIX VIIII XI XII XIII XIV in castella viginti: ex quibus nomine Caesaris qui-

---

[1] ducentinarie *C*; XX quinariae *Schultze*; X et VIIII quinariae *Poleni*.
[2] humilior turetia meatoribus *CB*; oritur et a metitoribus *Cod. Vat.*

---

[1] *i.e.* subject to the disposition of the Emperor.
[2] The manuscript reading here is unintelligible.

AQUEDUCTS OF ROME, II. 78-79

from certain aqueducts to supplement others and are set down twice in showing the distribution, figure only once in reckoning. Of this quantity there are delivered outside the City, 4,063 *quinariae*, 1,718 *quinariae* in the name of Caesar,[1] to private parties, 2,345. The remaining 9,955 were distributed within the City to 247 reservoirs; of these there were delivered in the name of Caesar 1,707½ *quinariae*, to private parties 3,847 *quinariae*, for public uses 4,401 *quinariae*,—namely to . . .[2] camps 279 *quinariae*, to seventy-five public structures 2,301 *quinariae*, to thirty-nine ornamental fountains 386 *quinariae*, to five hundred and ninety-one water-basins 1,335 *quinariae*. But the schedule must be made to apply also to the several aqueducts and to the several wards of the City.[3]

Of the 14,018 *quinariae*, then, which we set down as the total discharge of all the aqueducts, only 5 *quinariae* are given from Appia outside the City because [its source is so low].[4] The remaining 699 *quinariae* were distributed within the City throughout the second, eighth, ninth, eleventh, twelfth, thirteenth, and fourteenth wards, among twenty reservoirs. Of these there were furnished in the

---

[3] The original MS. of the De Aquis is a hopeless confusion of figures as regards the statistics given in chapters 78-86. The numbers, as they have come down to us in the Montecassino manuscript exhibit a great many errors, and it would be impossible to affirm what are errors, and what is the truth. Poleni endeavoured to reconcile these figures, so as to make them mathematically consistent. Tables showing his adjustments, and a second set of tables, prepared by Herschel on the basis of Bücheler's figures, are given at the end of the book.

[4] The manuscript reading here is hopeless except for *humilior*.

SEXTUS JULIUS FRONTINUS

nariae centum quinquaginta una, privatis quinariae centum nonaginta quattuor, usibus publicis quinariae trecentae quinquaginta quattuor: ex eo castris I quinariae quattuor, operibus publicis quattuordecim quinariae centum viginti tres, muneri uni quinariae duae, lacibus nonaginta duobus quinariae ducentae viginti sex.

80. Anionis Veteris erogabantur extra urbem nomine Caesaris quinariae centum sexaginta novem, privatis quinariae CCCCIIII. Reliquae quinariae mille quingentae octo semis intra urbem dividebantur per regiones primam III IIII V VI VII VIII VIIII XII XIIII in castella triginta quinque: ex quibus nomine Caesaris quinariae sexaginta VI S, privatis quinariae CCCCXC, usibus publicis quinariae quingentae tres: ex eo castris unis quinariae quinquaginta, operibus publicis XIX quinariae centum nonaginta sex, muneribus novem quinariae octoginta octo, lacibus nonaginta quattuor quinariae ducentae decem et octo.

81. Marciae erogabantur extra urbem nomine Caesaris quinariae CCLXI S. Reliquae quinariae mille quadringentae septuaginta duae intra urbem dividebantur per regiones primam tertiam quartam V VI VII VIII VIIII X XIIII in castella quinquaginta unum: ex quibus nomine Caesaris quinariae CXVI, privatis quinariae quingentae quadraginta tres,[1] usibus publicis quinariae CCCCXXXVIIII: ex eo castris IIII quinariae XLIIS, operibus publicis quindecim quinariae XLI,[2] muneribus XII quinariae CIIII, lacibus CXIII quinariae CCLVI.

---

[1] *no gap in C between* tres *and* castris; usibus publicis quinariae ccccxxxviiii *Poleni*; ex eo *B*.   [2] *C*; xv *B*.

---

[1] The MS. omits the general statement usually given of

410

AQUEDUCTS OF ROME, II. 79-81

name of Caesar 151 *quinariae*, to private parties 194 *quinariae*, for public uses 354 *quinariae*,—namely, to one camp 4 *quinariae*, to fourteen public structures 123 *quinariae*, to one ornamental fountain 2 *quinariae*, to ninety-two water-basins 226 *quinariae*.

Out of Old Anio were delivered outside the City in the name of Caesar 169 *quinariae*, to private parties 404 *quinariae*. The remaining 1,508½ *quinariae* were distributed inside the City through the first, third, fourth, fifth, sixth, seventh, eighth, ninth, twelfth, and fourteenth wards, among thirty-five reservoirs. Of these there were furnished in the name of Caesar 66½ *quinariae*, for the use of private parties 490 *quinariae*, for public uses 503 *quinariae*,—namely, to one camp 50 *quinariae*, to nineteen public structures 196 *quinariae*, to nine ornamental fountains 88 *quinariae*, to ninety-four water-basins 218 *quinariae*.

Out of Marcia were delivered outside the City in the name of Caesar 261½ *quinariae*. The remaining 1,472 *quinariae* were distributed inside the City through the first, third, fourth, fifth, sixth, seventh, eighth, ninth, tenth, and fourteenth wards, among fifty-one reservoirs. Of these there were furnished in the name of Caesar 116 *quinariae*, to private parties 543 *quinariae*, for public uses 439 [1] *quinariae*,—namely, to four camps 42½ *quinariae*, to fifteen public structures 41 *quinariae*, to twelve ornamental fountains 104 *quinariae*, to one hundred and thirteen water-basins 256 *quinariae*.[2]

the amount distributed for public uses, and mentions only the separate items which would fall under this general head. Poleni would insert a general statement, deriving his total from the addition of the separate amounts.

[2] A branch of Marcia supplied the baths of Caracalla (built 206 A.D.)

SEXTUS JULIUS FRONTINUS

82. Tepulae erogabantur extra urbem nomine Caesaris quinariae LVIII, privatis quinariae quinquaginta sex. Reliquae quinariae CCCXXXI intra urbem dividebantur per regiones quartam V VI VII in castella XIIII: ex quibus nomine Caesaris quinariae XXXIIII,[1] privatis quinariae CCXXXVII, usibus publicis quinariae quinquaginta: ex eo castris I quinariae duodecim, operibus publicis III quinariae septem, lacibus XIII quinariae XXXII.

83. Iuliae fluebant extra urbem nomine Caesaris quinariae LXXX quinque, privatis quinariae CXXI. Reliquae quinariae quingentae quadraginta octo intra urbem dividebantur per regiones secundam III V VI VIII X XII in castella decem et septem: ex quibus nomine Caesaris quinariae decem et octo, privatis quinariae, CXCVI,[2] usibus publicis quinariae CCCLXXXIII: ex eo castris . . .[3] quinariae sexaginta novem, operibus publicis . . .[4] quinariae CXXXI, muneribus III quinariae sexaginta septem, lacibus viginti octo quinariae sexaginta quinque.

84. Virginis nomine exibant extra urbem quinariae ducentae. Reliquae quinariae duo milia trecentae quattuor intra urbem dividebantur per regiones septimam nonam quartam decimam in castella decem et octo: ex quibus nomine Caesaris quinariae quingentae novem, privatis quinariae CCCXXXVIII, usibus publicis ∞ centum sexaginta septem: ex eo muneribus II quinariae XXVI, lacibus viginti quin-

---

[1] *C*; xxxxii *B*.
[2] octo usibus publicis *C*; privatis cxcvi *added by Poleni*, quinariae *Schultze*.
[3] *No gap in C*; iv *Schultze*; iii *Poleni*.
[4] *No gap in C*; xi *Schultze*; x *Poleni*.

AQUEDUCTS OF ROME, II. 82-84

Out of Tepula there were delivered outside the City in the name of Caesar 58 *quinariae*, to private parties 56 *quinariae*. The remaining 331 *quinariae* were distributed within the City through the fourth, fifth, sixth, and seventh wards among fourteen reservoirs. Of these there were furnished in the name of Caesar 34 *quinariae*, to private parties 237 *quinariae*, for public uses 50 *quinariae*,—namely, to one camp 12 *quinariae*, to three public structures 7 *quinariae*, to thirteen basins 32 *quinariae*.

Out of Julia there flowed outside the City in the name of Caesar 85 *quinariae*, to private parties 121 *quinariae*. The remaining 548 *quinariae* were distributed within the City to the second, third, fifth, sixth, eighth, tenth, and twelfth wards, among seventeen reservoirs. Of these there were furnished in the name of Caesar 18 *quinariae*, to private parties 196 *quinariae*,[1] for public uses 383 *quinariae*,—namely, to . . .[2] camps 69 *quinariae*, to . . .[2] public structures 181 *quinariae*, to three ornamental fountains 67 *quinariae*, to twenty-eight basins 65 *quinariae*.

Virgo delivered outside the City 200 *quinariae*. The remaining 2,304 *quinariae* were distributed within the City to the seventh, ninth, and fourteenth wards, among eighteen reservoirs. Of these there were furnished in the name of Caesar 509 *quinariae*, to private parties 338 *quinariae*, for public uses 1,167 *quinariae*,—namely, to two ornamental fountains 26 *quinariae*, to twenty-five basins 51 *quinariae*, to

---

[1] The MS. says nothing of deliveries to private persons. Poleni, for the sake of arithmetic, would amend the text here.

[2] The MS. shows no gap, though a numeral is clearly called for.

SEXTUS JULIUS FRONTINUS

que quinariae quinquaginta una, operibus publicis sedecim quinariae ∞ CCCLXXX, in quibus per se Euripo, cui ipsa nomen dedit, quinariae CCCCLX.

85. Alsietinae quinariae trecentae nonaginta duae. Haec tota extra urbem consumitur, nomine Caesaris quinariae trecentae quinquaginta quattuor, privatis quinariae centum triginta octo.

86. Claudia et Anio Novus extra urbem proprio quaeque rivo erogabantur, intra urbem confundebantur. Et Claudia quidem extra urbem dabat nomine Caesaris quinarias CCXIVI,[1] privatis quinarias CCCCXXX novem; Anio Novus nomine Caesaris quinarias septingentas viginti octo. Reliquae utriusque quinariae tria milia quadringentae nonaginta octo intra urbem dividebantur per regiones urbis XIIII in castella nonaginta duo; ex quibus nomine Caesaris quinariae octingentae XVV,[2] privatis quinariae ∞ sexaginta septem, usibus publicis quinariae ∞ XIV[3]: ex eo castris novem quinariae centum quadraginta novem, operibus publicis decem et octo quinariae CCCLXXIIII, muneribus XII quinariae centum septem, lacibus CC viginti sex quinariae CCCCXXCII.

87. Haec copia aquarum ad Nervam imperatorem usque computata ad hunc modum discribebatur. Nunc providentia diligentissimi principis quicquid aut fraudibus aquariorum intercipiebatur aut inertia

---

[1] *C*; *B has* ccxlvi. *The figures are for* 200 + 11 + 6, *according to Petschenig.*

[2] *C*; *B has* xviiii. *The figures, according to Petschenig, represent* 15 + 5.   [3] *C*; xii *B*.

---

[1] The most important of these were the Thermae of Agrippa.

[2] The name given to an artificial channel running through

AQUEDUCTS OF ROME, II. 84–87

sixteen public structures [1] 1,380 *quinariae*. In the amount delivered to public structures are included 460 *quinariae* for the Euripus [2] alone, to which Virgo itself gave its name.[3]

Alsietina has 392 *quinariae*. These are all used outside the City, 354 *quinariae* being furnished in the name of Caesar, and to private parties 138 *quinariae*.

Outside the City, Claudia and New Anio delivered each from its own channel; inside the City they were mixed together. Claudia discharged outside the City in the name of Caesar 217 *quinariae*, to private parties 439 *quinariae*; New Anio delivered in the name of Caesar 728 *quinariae*. The remaining 3,498 *quinariae* belonging to these two were distributed inside the City through all the fourteen wards, among ninety-two reservoirs. Of these, there were furnished in the name of Caesar 820 *quinariae*, to private parties 1,067 *quinariae*, for public uses 1,014 *quinariae*,—namely, to nine camps 149 *quinariae*, to eighteen public structures [4] 374 *quinariae*, to twelve ornamental fountains 107 *quinariae*, to two hundred and twenty-six basins 482 *quinariae*.

This is the schedule of the amount of water as reckoned up to the time of the Emperor Nerva[5] and this is the way in which it was distributed. But now, by the foresight of the most painstaking of sovereigns, whatever was unlawfully drawn by the water-men, or was wasted as the result of negligence, has been added to our supply;

---

the Gardens of Agrippa on the Campus Martius and emptying into the Tiber.

[3] This allusion is not clear.

[4] Among them the Baths of Nero.

[5] The Emperor Trajan is meant.

pervertebatur, quasi nova inventione fontium accrevit. Ac prope duplicata ubertas est et tam sedula deinde partitione distributa, ut regionibus quibus singulae serviebant aquae plures darentur, tamquam Caelio et Aventino in quos sola Claudia per arcus Neronianos ducebatur, quo fiebat ut quotiens refectio aliqua intervenisset, celeberrimi colles sitirent. Quibus nunc plures aquae et in primis Marcia reddita amplo opere a Spe in Aventinum usque perducitur. Atque etiam omni parte urbis lacus tam novi quam veteres plerique binos salientes diversarum aquarum acceperunt, ut si casus alterutram impedisset, altera sufficiente non destitueretur usus.

88. Sentit hanc curam imperatoris piissimi Nervae principis sui regina et domina orbis in dies et magis sentiet salubritas eiusdem aucto castellorum, operum, munerum et lacuum numero. Nec minus ad privatos commodum ex incremento beneficiorum eius diffunditur; illi quoque qui timidi inlicitam aquam ducebant, securi nunc ex beneficiis fruuntur. Ne pereuntes quidem aquae otiosae sunt: alia munditiarum facies, purior spiritus, et causae gravioris caeli quibus apud veteres urbis infamis aer fuit, sunt

AQUEDUCTS OF ROME, II. 87–88

just as though new sources had been discovered. And in fact the supply has been almost doubled, and has been distributed with such careful allotment that wards which were previously supplied by only one aqueduct now receive the water of several. Take for example the Caelian and the Aventine Hills, to which Claudia alone used to run on the arches of Nero. The result was, that whenever any repairs caused interruptions, these densely inhabited hills suffered a drought. They are now supplied by several aqueducts, above all, by Marcia, which has been rebuilt on a substantial structure and carried from Spes Vetus to the Aventine. In all parts of the City also, the basins, new and old alike, have for the most part been connected with the different aqueducts by two pipes each, so that if accident should put either of the two out of commission, the other may serve and the service may not be interrupted.

The effect of this care displayed by the Emperor Nerva, most patriotic of rulers, is felt from day to day by the present queen and empress of the world; and will be felt still more in the improved health of the city, as a result of the increase in the number of the works, reservoirs, fountains, and water-basins. No less advantage accrues also to private consumers from the increase in number of the Emperor's private grants; those also who with fear drew water unlawfully, now free from care, draw their supply by grant from the sovereign. Not even the waste water is lost; the appearance of the City is clean and altered; the air is purer; and the causes of the unwholesome atmosphere, which gave the air of the City so bad a name with the ancients, are now

SEXTUS JULIUS FRONTINUS

remotae. Non praeterit me, deberi operi novae erogationis ordinationem; sed haec cum incremento adiunxerimus; intellegi oportet, non esse ea ponenda nisi consummata fuerint.

89. Quid quod nec hoc diligentiae principis, quam exactissimam civibus suis praestat, sufficit, parum praesidii usibus ac voluptatibus nostris contulisse sese credentis, quod tantam copiam adiecit, nisi eam ipsam sinceriorem iucundioremque faciat? Operae pretium est ire per singula, per quae ille occurrendo vitiis quarundam universis adiecit utilitatem. Etenim quando civitas nostra, cum vel exigui imbres supervenerant, non turbulentas limosasque aquas habuit? Nec quia haec universis ab origine natura est, aut quia[1] istud incommodum sentire debeant quae[2] capiuntur ex fontibus, in primis Marcia et Claudia ac reliquae, quarum splendor a capite integer nihil aut minimum pluvia inquinatur, si putea exstructa et obtecta sint.

90. Duae Anienses minus permanent limpidae, nam sumuntur ex flumine ac saepe etiam sereno turbantur, quoniam Anio quamvis purissimo defluens lacu mollibus tamen cedentibus ripis aufert aliquid quo turbetur, priusquam deveniat in rivos. Quod incommodum non solum hibernis ac vernis, sed etiam aestivis imbribus sentit, quo tempore gratior aquarum sinceritas exigitur.

---

[1] quia *C cod. Urb.*; quasi *B.*   [2] quae *C*; quot *B.*

---

[1] *e. g.* an artificial basin made to receive the water at its source.

AQUEDUCTS OF ROME, II. 88-90

removed. I am well aware that I ought to indicate in detail the manner of the new distribution; but this I will add when the additions are made; it ought to be understood that no account should be given until they are completed.

What shall we say of the fact that the painstaking interest which our Emperor evinces for his subjects does not rest satisfied with what I have already described, but that he deems he has contributed too little to our needs and gratification merely by such increase in the water supply, unless he should also increase its purity and its palatableness? It is worth while to examine in detail how, by correcting the defects of certain waters, he has enhanced the usefulness of all of them. For when has our City not had muddy and turbid water, whenever there have been only moderate rain-storms? And this is not because all the waters are thus affected at their sources, or because those which are taken from springs ought to be subject to such pollution. This is especially true of Marcia and Claudia and the rest, whose purity is perfect at their sources, and which would be not at all, or but very slightly, made turbid by rains, if well-basins should be built and covered over.

The two Anios are less limpid, for they are drawn from a river, and are often muddy even in good weather, because the Anio, although flowing from a lake whose waters are very pure, is nevertheless made turbid by carrying away portions of its loose crumbling banks, before it enters the conduits—a pollution to which it is subject not only in the rain-storms of winter and spring, but also in the showers of summer, at which time of the year a more refreshing purity of the water is demanded.

SEXTUS JULIUS FRONTINUS

91. Et alter quidem ex his, id est Anio Vetus, cum plerisque libra sit inferior, incommodum intra se tenet. Novus autem Anio vitiabat ceteras, nam cum editissimus veniat et in primis abundans, defectioni aliarum succurrit. Imperitia vero aquariorum deducentium in alienos eum specus frequentius, quam explemento opus erat, etiam sufficientes aquas inquinabat, maxime Claudiam, quae per multa milia passuum proprio ducta rivo, Romae demum cum Anione permixta in hoc tempus perdebat proprietatem. Adeoque obvenientibus non succurrebatur, ut pleraeque accerserentur per imprudentiam non uti dignum erat aquas partientium. Marciam ipsam splendore et frigore gratissimam balneis ac fullonibus et relatu quoque foedis ministeriis deprehendimus servientem.

92. Omnes ergo discerni placuit, tum singulas ita ordinari ut in primis Marcia potui tota serviret et deinceps reliquae secundum suam quaeque qualitatem aptis usibus assignarentur sic ut Anio Vetus pluribus ex causis (quo inferior excipitur minus salubris) in hortorum rigationem atque in ipsius urbis sordidiora exiret ministeria.

93. Nec satis fuit principi nostro ceterarum restituisse copiam et gratiam; Anionis quoque Novi vitia excludi posse vidit. Omisso enim flumine

One of the Anios, namely Old Anio, running at a lower level than most of the others, keeps this pollution to itself. But New Anio contaminated all the others, because, coming from a higher altitude and flowing very abundantly, it helps to make up the shortage of the others; but by the unskilfulness of the water-men, who diverted it into the other conduits oftener than there was any need of an augmented supply, it spoiled also the waters of those aqueducts that had a plentiful supply, especially Claudia, which, after flowing in its own conduit for many miles, finally at Rome, as a result of its mixture with Anio, lost till recently its own qualities. And so far was New Anio from being an advantage to the waters it supplemented that many of these were then called upon improperly through the heedlessness of those who allotted the waters. We have found even Marcia, so charming in its brilliancy and coldness, serving baths, fullers, and even purposes too vile to mention.

It was therefore determined to separate them all and then to allot their separate functions so that first of all Marcia should serve wholly for drinking purposes, and then that the others should each be assigned to suitable purposes according to their special qualities, as for example, that Old Anio, for several reasons (because the farther from its source it is drawn, the less wholesome a water is), should be used for watering the gardens, and for the meaner uses of the City itself.

But it was not sufficient for our ruler to have restored the volume and pleasant qualities of the other waters; he also recognized the possibility of remedying the defects of New Anio, for he gave orders to

repeti ex lacu qui est super villam Neronianam Sublacensem, ubi limpidissimus est, iussit. Nam cum oriatur Anio supra Trebam Augustam, seu quia per saxosos montes decurrit, paucis circa ipsum oppidum obiacentibus cultis, seu quia lacuum altitudine in quos excipitur velut defaecatur, imminentium quoque nemorum opacitate inumbratus, frigidissimus simul ac splendidissimus eo pervenit. Haec tam felix proprietas aquae omnibus dotibus aequatura Marciam, copia vero superatura, veniet in locum deformis illius ac turbidae, novum auctorem imperatorem Caesarem Nervam Traianum Augustum praescribente titulo.

94. Sequitur ut indicemus quod ius ducendae tuendaeque sit aquae, quorum alterum ad cohibendos intra modum impetrati beneficii privatos, alterum ad ipsorum ductuum pertinet tutelam. In quibus dum altius repeto leges de singulis aquis latas, quaedam apud veteres aliter observata inveni. Apud quos omnis aqua in usus publicos erogabatur et cautum ita fuit: "Ne quis privatus aliam aquam ducat, quam quae ex lacu humum accidit"—haec enim sunt verba legis—id est quae ex lacu abundavit; eam nos caducam vocamus. Et haec ipsa non in alium usum quam in balnearum aut fullonicarum dabatur, eratque vectigalis, statuta mercede quae in publicum pende-

---

[1] That is, water the private right to which has become void.

stop drawing directly from the river and to take from the lake lying above the Sublacensian Villa of Nero, at the point where the Anio is the clearest; for inasmuch as the source of Anio is above Treba Augusta, it reaches this lake in a very cold and clear condition, be it because it runs between rocky hills and because there is but little cultivated land even around that hamlet, or because it drops its sediment in the deep lakes into which it is taken, being shaded also by the dense woods that surround it. These so excellent qualities of the water, which bids fair to equal Marcia in all points, and in quantity even to exceed it, are now to supersede its former unsightliness and impurity; and the inscription will proclaim as its new founder, "Imperator Caesar Nerva Trajanus Augustus."

We have further to indicate what is the law with regard to conducting and safeguarding the waters, the first of which treats of the limitation of private parties to the measure of their grants, and the second has reference to the upkeep of the conduits themselves. In this connection, in going back to ancient laws enacted with regard to individual aqueducts, I found certain points wherein the practice of our forefathers differed from ours. With them all water was delivered for the public use, and the law was as follows: "No private person shall conduct other water than that which flows from the basins to the ground" (for these are the words of the law); that is, water which overflows from the troughs; we call it "lapsed" water;[1] and even this was not granted for any other use than for baths or fulling establishments; and it was subject to a tax, for a fee was fixed, to be paid into the

retur. Aliquid et in domos principum civitatis dabatur, concedentibus reliquis.

95. Ad quem autem magistratum ius dandae vendendaeve aquae pertinuerit, in eis ipsis legibus variatur. Interdum enim ab aedilibus, interdum a censoribus permissum invenio; sed apparet, quotiens in re publica censores erant, ab illis potissimum petitum, cum ei non erant, aedilium eam potestatem fuisse. Ex quo manifestum est quanto potior cura maioribus communium utilitatium quam privatarum voluptatium fuerit, cum etiam ea aqua quam privati ducebant ad usum publicum pertineret.

96. Tutelam autem singularum aquarum locari solitam invenio positamque redemptoribus necessitatem certum numerum circa ductus extra urbem, certum in urbe servorum opificum habendi, et quidem ita ut nomina quoque eorum, quos habituri essent in ministerio per quasque regiones, in tabulas publicas deferrent; eorumque operum probandorum curam fuisse penes censores aliquando et aediles, interdum etiam quaestoribus eam provinciam obvenisse, ut apparet ex S. C.[1] quod factum est C. Licinio et Q. Fabio cos.[2]

---

[1] *i.e.* senatus consulto. *Suggested by Pithou*; ex eo *C*.
[2] *Poleni*; clycynio consule fabio censoribus *C*.

---

[1] Since censors were chosen only every five years, and held office for but eighteen months, their duties in the intervening periods devolved upon other officials.
[2] Since these grants were made only for the sake of public utilities, such as baths and fulleries, and since moreover the State profited from the tax paid into the public treasury.
[3] The manuscript reading has been amended as unintelligible, but the amended statement is not quite in agreement with recorded facts. Livy xxxix. 32 says that in 185 B.C. it

AQUEDUCTS OF ROME, II. 94-96

public treasury. Some water also was conceded to the houses of the principal citizens, with the consent of the others.

To which authorities belonged the right to grant water or to sell it, is variously given even in the laws, for at times I find that the grant was made by the aediles, at other times by the censors; but it is apparent that as often as there were censors in the government[1] these grants were sought chiefly from them. If there were none, then the aediles had the power referred to. It is plain from this how much more our forefathers cared for the general good than for private luxury, inasmuch as even the water which private parties conducted was made to subserve the public interest.[2]

The care of the several aqueducts I find was regularly let out to contractors, and the obligation was imposed upon these of having a fixed number of slave workmen on the aqueducts outside the City, and another fixed number within the City; and of entering in the public records the names also of those whom they intended to employ in the service for each ward of the City. I find also that the duty of inspecting their work devolved at times on the aediles and censors, and at times on the quaestors, as may be seen from the resolution of the Senate which was passed in the consulate of Gaius Licinius and Quintus Fabius.[3]

---

was certainly expected that Quintus Fabius Labeo and Lucius Porcius Licinus would be elected consuls for the next year. The elections, however, resulted otherwise: Quintus Fabius Labeo and Marcius Claudius Marcellus were consuls in 183, Lucius Porcius Licinus and Publius Claudius Pulcher in 184.

SEXTUS JULIUS FRONTINUS

97. Quanto opere autem curae fuerit ne quis violare ductus aquamve non concessam derivare auderet, cum ex multis apparere potest, tum et ex hoc quod Circus Maximus ne diebus quidem ludorum circensium nisi aedilium aut censorum permissu inrigabatur, quod durasse etiam postquam res ad curatores transiit sub Augusto, apud Ateium Capitonem legimus.   Agri vero, qui aqua publica contra legem essent inrigati, publicabantur   Mancipi etiam si clam [1] eo quem adversus legem fecisset, multa dicebatur.   In eisdem legibus adiectum est ita : " Ne quis aquam oletato dolo malo, ubi publice saliet.   Si quis oletarit, sestertiorum decem milium multa esto." Cuius rei causa aediles curules iubebantur per vicos singulos ex eis qui in unoquoque vico habitarent praediave haberent binos praeficere, quorum arbitratu aqua in publico saliret.

98. Primus M. Agrippa post aedilitatem, quam gessit consularis, operum suorum et munerum velut perpetuus curator fuit.   Qui iam copia permittente discripsit, quid aquarum publicis operibus, quid lacibus, quid privatis daretur.   Habuit et familiam

---

[1] *B* ; cum *C. legendum* mancipium *vel tale quid B.*

---

[1] An eminent Roman jurist under Augustus and Tiberius.
[2] The meaning of this sentence is clear, though not its construction.
[3] A very large sum for those days, approximately £85.
[4] The aedileship usually preceded the consulship by six years.   In 33 B.C. Octavian was anxious that the people should be amused by shows and buildings of more than usual splendour, and Agrippa, in his loyalty to the Triumvir, descended from the rank which he had attained as a consular to serve in the inferior office of aedile.

How much care was taken that no one should venture to injure the conduits, or draw water that had not been granted, may be seen not only from many other things, but especially from the fact that the Circus Maximus could not be watered, even on the days of the Circensian Games, except with permission of the aediles or censors, a regulation which, as we read in the writings of Ateius Capito,[1] was still in force even after the care of the waters had passed, under Augustus, to commissioners. Indeed, lands which had been irrigated unlawfully from the public supply were confiscated. Whenever a slave infringed the law, even without the knowledge of his master, a fine was imposed.[2] By the same laws it is also enacted as follows: "No one shall with malice pollute the waters where they issue publicly. Should any one pollute them, his fine shall be ten thousand sestertii."[3] Therefore the order was given to the Curule Aediles to appoint two men in each district from the number of those who lived in it, or owned property in it, in whose care the public fountains should be placed.

Marcus Agrippa, after his aedileship (which he held after his consulship)[4] was the first man to become the permanent incumbent[5] of this office, so to speak—a commissioner charged with the supervision of works which he himself had created. Inasmuch as the amount of water now available warranted it, he determined how much should be allotted to the public structures, how much to the basins, and how much to private parties. He also

---

[5] Under the Republic, these special offices had been of a temporary nature; now Agrippa, and certain others following him, held this office for life.

SEXTUS JULIUS FRONTINUS

propriam aquarum, quae tueretur ductus atque castella et lacus. Hanc Augustus hereditate ab eo sibi relictam publicavit.

99. Post eum Q. Aelio Tuberone Paulo Fabio Maximo cos. cum res usque in id tempus quasi potestate acta certo iure eguisset, senatus consulta facta sunt ac lex promulgata. Augustus quoque edicto complexus est, quo iure uterentur qui ex commentariis Agrippae aquas haberent, tota re in sua beneficia translata. Modulos etiam, de quibus dictum est, constituit et rei continendae exercendaeque curatorem fecit Messalam Corvinum, cui adiutores dati Postumius Sulpicius praetorius et Lucius Cominius pedarius. Insignia eis quasi magistratibus concessa, deque eorum officio senatus consultum factum, quod infra scriptum est.

100. "Quod Q. Aelius Tubero Paulus Fabius Maximus cos. V. F.[1] de eis qui curatores aquarum publicarum ex consensu senatus a Caesare Augusto nominati essent ornandis,[2] D. E. R. Q. F. P. D. E. R. I. C.[3] placere huic ordini, eos qui aquis publicis praeessent, cum eius rei causa extra urbem essent, lictores binos et servos publicos ternos, architectos

---

[1] *i.e.* verba fecerunt.   [2] *Bergk*; ordinandis *CB*.
[3] *i.e.* de ea re quid facere placeret, de ea re ita censuerunt.

---

[1] In 11 B.C.   [2] *Cf.* 23.
[3] The *senatores pedarii* did not enjoy full senatorial rights.

AQUEDUCTS OF ROME, II. 98-100

kept his own private gang of slaves for the maintenance of the aqueducts and reservoirs and basins. This gang was given to the State as its property by Augustus, who had received it in inheritance from Agrippa.

Following him, under the consulate of Quintus Aelius Tubero and Paulus Fabius Maximus,[1] resolutions of the Senate were passed and a law was promulgated in these matters, which until that time had been managed at the option of officials, and had lacked definite control. Augustus also determined by an edict what rights those should possess who were enjoying the use of water according to Agrippa's records, thus making the entire supply dependent upon his own grants. The ajutages, also, of which I have spoken above,[2] were established by him; and for the maintenance and operation of the whole system he named Messala Corvinus commissioner, and gave him as assistants Postumius Sulpicius, ex-praetor, and Lucius Cominius, a junior[3] senator. They were allowed to wear regalia as though magistrates; and concerning their duties a resolution of the Senate was passed, which is here given:—

"The consuls, Quintus Aelius Tubero and Paulus Fabius Maximus, having made a report relating to the duties and privileges of the water-commissioners appointed with the approval of the Senate by Caesar Augustus, and inquiring of the Senate what it would please to order upon the subject, it has been RESOLVED that it is the sense of this body: That those who have the care of the administration of the public waters, when they go outside the City in the discharge of their duties, shall have two lictors, three public servants, and an architect for

singulos et scribas, librarios, accensos praeconesque totidem habere, quot habent ei per quos frumentum plebei datur. Cum autem in urbe eiusdem rei causa aliquid agerent, ceteris apparitoribus eisdem praeterquam lictoribus uti. Utique quibus apparitoribus ex hoc senatus consulto curatoribus aquarum uti liceret, eos diebus decem proximis, quibus senatus consultum factum esset, ad aerarium deferrent; quique ita delati essent, eis praetores aerarii mercedem cibaria, quanta praefecti frumento dando dare deferreque solent, annua darent et adtribuerent; eisque eas pecunias sine fraude sua capere liceret. Utique tabulas, chartas ceteraque quae eius curationis causa opus essent eis curatoribus Q. Aelius Paulus Fabius cos. ambo alterve, si eis videbitur, adhibitis praetoribus qui aerario praesint, praebenda locent.

101. "Itemque cum viarum curatores frumentique parte quarta anni publico fungantur ministerio, ut curatores aquarum iudiciis vacent privatis publicisque." Apparitores et ministeria, quamvis perseveret adhuc aerarium in eos erogare, tamen esse curatorum videntur desisse inertia ac segnitia non agentium officium. Egressis autem urbem dumtaxat agendae

---

[1] The resolution of the senate ends at this point. The rest of the section is taken up with some comments of Frontinus on various provisions of the bill.

each of them, and the same number of secretaries, clerks, assistants, and criers as those have who distribute wheat among the people; and when they have business inside the City on the same duties, they shall make use of all the same attendants, omitting the lictors; and, further, that the list of attendants granted to the water-commissioner by this resolution of the Senate shall be by them presented to the public treasurer within ten days from its promulgation, and to those whose names shall be thus reported the praetors of the treasury shall grant and give, as compensation, food by the year, as much as the food-commissioners are wont to give and allot, and they shall be authorized to take money for that purpose without prejudice to themselves. Further, there shall be furnished to the commissioners tablets, paper, and everything else necessary for the exercise of their functions. To this effect, the consuls, Quintus Aelius and Paulus Fabius, are ordered, both or either one, as may seem best to them, to consult with the praetors of the treasury in contracting for these supplies.

"Furthermore, inasmuch as the superintendents of streets and those in charge of the distribution of grain occupy a fourth part of the year in fulfilling their State duties, the water-commissioners likewise shall adjudicate (for a like period) in private and State causes."[1] Although the treasury has continued down to the present to pay for these attendants and servants, they have, as far as appearance goes, ceased to belong to the commissioners, who through laziness and indolence neglect their duties. Moreover, when the commissioners went out of the City, provided it was

SEXTUS JULIUS FRONTINUS

rei causa senatus praesto esse lictores iusserat. Nobis circumeuntibus rivos fides nostra et auctoritas a principe data pro lictoribus erit.

102. Cum perduxerimus rem ad initium curatorum, non est alienum subiungere qui post Messalam huic officio ad nos usque praefuerint. Messalae successit Planco et Silio cos. Ateius Capito. Capitoni C. Asinio Pollione C. Antistio Vetere cos. Tarius Rufus. Tario Servio Cornelio Cethego L. Visellio Varrone consulibus M. Cocceius Nerva, divi Nervae avus, scientia etiam iuris inlustris. Huic successit Fabio Persico L. Vitellio cos. C. Octavius Laenas. Laenati Aquila Iuliano et Nonio Asprenate consulibus M. Porcius Cato. Huic successit post mensem Ser. Asinio Celere A. Nonio[1] Quintiliano consulibus A. Didius Gallus. Gallo Q. Veranio et Pompeio Longo cos. Cn. Domitius Afer. Afro Nerone Claudio Caesare IIII et Cosso Cossi f. consulibus L. Piso. Pisoni Verginio Rufo et Memmio Regulo consulibus Petronius Turpilianus. Turpiliano Crasso Frugi et Laecanio Basso consulibus P. Marius. Mario Luccio Telesino et Suetonio Paulino cos. Fonteius Agrippa.

---

[1] *Nipperdey*; huic successit post quem serasinius celera . . . tonio *C*.

---

[1] Messala succeeded to Agrippa, under the consulate of Q. Aelius Tubero and Paulus Fabius Maximus, according to Varro.
[2] A.D. 13.   [3] A.D. 23.   [4] A.D. 24.
[5] A.D. 34.   [6] A.D. 38.
[7] *Consules suffecti* in 38 A.D., according to the conjecture of Nipperdey.
[8] A.D. 49.   [9] A.D. 60.   [10] A.D. 63.
[11] A.D. 64.   [12] A.D. 66.

on official business, the Senate had commanded the lictors to accompany them. For myself, when I go about to examine the aqueducts, my self-reliance and the authority given me by the sovereign will stand in place of the lictors.

As I have followed the matter down to the introduction of the commissioners, it will not be out of place now to subjoin the names of those who followed Messala[1] in this office up to my incumbency:—To Messala succeeded, under the consulate of Silius and Plancus,[2] Ateius Capito; to Capito, under the consulate of Gaius Asinius Pollio and Gaius Antistius Vetus,[3] Tarius Rufus; to Tarius, under the consulate of Servius Cornelius Cethegus and Lucius Visellius Varro,[4] Marcus Cocceius Nerva, the grandfather of the Deified Nerva, who was also noted as learned in the science of law. To him succeeded, under the consulate of Fabius Persicus and Lucius Vitellius,[5] Gaius Octavius Laenas; to Laenas, under the consulate of Aquila Julianus and Nonius Asprenas,[6] Marcus Porcius Cato. To him succeeded, after a month, under the consulate of Servius Asinius Celer and Aulus Nonius Quintilianus,[7] Aulus Didius Gallus; to Gallus, under the consulate of Quintus Veranius and Pompeius Longus,[8] Gnaeus Domitius Afer; to Afer, under the fourth consulate of Nero Claudius Caesar, and that of Cossus, the son of Cossus,[9] Lucius Piso; to Piso, under the consulate of Verginius Rufus and Memmius Regulus,[10] Petronius Turpilianus; to Turpilianus, under the consulate of Crassus Frugi and Lecanius Bassus,[11] Publius Marius; to Marius, under the consulate of Lucius Telesinus and Suetonius Paulinus,[12] Fonteius Agrippa; to

SEXTUS JULIUS FRONTINUS

Agrippae Silio et Galerio Trachalo cos. Albius Crispus. Crispo Vespasiano III et Cocceio Nerva cos. Pompeius Silvanus. Silvano Domitiano II Valerio Messalino consulibus Tampius Flavianus. Flaviano Vespasiano V Tito III consulibus Acilius Aviola. Post quem imperatore Nerva III et Verginio Rufo III consulibus ad nos cura translata est.

103. Nunc quae observare curator aquarum debeat et legem senatusque consulta ad instruendum actum pertinentia subiungam. Circa ius ducendae aquae in privatis observanda sunt, ne quis sine litteris Caesaris, id est ne quis aquam publicam non impetratam, et ne quis amplius quam impetravit ducat. Ita enim efficiemus ut modus, quem adquiri diximus, possit ad novos salientes et ad nova beneficia principis pertinere. In utroque autem magna cura multiplici opponenda fraudi est: sollicite subinde ductus extra urbem circumeundi ad recognoscenda beneficia; idem in castellis et salientibus publicis faciendum, ut sine intermissione diebus noctibusque aqua fluat. Quod senatus quoque consulto facere curator iubetur, cuius haec verba sunt:

104. "Quod Q. Aelius Tubero Paulus Fabius Maxi-

---

[1] A.D. 68. [2] A.D. 71. [3] A.D. 73.
[4] A.D. 74. [5] A.D. 97.

Agrippa, under the consulate of Silius and Galerius Trachalus,[1] Albius Crispus; to Crispus, under the third consulate of Vespasian, and that of Cocceius Nerva,[2] Pompeius Silvanus; to Silvanus, under the second consulate of Domitian and that of Valerius Messalinus,[3] Tampius Flavianus; to Flavianus, under the fifth consulate of Vespasian, and the third of Titus,[4] Acilius Aviola. After Aviola, under the third consulate of the Emperor Nerva, and the third of Verginius Rufus,[5] the office was transferred to me.

I will now set down what the water-commissioner must observe, being the laws and Senate enactments which serve to determine his procedure. As concerns the draft of water by private consumers, it is to be noted: No one shall draw water without an authorisation from Caesar, that is, no one shall draw water from the public supply without a licence, and no one shall draw more than has been granted. By this means, we shall make it possible that the quantity of water, which has been regained, as we have said, may be distributed to new fountains and may be used for new grants from the sovereign. But in both cases it will be necessary to exert great resistance to manifold forms of fraud. Frequent rounds must be made of channels of the aqueducts outside the City, and with great care, to check up the granted quantities. The same must be done in case of the reservoirs and public fountains, that the water may flow without interruption, day and night. For this the commissioner has been directed to provide, by a resolution of the Senate, the language of which is as follows:

"The consuls, Quintus Aelius Tubero and Paulus Fabius Maximus, having made a report upon the

SEXTUS JULIUS FRONTINUS

mus cos. V. F. de numero publicorum salientium qui in urbe essent intraque aedificia urbi coniuncta, quos M. Agrippa fecisset, Q. F. P. D. E. R. I. C. neque augeri placere nec minui numerum publicorum salientium, quos nunc esse rettulerunt ei, quibus negotium a senatu est imperatum ut inspicerent aquas publicas inirentque numerum salientium publicorum. Itemque placere curatores aquarum, quos Caesar Augustus ex senatus auctoritate nominavit, dare operam uti salientes publici quam adsiduissime interdiu et noctu aquam in usum populi funderent." In hoc senatus consulto crediderim adnotandum quod senatus tam augeri quam minui salientium publicorum numerum vetuerit. Id factum existimo, quia modus aquarum quae eis temporibus in urbem veniebant, antequam Claudia et Anio Novus perducerentur, maiorem erogationem capere non videbatur.

105. Qui aquam in usus privatos deducere volet, impetrare eam debebit et a principe epistulam ad curatorem adferre; curator deinde beneficio Caesaris praestare maturitatem et procuratorem eiusdem officii libertum Caesaris protinus scribere. Procuratorem autem primus Ti. Claudius videtur admovisse, postquam Anionem Novum et Claudiam induxit Quid contineat epistula, vilicis quoque fieri notum debet, ne quando neglegentiam aut fraudem suam ignorantiae colore defendant. Procurator calicem eius

---

[1] *Cf.* 116, 117.

number of public fountains established by Marcus Agrippa in the City and within structures adjacent to the City, and having inquired of the Senate what it would please to order upon the subject, it has been RESOLVED that it is the sense of this body: That the number of public fountains which exist at present, according to the report of those who were ordered by the Senate to examine the public aqueducts and to inventory the number of public fountains, shall be neither increased nor diminished. Further, that the water-commissioners, who have been appointed by Caesar Augustus, with the endorsement of the Senate, shall take pains that the public fountains may deliver water as continuously as possible for the use of the people day and night." In this resolution of the Senate, I think it should be noted that the Senate forbade any increase as well as any decrease in the number of public fountains. I think this was done because the quantity of water, which at that time came into the City, before Claudia and New Anio had been brought in, did not seem to permit of a greater distribution.

Whoever wishes to draw water for private use must seek for a grant and bring to the commissioner a writing from the sovereign; the commissioner must then immediately expedite the grant of Caesar, and appoint one of Caesar's freedmen as his deputy for this service. Tiberius Claudius appears to have been the first man to appoint such a deputy after he introduced Claudia and New Anio. The overseers[1] must also be made acquainted with the contents of the writing, that they may not excuse their negligence or fraud on the plea of ignorance. The

SEXTUS JULIUS FRONTINUS

moduli, qui fuerit impetratus, adhibitis libratoribus signari cogitet, diligenter intendat mensurarum quas supra diximus modum et positionis notitiam habeat, ne sit in arbitrio libratorum, interdum maioris luminis, interdum minoris pro gratia personarum calicem probare. Sed nec statim ab hoc liberum subiciendi qualemcumque plumbeam fistulam permittatur arbitrium, verum eiusdem luminis quo calix signatus est per pedes quinquaginta, sicut senatus consulto quod subiectum est cavetur.

106. "Quod Q. Aelius Tubero Paulus Fabius Maximus cos. V. F. quosdam privatos ex rivis publicis aquam ducere, Q. D. E. R. F. P. D. E. R. I. C. ne cui privato aquam ducere ex rivis publicis liceret, utique omnes ei quibus aquae ducendae ius esset datum ex castellis ducerent, animadverterentque curatores aquarum, quibus locis intra urbem apte castella privati facere possent, ex quibus aquam ducerent quam ex castello communi[1] accepissent a curatoribus aquarum. Neve cui eorum quibus aqua daretur publica ius esset, intra quinquaginta pedes eius castelli, ex quo aquam ducerent, laxiorem

---

[1] *Bennett*; communem *CB*.

---

[1] *Cf.* 36.   [2] *Cf.* 24 ff.   [3] *Cf.* 26 ff. and 36.
[4] *Cf.* 26, and footnote 2 on p. 364.   [5] *Cf.* 25 and footnote.

deputy must call in the levellers and provide that the *calix*[1] is stamped as conforming to the deeded quantity, and must study the size of the ajutages we have enumerated above,[2] as well as have knowledge of their location,[3] lest it rest with the caprice of the levellers to approve a *calix* of sometimes greater, or sometimes smaller, interior area,[4] according as they interest themselves in the parties. Neither must the deputy permit the free option of connecting directly to the ajutages any sort of lead pipe, but there must rather be attached for a length of fifty feet one of the same interior area as that which the ajutage has been certified to have, as has been ordained by a vote of the Senate which follows:

"The consuls, Quintus Aelius Tubero and Paulus Fabius Maximus, having made a report that some private parties take water directly from the public conduits, and having inquired of the Senate what it would please to order upon the subject, it has been RESOLVED that it is the sense of this body: That it shall not be permitted to any private party to draw water from the public conduits; and all those to whom the right to draw water has been granted shall draw it from the reservoirs, the water-commissioners to direct at what points, within the City, private parties may suitably erect reservoirs for the purpose of drawing from them the water which they had received at the hands of the water-commissioner from some public reservoir; and no one of those to whom a right to draw water from the public conduits has been granted shall have the right to use a larger pipe than a *quinaria*[5] for a space of fifty feet from the reservoir

SEXTUS JULIUS FRONTINUS

fistulam subicere quam quinariam." In hoc S. C. dignum adnotatione est, quod aquam non nisi ex castello duci permittit, ne aut rivi aut fistulae publicae frequenter lacerentur.

107. Ius impetratae aquae neque heredem neque emptorem neque ullum novum dominum praediorum sequitur. Balneis quae publice lavarent privilegium antiquitus concedebatur, ut semel data aqua perpetuo maneret. Sic ex veteribus senatus consultis cognoscimus, ex quibus unum subieci; nunc omnis aquae cum possessore instauratur beneficium.

108. "Quod Q. Aelius Tubero Paulus Fabius Maximus cos. V. F. constitui oportere, quo iure extra intraque urbem ducerent aquas, quibus adtributae essent, Q. D. E. R. F. P. D. E. R. I. C. uti usque eo maneret adtributio aquarum, exceptis quae in usum balinearum essent datae aut Augusti[1] nomine, quoad eidem domini possiderent id solum, in quod accepissent aquam."

109. Cum vacare aliquae coeperunt aquae, adnuntiatur et in commentarios redigitur, qui respiciuntur ut petitoribus ex vacuis dari possint. Has aquas statim intercipere solebant, ut medio tempore venderent aut possessoribus praediorum aut aliis

[1] *Herschel*; haustus *CB*.

---

[1] *i.e.* through the death of the grantee or the transfer of this property.

AQUEDUCTS OF ROME, II 106-109

out of which he is to draw the water." In this resolution of the Senate it is worthy of note that the resolution does not permit water to be drawn except from reservoirs, in order that the conduits or the public pipes may not be frequently cut into.

The right to granted water does not pass either to the heirs, or to the buyer, or to any new proprietor of the land. The public bathing establishments had from old times the privilege that water once granted to them should remain theirs for ever. We know this from old resolutions of the Senate, of which I give one below:—(Nowadays every grant of water is renewed to the new owner.)

"The consuls, Quintus Aelius Tubero and Paulus Fabius Maximus, having made a report upon the necessity of determining in accordance with what law those persons, to whom water had been granted, should draw water inside and outside the City, and having inquired of the Senate what it would please to order upon the subject, it has been RESOLVED that it is the sense of this body: That a grant of water, with the exception of those supplies which have been granted for the use of bathing establishments, or in the name of Augustus, shall remain in force as long as the same proprietors continue to hold the ground for which they received the grant of the water."

As soon as any water-rights are vacated,[1] this is announced, and entered in the records, which are consulted, in order that vacant water-rights may be given to applicants. These waters they formerly used to cut off immediately, in order that between times they might sell them either to the occupants of the land, or to outsiders even. It seemed

SEXTUS JULIUS FRONTINUS

etiam. Humanius visum est principi nostro, ne praedia subito destituerentur, triginta dierum spatium indulgeri, intra quod ei ad quos res pertineret . . . .[1] De aqua in praedia sociorum data nihil constitutum invenio. Perinde tamen observatur ac iure cautum, ut dum quis ex eis qui communiter impetraverunt superesset, totus modus praediis adsignatus flueret et tunc demum renovaretur beneficium, cum desisset quisque ex eis quibus datum erat possidere. Impetratam aquam alio, quam in ea praedia in quae data erit, aut ex alio castello, quam ex quo epstula principiis continebit, duci palamst non oportere; sed et mandatis prohibetur.

110. Impetrantur autem et eae aquae quae caducae vocantur, id est quae aut ex castellis aut ex manationibus fistularum,[2] quod beneficium a principibus parcissime tribui solitum. Sed fraudibus aquariorum obnoxium est, quibus prohibendis quanta cura debeatur, ex capite mandatorum manifestum erit quod subieci.[3]

111. "Caducam neminem volo ducere nisi qui meo[4] beneficio aut priorum principum habent. Nam necesse est ex castellis aliquam partem aquae effluere, cum hoc pertineat non solum ad urbis nostrae salubritatem, sed etiam ad utilitatem cloacarum abluendarum."

112. Explicitis quae ad ordinationem aquarum privati usus pertinebant, non ab re est quaedam

---

[1] *Addendum cum Iordansis petere possent beneficii instaurationem B*; *no gap in C.*
[2] effluunt *add. Iocundus B*; *no gap in C.*
[3] *Between 110 and 111 is a gap of about twenty letters in C.*
[4] mō *C*; nostro *B.*

---

[1] *i.e.* Imperial edict.

less harsh to our ruler, in order not to deprive estates of water suddenly, to give thirty days' grace, within which those whose interests were involved [might make suitable arrangements]. I did not find anything set down about the water granted to an estate belonging to a syndicate. Nevertheless, the following practice is observed, just as though prescribed by law, "that as long as one of those who have received a common grant of water survives, the full amount of granted water shall flow upon the land, and the grant shall be renewed only when every one of those who received it has ceased to remain in possession of the property." That granted water must not be carried elsewhere than upon the premises to which it has been made appurtenant, or taken from another reservoir than the one designated in the writing of the sovereign, is self-evident, but is forbidden also by ordinance.[1]

Those waters also that are called "lapsed," namely, those that come from the overflow of the reservoirs or from leakage of the pipes, are subject to grants; which are wont to be given very sparingly, however, by the sovereign. But this offers opportunity for thefts by the water-men; and how much care should be devoted to preventing these, may be seen from a paragraph of an ordinance, which I append:

"I desire that no one shall draw 'lapsed' water except those who have permission to do so by grants from me or preceding sovereigns; for there must necessarily be some overflow from the reservoirs, this being proper not only for the health of our City, but also for use in the flushing of the sewers."

Having now explained those things that relate to the administration of water for the use of private

SEXTUS JULIUS FRONTINUS

ex eis, quibus circumscribi saluberrimas constitutiones in ipso actu deprehendimus, exempli causa attingere. Ampliores quosdam calices, quam impetrati erant, positos in plerisque castellis inveni et ex eis aliquos ne signatos quidem. Quotiens autem signatus calix excedit legitimam mensuram, ambitio procuratoris qui eum signavit detegitur. Cum vero ne signatus quidem est, manifesta culpa omnium, maxime accipientis, deprehenditur, deinde vilici. In quibusdam, cum calices legitimae mensurae signati essent, statim amplioris moduli fistulae subiectae fuerunt, unde acciderat ut aqua non per legitimum spatium coercita, sed per brevis angustias expressa facile laxiorem in proximo fistulam impleret. Ideoque illud adhuc, quotiens signatur calix, diligentiae adiciendum est, ut fistulae quoque proximae per spatium, quod S. C. comprehensum diximus, signentur. Ita demum enim vilicus cum scierit non aliter quam signatas conlocari debere, omni carebit excusatione.

113. Circa conlocandos quoque calices observari oportet ut ad lineam ordinentur nec alterius inferior calix, alterius superior ponatur. Inferior plus trahit; superior, quia cursus aquae ab inferiore rapitur, minus ducit. In quorundam fistulis ne calices quidem positi fuerunt. Hae fistulae solutae vocantur et ut aquario libuit, laxantur vel coartantur.

---

[1] *i.e.* fifty feet; *cf.* 105.
[2] "The ajutage was not less than about nine inches long." —*Herschel.*
[3] *Cf.* 36.
[4] *i. e.* the amount of water flowing through them.

parties, it will not be foreign to the subject to touch upon certain practices, by way of illustration, whereby we have caught these most wholesome ordinances in the very act of being defeated. In a great number of reservoirs I found certain ajutages of a larger size than had been granted, and among them some that had not even been stamped. Now whenever a stamped ajutage is larger than its legitimate measure it reveals designing dishonesty on the part of the deputy who stamped it; but when it is not even stamped, it clearly reveals the fault of all, especially of the grantee, also of the overseer. In some of the reservoirs, though their ajutages were stamped in conformity with their lawful admeasurements, pipes of a greater diameter [than the ajutages] were at once attached to them. As a consequence, the water not being held together for the lawful distance,[1] and being on the contrary forced through the short restricted distance,[2] easily filled the adjoining larger pipes. Care should therefore be taken, as often as an ajutage is stamped, to stamp also the adjoining pipe over the length which we stated was prescribed by the resolution of the Senate. For then and then only can the overseer be held to his full responsibility, when he understands that none but stamped pipes must be set in place.

In setting[3] ajutages also, care must be taken to set them on the level, and not place the one higher and the other lower down. The lower one will take in more; the higher one will suck in less, because the current of water is drawn in by the lower one. To some pipes no ajutages were even attached. Such pipes are called "uncontrolled," and are enlarged or diminished[4] as pleases the water-men.

SEXTUS JULIUS FRONTINUS

114. Adhuc illa aquariorum intolerabilis fraus est: translata in novum possessorem aqua foramen novum castello imponunt, vetus relinquunt quo venalem extrahunt aquam. In primis ergo hoc quoque emendandum curatori crediderim. Non enim solum ad ipsarum aquarum custodiam, sed etiam ad castelli tutelam pertinet, quod subinde et sine causa foratum vitiatur.

115. Etiam ille aquariorum tollendus est reditus, quem vocant puncta. Longa ac diversa sunt spatia, per quae fistulae tota meant urbe latentes sub silice. Has comperi per eum qui appellabatur a punctis passim convulneratas omnibus in transitu negotiationibus praebuisse peculiaribus fistulis aquam, quo efficiebatur ut exiguus modus ad usus publicos perveniret. Quantum ex hoc modo aquae surreptum sit, aestimo ex eo quod aliquantum plumbi sublatis eiusmodi ramis redactum est.

116. Superest tutela ductuum, de qua priusquam dicere incipiam, pauca de familia quae huius rei causa parata est explicanda sunt.[1] Familiae sunt duae, altera publica, altera Caesaris. Publica est antiquior, quam ab Agrippa relictam Augusto et ab eo publicatam diximus; habet homines circiter

---

[1] *At this point there is a gap of about twenty letters in C.*

There is, besides, this intolerable method of cheating practised by the water-men: When a water-right is transferred to a new owner, they will insert a new ajutage in the reservoir; the old one they leave in the tank and draw from it water, which they sell. This practice especially, therefore, as I believe, should be corrected by the Commissioner; for this concerns not only the protection of the water itself, but also the maintenance of the reservoirs, which get to be leaky when they are often and unnecessarily tapped into.

The following mode of gaining money, practised by the water-men, is also to be abolished: the one called "puncturing." There are extensive areas in various places where secret pipes run under the pavements all over the City. I discovered that these pipes were furnishing water by special branches to all those engaged in business in those localities through which the pipes ran, being bored for that purpose here and there by the so-called "puncturers"; whence it came to pass that only a small quantity of water reached the places of public supply. How large an amount of water has been stolen in this manner, I estimate by means of the fact that a considerable quantity of lead has been brought in by the removal of that kind of branch pipes.

It remains to speak of the maintenance of the conduits; but before I say anything about this, a little explanation should be given about the gangs of slaves established for this purpose. There are two of those gangs, one belonging to the State, the other to Caesar. The one belonging to the State is the older, which, as we have said, was left by Agrippa to Augustus, and was by him made over to the

SEXTUS JULIUS FRONTINUS

ducentos quadraginta. Caesaris familiae numerus est quadringentorum sexaginta, quam Claudius cum aquas in urbem perduceret constituit.

117. Utraque autem familia in aliquot ministeriorum species diducitur, vilicos, castellarios, circitores, silicarios, tectores aliosque opifices. Ex his aliquos extra urbem esse oportet ad ea quae non sunt magnae molitionis, maturum tamen auxilium videntur exigere. Homines in urbe circa castellorum et munerum stationes opera quaeque urgebunt, in primis ad subitos casus, ut ex compluribus regionibus, in quam necessitas incubuerit, converti possit praesidium aquarum abundantium. Tam amplum numerum utriusque familiae solitum ambitione aut neglegentia praepositorum in privata opera diduci revocare ad aliquam disciplinam et publica ministeria ita instituimus, ut pridie quid esset actura dictaremus et quid quoque die egisset actis comprehenderetur.

118. Commoda publicae familiae ex aerario dantur, quod impendium exoneratur vectigalium reditu ad ius aquarum pertinentium. Ea constant ex locis aedificiisve quae sunt circa ductus et castella aut munera aut lacus. Quem reditum prope sestertiorum ducentorum quinquaginta milium alienatum ac vagum, proximis vero temporibus in Domitiani loculos conversum iustitia Divi Nervae populo restituit, nostra sedulitas ad certam regulam redegit, ut constaret quae essent ad hoc vectigal pertinentia loca. Caesaris familia ex fisco accipit

---

[1] *Cf.* 98.
[2] Equivalent at this time to about £2,125 or from $10,000 to $12,000.

State.[1] It numbers about 240 men. The number in Caesar's gang is 460; it was organized by Claudius at the time he brought his aqueduct into the City.

Both gangs are divided into several classes of workmen: overseers, reservoir-keepers, inspectors, pavers, plasterers, and other workmen; of these, some must be outside the city for purposes which do not seem to require any great amount of work, but yet demand prompt attention; the men inside the city at their stations at the reservoirs and fountains will devote their energies to the several works, especially in case of sudden emergencies, in order that a plentiful reserve supply of water may be turned from several wards of the city to the one afflicted by an emergency. Both of these large gangs, which regularly were diverted by exercise of favouritism, or by negligence of their foremen, to employment on private work, I resolved to bring back to some discipline and to the service of the State, by writing down the day before what each gang was to do, and by putting in the records what it had done each day.

The wages of the State gang are paid from the State treasury, an expense which is lightened by the receipt of rentals from water-rights, which are received from places or buildings situated near the conduits, reservoirs, public fountains, or water-basins. This income of nearly 250,000 *sestertii*[2] formerly lost through loose management, was turned in recent times into the coffers of Domitian; but with a due sense of right the Deified Nerva restored it to the people. I took pains to bring it under fixed rules, in order that it might be clear what were the places which fell under this tax. The gang of Caesar gets its wages from the emperor's privy purse,

commoda, unde et omne plumbum et omnes impensae ad ductus et castella et lacus pertinentes erogantur.

119. Quoniam quae videbantur ad familiam pertinere exposuimus, ad tutelam ductuum sicut promiseram divertemus, rem enixiore cura dignam, cum magnitudinis Romani imperii vel praecipuum sit indicium. Multa atque ampla opera subinde dilabuntur, quibus ante succurri debet quam magno auxilio egere incipiant, plerumque tamen prudenti temperamento sustinenda, quia non semper opus aut facere aut ampliare quaerentibus credendum est. Ideoque non solum scientia peritorum sed et proprio usu curator instructus esse debet, nec suae tantum stationis architectis uti, sed plurium advocare non minus fidem quam subtilitatem, ut aestimet quae repraesentanda, quae differenda sint, et rursus quae per redemptores effici debeant, quae per domesticos artifices.

120. Nascuntur opera ex his causis : aut impotentia possessorum quid corrumpitur aut vetustate aut vi tempestatium aut culpa male facti operis, quod saepius accidit in recentibus.

121. Fere aut vetustate aut vi tempestatium eae partes ductuum laborant quae arcuationibus sustinentur aut montium lateribus applicatae sunt, et ex arcuationibus eae quae per flumen traiciuntur.

from which are also drawn all expenses for lead and for conduits, reservoirs, and basins.

As I have now explained all, I think, that has to do with slave-gangs, I will now, as I promised, come back to the maintenance of the conduits, a thing which is worthy of more special care, as it gives the best testimony to the greatness of the Roman Empire. The numerous and extensive works are continually falling into decay, and they must be attended to before they begin to demand extensive repair. Very often, however, it is best to exercise a wise restraint in attending to their upkeep, since those who urge the construction or extension of the works cannot always be trusted. The water-commissioner, therefore, not only ought to be provided with competent advisers, but ought also to be equipped with practical experience of his own. He must consult not only the architects of his own office, but must also seek aid from the trustworthy and thorough knowledge of numerous other persons, in order to judge what must be taken in hand forthwith, and what postponed, and, again, what is to be carried out by public contractors and what by his own regular workmen.

The necessity of repairs arises from the following reasons: damage is done either by the lawlessness of abutting proprietors, by age, by violent storms, or by defects in the original construction, which has happened quite frequently in the case of recent works.

As a rule, those parts of the aqueducts which are carried on arches or are placed on side-hills and, of those on arches, the parts that cross rivers suffer most from the effects of age or of violent storms.

SEXTUS JULIUS FRONTINUS

Ideoque haec opera sollicita festinatione explicanda sunt. Minus iniuriae subiacent subterranea nec gelicidiis nec caloribus exposita. Vitia autem eiusmodi sunt, ut aut non interpellato cursu subveniatur eis, aut emendari nisi averso non possint, sicut ea quae in ipso alveo fieri necesse est.

122. Haec duplici ex causa nascuntur: aut enim limo concrescente, qui interdum in crustam indurescit, iter aquae coartatur, aut tectoria corrumpuntur, unde fiunt manationes quibus necesse est latera rivorum et substructiones vitiari. Pilae quoque ipsae tofo exstructae sub tam magno onere labuntur. Refici quae circa alveos rivorum sunt aestate non debent, ne intermittatur usus tempore quo praecipue desideratur, sed vere vel autumno et maxima cum festinatione, ut scilicet ante praeparatis omnibus quam paucissimis diebus rivi cessent. Neminem fugit, per singulos ductus hoc esse faciendum, ne si plures pariter avertantur, desit aqua civitati.

123. Ea quae non interpellato aquae cursu effici debent maxime structura constant, quam et suis temporibus et fidelem fieri oportet. Idoneum structurae tempus est a Kalendis Aprilibus in Kalendas Novembres ita ut optimum sit intermittere eam partem aestatis quae nimiis caloribus incandescit,

These, therefore, must be put in order with care and despatch. The underground portions, not being subjected to either heat or frost, are less liable to injury. Defects are either of the sort that can be remedied without stopping the flow of the water, or such as cannot be made without diverting the flow, as, for example, those which have to be made in the channel itself.

These latter become necessary from two causes: either the accumulation of deposit, which sometimes hardens into a crust, contracts the channel of the water; or else the concrete lining is damaged, causing leaks, whereby the sides of the conduits and the substructures are necessarily injured. Sometimes even the piers, which are built of tufa, give way under the great load. Repairs to the channel itself should not be made in the summer time, in order not to stop the flow of water at a time when the demand for it is the greatest, but should be made in the spring or autumn, and with the greatest speed possible, and of course with all preparations made in advance, in order that the conduits may be out of commission as few days as possible. As is obvious to every one, a single aqueduct must be taken at a time, for if several were cut off at once, the supply would prove inadequate for the City's needs.

Repairs that should be executed without cutting off the water consist principally of masonry work, which should be constructed at the right time, and conscientiously. The suitable time for masonry work is from April 1 to November 1, but with this restriction, that the work would best be interrupted during the hottest part of the summer, because

SEXTUS JULIUS FRONTINUS

quia temperamento caeli opus est, ut ex commodo [1] structura combibat et in unitatem corroboretur; non minus autem sol acrior quam gelatio praecipit materiam. Nec ullum opus diligentiorem poscit curam quam quod aquae obstaturum est; fides itaque eius per singula secundum legem notam omnibus sed a paucis observatam exigenda est.

124. Illud nulli dubium esse crediderim, proximos ductus, id est qui a septimo miliario lapide quadrato consistunt, maxime custodiendos, quoniam et amplissimi operis sunt et plures aquas singuli sustinent. Quos si necesse fuerit interrumpere, maiorem partem aquarum urbis destituent. Remedia tamen sunt et huius difficultatis: opus incohatum excitatur ad libram deficientis, alveus vero plumbatis canalibus per spatium interrupti ductus rursus continuatur. Porro quoniam fere omnes specus per privatorum agros derecti erant et difficilis videbatur futurae impensae praeparatio, nisi et aliqua iuris constitutione succurreretur, simul ne accessu ad reficiendos rivos redemptores a possessoribus prohiberentur, S. C. factum est quod subieci.

125. "Quod Q. Aelius Tubero Paulus Fabius Maximus cos. V. F. de rivis, specibus, fornicibus

---

[1] *Schöne*; ex commodi *C*; et humorem commode *B*.

moderate weather is necessary for the masonry properly to absorb the mortar, and to solidify into one compact mass; for excessive heat of the sun is no less destructive than frost to masonry. Nor is greater care required upon any works than upon such as are to withstand the action of water; for this reason, in accordance with principles which all know but few observe, honesty in all details of the work must be insisted upon.

I think no one will doubt that the greatest care should be taken with the aqueducts nearest to the City (I mean those inside the seventh mile-stone, which consist of block-stone masonry), both because they are structures of the greatest magnitude, and because each one carries several conduits; for should it once be necessary to interrupt these, the City would be deprived of the greater part of its water-supply. But there are methods for meeting even these difficulties: provisional works are built up to the level of the conduit which is being put out of use, and a channel, formed of leaden troughs, running along the course of the portion that has been cut off, again provides a continuous passage. Furthermore, since almost all the aqueducts ran through the fields of private parties and it seemed difficult to provide for future outlays without the help of some constituted law; in order, also, that contractors should not be prevented by proprietors from access to the conduits for the purpose of making repairs, a resolution of the Senate was passed, which I give below:—

"The consuls, Quintus Aelius Tubero and Paulus Fabius Maximus, having made a report relating to the restoration of the canals, conduits, and arches of

SEXTUS JULIUS FRONTINUS

aquae Iuliae, Marciae, Appiae, Tepulae, Anienis reficiendis, Q. D. E. R. F. P. D. E. R. I. C. uti cum ei rivi, specus, fornices, quos Augustus Caesar se refecturum impensa sua pollicitus senatui est, reficerentur,[1] ex agris privatorum terra, limus, lapides, testa, harena, ligna ceteraque quibus ad eam rem opus esset, unde quaeque eorum proxime sine iniuria privatorum tolli, sumi, portari possint, viri boni arbitratu aestimata darentur, tollerentur, sumerentur, exportarentur; et ad eas res omnes exportandas earumque rerum reficiendarum causa, quotiens opus esset, per agros privatorum sine iniuria eorum itinera, actus paterent, darentur."

126. Plerumque autem vitia oriuntur ex impotentia possessorum, qui pluribus rivos violant. Primum enim spatia, quae circa ductus aquarum ex S. C. vacare debent, aut aedificiis aut arboribus occupant. Arbores magis nocent, quarum radicibus concamerationes et latera solvuntur. Dein vicinales vias agrestesque per ipsas formas derigunt. Novissime aditus ad tutelam praecludunt. Quae omnia S. C. quod subieci provisa sunt.

127. "Quod Q. Aelius Tubero Paulus Fabius Maximus cos. V. F. aquarum, quae in urbem veni-

---

[1] *Gap of about ten letters in C.*

AQUEDUCTS OF ROME, II. 125-127

Julia, Marcia, Appia, Tepula, and Anio, and having inquired of the Senate what it would please to order upon the subject, it has been RESOLVED: That when those canals, conduits, and arches, which Augustus Caesar promised the Senate to repair at his own cost, shall be repaired, the earth, clay, stone, potsherds, sand, wood, etc., which are necessary for the work in hand, shall be granted, removed, taken, and brought from the lands of private parties, their value to be appraised by some honest man, and each of these to be taken from whatever source it may most conveniently and, without injury to private parties, be removed, taken, and brought; and that thoroughfares and roads through the lands of private parties shall, without injury to them, remain open and their use be permitted, as often as it is necessary for the transportation of all these things for the purposes of repairing these works."

But very often damages occur by reason of the lawlessness of private owners, who injure the conduits in numerous ways. In the first place, they occupy with buildings or with trees the space around the aqueducts, which according to a resolution of the Senate should remain open. The trees do the most damage, because their roots burst asunder the top coverings as well as the sides. They also lay out village and country roads over the aqueducts themselves. Finally, they shut off access to those coming to make repairs. All these offences have been provided against in the resolution of the Senate, which I append:—

"The consuls, Quintus Aelius Tubero and Paulus Fabius Maximus, having made a report that the routes of the aqueducts coming to the city are being

SEXTUS JULIUS FRONTINUS

rent, itinera occuparì monumentis et aedificiis et arboribus conseri, Q. F. P. D. E. R. I. C. cum ad reficiendos rivos specusque per quae opera publica corrumpantur,[1] placere circa fontes et fornices et muros utraque ex parte quinos denos pedes patere, et circa rivos qui sub terra essent et specus intra urbem et urbi continentia aedificia utraque ex parte quinos pedes vacuos relinqui ita ut neque monumentum in eis locis neque aedificium post hoc tempus ponere neque conserere arbores liceret, sique nunc essent arbores intra id spatium, exciderentur praeterquam si quae villae continentes et inclusae aedificiis essent. Si quis adversus ea commiserit, in singulas res poena HS [2] dena milia essent, ex quibus pars dimidia praemium accusatori daretur, cuius opera maxime convictus esset qui adversus hoc S. C. commisisset, pars autem dimidia in aerarium redigeretur. Deque ea re iudicarent cognoscerentque curatores aquarum."

128. Posset hoc S. C. aequissimum videri, etiam si ex re tantum publicae utilitatis [3] ea spatia vindicarentur. Multo magis autem maiores nostri admirabili aequitate ne ea quidem eripuerunt privatis quae ad modum publicum pertinebant, sed cum aquas perducerent, si difficilior possessor in parte

---

[1] *Various emendations have been suggested for the reading here given, which is that of C, but none is thoroughly satisfactory.*

[2] Sestertium.   [3] *C*; utilitate *B*.

---

[1] About £85.

encumbered with tombs and edifices and planted with trees, and having inquired of the Senate what it would please to order upon the subject, it has been RESOLVED: That since, for the purpose of repairing the channels and conduits [obstructions must be removed] by which public structures are damaged, it is decreed that there shall be kept clear a space of fifteen feet on each side of the springs, arches, and walls; and that about the subterranean conduits and channels, inside the City, and inside buildings adjoining the City, there shall be left a vacant space of five feet on each side; and it shall not be permitted to erect a tomb at these places after this time, nor any structures, nor to plant trees. If there be any trees within this space at the present time they shall be taken out by the roots except when they are connected with country seats or enclosed in buildings. Whoever shall contravene these provisions, shall pay as penalty, for each contravention, 10,000 *sestertii*,[1] of which one-half shall be given as a reward to the accuser whose efforts have been chiefly responsible for the conviction of the violator of this vote of the Senate. The other half shall be paid into the public treasury. About these matters the water-commissioners shall judge and take cognizance."

This resolution of the Senate would appear perfectly just, even if this ground were claimed solely in view of the public advantage; but with much more admirable justice, our forefathers did not seize from private parties even those lands which were necessary for public purposes but, in the construction of water-works, whenever a proprietor made any difficulty in the sale of a portion, they

vendunda fuerat, pro toto agro pecuniam intulerunt
et post determinata necessaria loca rursus eum agrum
vendiderunt, ut in suis finibus proprium ius res
publica privatique haberent. Plerique tamen non
contenti occupasse fines ipsis ductibus manus ad-
tulerunt per suffossa latera passim cursus aquarum
tam ei qui ius aquarum impetratum [1] habent, quam
ei qui quantulicumque beneficii occasione ad ex-
pugnandos rivos abutuntur. Quid porro fieret, si
non universa ista diligentissima lege prohiberentur
poenaque non mediocris contumacibus intentaretur?
Quare subscripsi verba legis.

129. " T. Quintius Crispinus consul . . . [2] populum
iure rogavit populusque iure scivit in foro pro rostris
aedis Divi Iulii pr. K.[3] Iulias. Tribus Sergia
principium fuit. Pro tribu Sex. L. f. Varro
primus scivit. Quicumque post hanc legem rogatam
rivos, specus, fornices, fistulas, tubulos, castella, lacus
aquarum publicarum, quae ad urbem ducuntur, sciens
dolo malo foraverit, ruperit, foranda rumpendave
curaverit peiorave fecerit, quo minus eae aquae
earumve quae pars in urbem Romam ire, cadere,
fluere, pervenire, duci possit, quove minus in urbe
Roma et in eis locis, aedificiis, quae loca, aedificia
urbi continentia sunt, erunt, in eis hortis, praediis,

---

[1] *B*; cursus us aquarum imperatum *C*; tam ei qui *Poleni*.
[2] *Gap of eight or ten letters in C.*
[3] *B*; P. R. Iulias *C*.

AQUEDUCTS OF ROME. II. 128-129

paid for the whole field, and after marking off the needed part, again sold the land with the understanding that the public as well as private parties should, each one within his boundaries, have his own full rights. But many have not been content to confine themselves to their limits, but have laid hands on the aqueducts themselves by puncturing, here and there, the side walls of the channels, not merely those who have secured a right to draw water, but also those who misuse the occasion of the least favour for attacking the walls of the conduits. What more would not be done, were all those things not prevented by a carefully drawn law, and were not the transgressors threatened with a serious penalty? Accordingly, I append the words of the law:—

"The consul Titus Quinctius Crispinus duly put the question to the people, and the people duly passed a vote in the Forum, before the Rostra of the temple of the Deified Julius on the thirtieth day of June. The Sergian tribe was to vote first. On their behalf, Sextus Varro, the son of Lucius, cast the first vote for the following measure: Whoever, after the passage of this law, shall maliciously and intentionally pierce, break, or countenance the attempt to pierce or break, the channels, conduits, arches, pipes, tubes, reservoirs, or basins of the public waters which are brought into the City, or who shall do damage with intent to prevent water-courses, or any portions of them from going, falling, flowing, reaching, or being conducted into the City of Rome; or so as to prevent the issue, distribution, allotment, or discharge into reservoirs or basins of any water at Rome or in those places or buildings which are now or shall hereafter

SEXTUS JULIUS FRONTINUS

locis, quorum hortorum, praediorum, locorum dominis possessoribusve aqua data vel adtributa est vel erit, saliat, distribuatur, dividatur, in castella, lacus immittatur, is populo Romano HS centum milia dare damnas esto. Et qui D. M.[1] quid eorum ita fecerit, id omne sarcire, reficere, restituere, aedificare, ponere et celere demolire damnas esto sine dolo malo atque omnia ita ut[2] quicumque curator aquarum est, erit, si curator aquarum nemo erit, tum is praetor qui inter cives et peregrinos ius licet, multa, pignoribus cogito, coercito, eique curatori aut si curator non erit, tum ei praetori eo nomine cogendi, coercendi, multae dicendae sive pignoris capiendi ius potestasque esto. Si quid eorum servus fecerit, dominus eius HS centum milia populo Romano D. D. E.[3] Si qui locus circa rivos, specus, fornices, fistulas, tubulos, castella, lacus aquarum publicarum, quae ad urbem Romam ducuntur et ducentur, terminatus est et erit, ne quis in eo loco post hanc legem rogatam quid opponito, molito, obsaepito, figito, statuito, ponito, conlocato, arato, serito, neve in eum quid immittito, praeterquam eorum faciendorum, reponendorum causa, quae hac lege licebit, oportebit. Qui adversus ea quid fecerit, siremps lex, ius causaque omnium rerum omnibusque esto, atque uti esset

---

[1] *i.e.* dolo malo.
[2] *lacunam indicavit Heinrichius, deest tale quid:* curator aquarum iusserit *vel* curator aquarum e re publica fideve sua videbitur *B* ; *no gap in C.*
[3] *i.e.* dare damnas esto.

---

[1] About £850.

be adjacent to the City, or in the gardens, properties, or estates of those owners or proprietors to whom the water is now or in future shall be given or granted, he shall be condemned to pay a fine of 100,000 *sestertii*[1] to the Roman people; and in addition, whoever shall maliciously do any of these things shall be condemned to repair, restore, re-establish, reconstruct, replace what he has damaged, and quickly demolish what he has built—all in good faith and in such manner [as the commissioners may determine]. Further, whoever is or shall be water-commissioner, or in default of such officer, that praetor who is charged with judging between the citizens and strangers, is authorized to fine, bind over by bail, or restrain the offender. For that purpose, the right and power to compel, restrain, fine, and bind over, shall belong to every water-commissioner, or if there be none, to the praetor. If a slave shall do any such damage, his master shall be condemned to pay 100,000 *sestertii* to the Roman people. If any enclosure has been made or shall be made near the channels, conduits, arches, pipes, tubes, reservoirs, or basins of the public waters, which now are or in future shall be conducted into the City of Rome, no one shall, after the passage of this law, put in the way, construct, enclose, plant, establish, set up, place, plough, sow anything, or admit anything in that space unless for the purpose of doing those things and making those repairs which shall be lawful and obligatory under this law. If any one contravenes these provisions, against him shall apply the same statute, the same law, and the same procedure in every particular as could apply and ought to apply against him who in contravention of this

SEXTUS JULIUS FRONTINUS

esseve oporteret, si is adversus hanc legem rivum, specum rupisset forassetve. Quo minus in eo loco pascere, herbam, fenum secare, sentes tollere liceat, eius hac lege nihilum rogatur.[1] Curatores aquarum, qui nunc sunt quique erunt, faciunto ut in eo loco, qui locus circa fontes et fornices et muros et rivos et specus terminatus est, arbores, vites, vepres, sentes, ripae, maceria, salicta, harundineta tollantur, excidantur, effodiantur, excodicentur, uti quod recte factum esse volent; eoque nomine eis pignoris capio, multae dictio coercitioque esto; idque eis sine fraude sua facere liceat, ius potestasque esto. Quo minus vites, arbores, quae villis, aedificiis maceriisve inclusae sunt, maceriae, quas curatores aquarum causa cognita ne demolirentur dominis permiserunt, quibus inscripta insculptaque essent ipsorum qui permisissent curatorum nomina, maneant, hac lege nihilum rogatur.[2] Quo minus ex eis fontibus, rivis, specibus, fornicibus aquam sumere, haurire eis, quibuscumque curatores aquarum permiserunt, permiserint, praeterquam rota calice, machina liceat, dum ne qui puteus neque foramen novum fiat, eius hac lege nihilum rogatur."[2]

130. Utilissimae legis contemptores non negaverim dignos poena quae intenditur, sed neglegentia longi temporis deceptos leniter revocari oportuit. Itaque sedulo laboravimus ut quantum in nobis fuit, etiam

---

[1] *Mommsen; gap of about twenty letters between* sentes *and* curatores *CB*.
[2] rogator *B; the MS. is uncertain*.

---

[1] *i.e.* by the neglect of the officials to enforce the law.

statute has broken into or pierced the channel or conduit of an aqueduct. Nothing of this law shall revoke the privilege of pasturing cattle, cutting grass or hay, or gathering brambles in this place. The water-commissioners, present or future, in any place which is now enclosed about any springs, arches, walls, channels, or conduits, are authorized to have removed, pulled out, dug up, or uprooted, any trees, vines, briars, brambles, banks, fences, willow-thickets, or beds of reeds, so far as they are ready to proceed with justice; and to that end they shall possess the right to bind over, to impose fines, or to restrain the offender; and it shall be their privilege, right, and power to do the same without prejudice. As for the vines and trees inside the enclosures of country-houses, structures or fences; as to the fences, which the commissioners after due process have exempted their owners from tearing down, and on which have been inscribed or carved the names of the commissioners who gave the permission—as to all these, nothing in this enactment prevents their remaining. Nor shall anything in this law revoke the permits that have been given by the water-commissioners to any one to take or draw water from springs, channels, conduits, or arches, and besides that to use wheel, *calix*, or machine, provided that no well be dug, and that no new tap be made."

I should call the transgressor of so beneficent a law worthy of the threatened punishment. But those who had been lulled into confidence by long-standing neglect[1] had to be brought back by gentle means to right conduct. I therefore endeavoured with diligence to have the erring ones remain unknown as far as possible. Those who sought

ignorarentur qui erraverant. Eis vero qui admoniti ad indulgentiam imperatoris decucurrerunt, possumus videri causa impetrati beneficii fuisse. In reliquum vero opto ne exsecutio legis necessaria sit, cum officii fidem etiam per offensas tueri praestiterit.

the Emperor's pardon, after due warning received, may thank me for the favour granted. But for the future, I hope that the execution of the law may not be necessary, since it will be advisable for me to maintain the honour of my office even at the risk of giving offence.

INDEX OF PROPER NAMES IN THE STRATAGEMS

(*The References are to Book, Chapter, and Section.*)

ABYDENI, II. vii. 6
Abydos, a town in Asia Minor on the Hellespont, I. iv. 7
Achaia, II. iv. 4
Acilius Glabrio, II. iv. 4
Adathas, II. v. 30
Aegospotami, a town and river in the Thracian Chersonese, II. i. 18
Aegyptii, Egyptians, I. i. 5; II. iii. 13; II. v. 6
Aelii, IV. v. 14
Aelius, IV. v. 14
Aemilian Way, II. v. 39
Aemilianus, *see* Cornelius
Aemilius: (1) Aemilius Paulus, I. ii. 7; I. iv. 1; (*cf.* notes on these passages)
(2) L. Aemilius Paulus, consul 216 B.C., IV. i. 4; IV. v. 5
(3) L. Aemilius Paulus Macedonicus, II. iii. 20; III. xvii. 2; IV. vii. 3
(4) Aemilius Rufus, IV. i. 28
(5) M. Aemilius Scaurus, IV. i. 13; IV. iii. 13
Aequi, a warlike people of ancient Italy, dwelling to the north-east of Latium, II. viii. 2; III. i. 1
Aesernia, a town in Samnium, I. v. 17
Aetolia, a division of Central Greece, II. v. 19
Afranians, followers of Afranius, II. v. 38
Afranius, one of Pompey's generals, I. v. 9; I. viii. 9; II. i. 11; II. v. 38; II. xiii. 6
Afri, Africans, I. viii. 11; II. ii. 11; II. iii. 1; II. v. 12
Africa, I. ii. 3; I. iii. ; I. vii. 7;
I. xii. 1; II. ii. 10; II. iii. 10; II. iii. 16; II. v. 40; II. vii. 4; III. vi. 1
Africanus, *see* Cornelius
Agathocles, I. xii. 9
Agesilaus, king of Sparta, I. iv. 2; I. iv. 3; I. viii. 12; I. x. 3; I. xi. 5; I. xi. 17; II. vi. 6; III. xi. 2
Agrigentini, III. ii. 6; III. x. 5
Agrigentum, a city of southern Sicily, III. x. 5
Agrippa, *see* Furius
Albani, inhabitants of Alba Longa, II. vii. 1
Albania, an Asian province on the Caspian Sea, II. iii. 14
Alcetas, IV. vii. 19
Alcibiades, an Athenian general at the time of the Peloponnesian War, II. v. 44; II. v. 45; II. vii. 6; III. ii. 6; III. vi. 6; III. ix. 6; III. xi. 3; III. xii. 1
Alexander of Macedon, I. i. 3; I. iii. 1; I. iv. 9; I. iv. 9a; I. v. 11; I. vii. 7; I. xi. 14; II. iii. 19; II. v. 17; II. xi. 3; II. xi. 4; II. xi. 6; III. vii. 4; IV. ii. 4; IV. iii. 10; IV. vi. 3; IV. vii. 34
Alexander, the Epirote, II. v. 10; III. iv. 5
Alexandrini Mores, Egyptian customs, I. i. 5
Alicius, II. xi. 5
Alyattes, king of Lydia, father of Croesus, III. xv. 6
Ambiorix, a Gallic chieftain, III. xvii. 6
Ambracia, a town in Epirus, II. vii. 14
Amphipolis, a town in Macedonia, I. v. 23; III. xvi. 5

469

INDEX OF PROPER NAMES

Anaxibius, a Spartan general, I. iv. 7; II. v. 42
Antigonus, the name of several Macedonian kings: (1) Father of Demetrius, IV. i. 10
(2) A. Gonatas, III. iv. 2
(3) A. Doson, II. vi. 5
Antiochus, the name of several Syrian kings: (1) III. ii. 9; III. ix. 10
(2) A., the Great, I. viii. 7; II. iv. 4; IV. vii. 10; IV. vii. 30
(3) A. Euergetes, II. xiii. 2
Antipater, a general of Alexander of Macedon, I. iv. 13; II. xi. 4
Antium, a town in Latium, III. i. 1
Antony, I. iii. 15; I. iii. 15; I. iv. 39; II. xiii. 7; III. xiii. 7; III. xiv. 3; III. xiv. 4; IV. i. 37
Apollonides, III. iii. 5
Apennines, II. iii. 3
Appius, see Claudius
Aquae Sextiae, a Roman colony in southern Gaul, II. iv. 6
Aquilius, IV. i. 36
Arabians, II. v. 16
Arbela, a city in Assyria, II. iii. 19
Arcadia, a province in the Peloponnesus, III. ii. 7
Arcadians, I. xi. 9; III. ii. 4
Archelaus, a general of Mithridates, I. v. 18; I. xi. 20; II. iii. 17; II. viii. 12
Archidamus, a king of Sparta, I. xi. 9
Ariovistus, a German king, I. xi. 3; II. i. 16; IV. v. 11
Aristides, an Athenian, renowned for his integrity, and surnamed "the Just," IV. iii. 5
Aristippus, III. ii. 8
Aristonicus, brother of Attalus III of Pergamum, IV. v. 16
Armenia, II. v. 33; IV. i. 21; IV. i. 28
Armenia Major, II. i. 14; II. ii. 4
Armenians, I. iv. 10; II. ix. 5
Arminius, a German chieftain, II. ix. 4
Arpi, a city of Apulia, III. ix. 2
Artaxerxes Mnemon, king of Persia, II. iii. 6; IV. ii. 7
Arusian Plains, in Samnium, IV. i. 14
Asculani, III. xvii. 8
Asculum, a town in Apulia, II. iii. 21
Asia, II. ix. 8; II. xi. 3; III. xvii. 5; IV. v. 16
Ategua, a city in Spain, III. xiv. 1

Atheas, II. iv. 20
Athenians, I. iii. 6; I. iii. 9; I. iv. 7; I. iv. 13; I. iv. 13a; I. v. 7; I. v. 23; I. xi. 10; II. i. 9; II. i. 18; II. ix. 8; II. ix. 9; II. ix. 10; III. ii. 6; III. iv. 2; III. vi. 6; III. xii. 1; III. xv. 2; IV. iii. 5; IV. vii. 13
Athens, I. i. 10; IV. vii. 13
Atilius Calatinus, I. v. 15; IV. v. 10
Atilius Regulus: (1) Consul 294 B.C., II. viii. 11; IV. i. 29
(2) Consul 267 B.C., II. iii. 10; IV. iii. 3
Atrebas, a member of a Belgian tribe, II. xiii. 11
Attalus, II. xiii. 1
Augustus, see Domitian and Vespasian
Aurelius, IV. i. 31
Aurelius Cotta, IV. i. 22; IV. i. 30; IV. i. 31
Aurunculeius Cotta, a lieutenant of Caesar, III. xvii. 6
Autophradates, a Persian general, I. iv. 5; II. vii. 9

Babylon, III. vii. 4
Babylonians, III. iii. 4; III. vii. 5
Baleares, inhabitants of the Balearic Islands, famous as slingers, II. iii. 16
Bantius, III. xvi. 1
Barca, see Hamilcar
Bardylis, II. v. 19
Boeotia, II. viii. 12
Boii, a Celtic people, who early migrated to Italy, I. ii. 7; I. vi. 4
Brasidas, a distinguished Spartan general during the Peloponnesian War, I. v. 23
Britannia, II. xiii. 11
Brundisium, a seaport of southern Italy, I. v. 5
Brutiani, followers of Brutus, IV. ii. 1
Brutti, the inhabitants of the southern point of Italy, II. iii. 21
Brutus, see Junius
Byzantii, the inhabitants of Byzantium, the modern Constantinople, I. iii. 4; I. iv. 13a; III. xi. 3

Cadurci, a people of southern Gaul, III. vii. 2
Caecilius Metellus: (1) L. Caecilius Metellus, consul 251 B.C., I. vii. 1; II. v. 4

470

IN THE STRATAGEMS

(2) Q. Caecilius Metellus Macedonicus, consul 143 B.C., III. vii. 3; IV. i. 11; IV. i. 23; IV. vii. 42

(3) Q. Caecilius Metellus Numidicus, I. viii. 8; IV. i. 2; IV. ii. 2; (*cf.* note to IV. i. 11)

(4) Q. Caecilius Metellus Pius, I. i. 12; II. i. 2; II. i. 3; II. iii. 5; II. xiii. 3

Caedicius, IV. vii. 8

(Q.) Caedicius, I. v. 15; IV. v. 10

Caelii, IV. v. 14

Caelius, IV. v. 14

Caesar, *see* Julius and Domitian

Caeso, *see* Fabius

Calatinus, *see* Atilius

Callidromus, II. iv. 4

Calpurnius Flamma, I. v. 15; IV. v. 10

Calpurnius Piso: (1) Cn. Calpurnius Piso, III. v. 6

(2) L. Calpurnius Piso, IV. i. 26

Calvinus, II. ii. 1

Camalatrum, II. iv. 7

Camertes, I. ii. 2

Camillus, *see* Furius

Campania, I. ix. 1

Campanians, III. iv. 1; III. xiii. 2; IV. vii. 29

Cannae, a village in Apulia, II. ii. 7; II. iii. 7; II. v. 27; III. xvi. 1; IV. i. 44; IV. v. 5; IV. v. 7; IV. vii. 39

Cannicus, II. iv. 7; II. v. 34

Cantenna, II. v. 34

Canusium, a town in Apulia, IV. v. 7

Capitol, III. xiii. 1; III. xv. 1

Capitolinus, *see* Quintius

Cappadocia, a country in Asia Minor, I. i. 6; I. v. 18; II. ii. 2; II. vii. 9; III. ii. 9

Capua, an ancient city of Campania, IV. xviii. 3

Cardianus, from Cardia, a town of the Thracian Chersonese, IV. vii. 34

Caria, a province of Asia Minor, I. viii. 12; III. ii. 5

Carpetani, a people of central Spain, II. vii. 7

Carthage, I. xi. 4

Carthago Nova, a seaport of southeastern Spain, III. ix. 1

Carthaginienses, *passim*: I. ii. 3; I. ii. 4; I. iii. 8; I. vii. 3; I. viii. 7; II. i. 4; II. ii. 11; II. v. 11; II. v. 12; II. v. 29; II. ix. 6; III. ii. 2

Casilini, IV. v. 20

Casilinum, a town in Campania, III. xiv. 2; III. xv. 3

Cassiani, followers of Cassius, IV. ii. 1

Cassius, II. v. 35; IV. ii. 1; IV. vii. 14

Castor and Pollux, I. xi. 8; I. xi. 9

Castus, II. iv. 7; II. v. 34

Catina, a city in Sicily, III. vi. 6

Catinenses, III. vi. 6

Cato, *see* Porcius

Catulus, *see* Lutatius

Caudine Forks, a mountain pass in Samnium, I. v. 16

Celtiberians, a people of Spain, II. v. 3; II. v. 7.

Ceres, the Italian goddess of agriculture, II. ix. 9

Chabrias, a celebrated Athenian general, I. iv. 14; I. xii. 2

Chaeronea, a town in Boeotia, II. i. 9

Chalcidians, the inhabitants of Chalcis in Euboea, III. xi. 1

Chares, an Athenian general, II. xii. 3; III. x. 8

Charmades, III. ii. 11

Chatti, a people of western Germany, II. iii. 23

Chaucenses, II. xi. 2

Chersonesus, the Thracian peninsula, west of the Hellespont, I. iv. 13a; II. v. 42

Chians, the inhabitants of Chios, an island in the Aegean Sea, I. iv. 13a

Cicero, *see* Tullius

Cimbrians, a people of northern Germany, I. ii. 6; I. v. 3; II. ii. 8; II. v. 8; II. vii. 12

Ciminian Forest, in Etruria, I. ii. 2

Cimon, a distinguished Athenian general, son of Miltiades, II. ix. 10; III. ii. 5

Cineas, an Epirote, a friend of Pyrrhus, IV. iii. 2

Civilis, IV. iii. 14

Claudius: (1) Appius Claudius Sabinus, IV. i. 34

(2) Appius Claudius Caecus, IV. i. 18

(3) Appius Claudius Caudex, I. iv. 11

(4) Appius Claudius Pulcher, IV. i. 44

(5) Publius Claudius, II. xiii. 9

(6) M. Claudius Marcellus, consul 222 B.C., II. ii. 6; II. iii. 9; II. iv. 8; III. iii. 2; III. xvi. 1;

471

INDEX OF PROPER NAMES

IV. i. 44; IV. v. 4; IV. vii. 26; IV. vii. 38
(7) Claudius Marcellus, officer under Marius, II. iv. 6
(8) C. Claudius Nero, I. i. 9; I. ii. 9; I. v. 19; II. iii. 8; II. ix. 2
(9) Ti. Claudius Nero, II. i. 15
Cleandridas, II. iii. 12
Clearchus, a Spartan general in the Peloponnesian War, III. v. 1; IV. i. 17
Cleomenes, king of Sparta, II. ii. 9
Cleonymus, III. vi. 7
Clisthenes, tyrant of Sicyon, III. vii. 6
Clodius, I. v. 21
Clusium, an important town in Etruria, I. viii. 3
Cocles, see Horatius
Cominius, III. xiii. 1
Commius, II. xiii. 11
Cononeus, III. iii. 6
Consabra, IV. v. 19
Corbulo, see Domitius
Corcyreans, the inhabitants of Corcyra, an island in the Ionian Sea, I. xii. 11
Corinth, III. xii. 2; IV. iii. 15
Coriolanus, an early Roman patrician condemned to exile by the plebeian assembly. Hero of Shakespeare's play of that name, I. viii. 1
Cornelius: (1) Cornelius Cossus, 426 B.C., II. viii. 9
(2) Cornelius Cossus, consul, 343 B.C., I. v. 14; IV. v. 9
(3) Cornelius Lentulus, IV. v. 5
(4) P. Cornelius Rufinus, III. vi. 4
(5) L. Cornelius Rufinus, III. ix.4; (*cf.* note)
(6) Cornelius Scipio, legatus 297 B.C., II. iv. 2
(7) L. Cornelius Scipio, consul 259 B.C., III. x. 2; (*cf.* III. ix. 4 and note)
(8) Cn. Cornelius Scipio, II. iii. 1; IV. iii. 4; IV. vii. 9
(9) P. Cornelius Scipio Africanus Major, I. ii. 1; I. iii. 5; I. iii. 8; I. viii. 10; I. xii. 1; II. i. 1; II. iii. 4; II. iii. 16; II. v. 29; II. vii. 4; II. xi. 5; III. vi. 1; III. ix. 1; IV. vii. 30; IV. vii. 39
(10) L. Cornelius Scipio Asiaticus, IV. vii. 30

(11) P. Cornelius Scipio Aemilianus Africanus Minor, II. viii. 7; IV. i. 1; IV. i. 5; IV. iii. 9; IV. vii. 4; IV. vii. 16; IV. vii. 27
(12) P. Cornelius Scipio Nasica, IV. i. 15
(13) P. Cornelius Scipio Nasica Corculum, III. vi. 2
(14) P. Cornelius Scipio Nasica Serapio, IV. i. 20
(15) L. Cornelius Sulla, I. v. 17; I. v. 18; I. ix. 2; I. xi. 11; I. xi. 20; II. iii. 17; II. vii. 2; II. vii. 3; II. viii. 12; II. ix. 3; IV. i. 27
Cossus, see Cornelius
Cotta, see Aurelius and Aurunculeius
Crassus, see Fonteius, Licinius, Otacilius
Craterus, III. vi. 7
Crispinus, see Quintius
Crissaei, the inhabitants of Crissa, a town in Phocis, III. vii. 6
Croesus, the last king of Lydia, I. v. 4; II. iv. 12; III. viii. 3
Crotona (Croton), a Greek town in southern Italy, III. vi. 4
Crotonienses, III. vi. 4
Cubii, II. xi. 5
Cunctator, see Fabius
Curio, see Scribonius
Curius, M'. Curius Dentatus, a hero of the Roman republic, I. viii. 4; II. ii. 1; IV. iii. 12
Cursor, see Papirius
Cyprus, the island, II. ix. 10
Cyrrhestes, a native of Cyrrhestica, a province of Syria, I. i. 6
Cyrus, the Elder, king of Persia, I. xi. 19; II. v. 5; III. iii. 4; III. viii. 3
Cyrus, the Younger, IV. ii. 5; IV. ii. 7
Cyzicenes, the inhabitants of Cyzicus, III. xiii. 6
Cyzicus, a town and island north of Asia Minor, III. ix. 6; IV. v. 21

Daci, the people of Dacia, on the north bank of the Danube, I. x. 4; II. iv. 3
Dardanicum bellum, IV. i. 43
Darius, king of Persia, I. v. 25
Datames, a distinguished Persian general, II. vii. 9
Decelea, a deme in Attica, north of Athens, I. iii. 9

IN THE STRATAGEMS

Decius: P. Decius Mus, (1) consul 340 B.C., I. v. 14; IV. v. 9; IV. v. 15
(2) Consul 295 B.C., I. viii. 3; IV. v. 15
Delminus, III. vi. 2
Delphi, seat of the oracle of Apollo in Phocis, I. xi. 11
Diana, the goddess, III. ii. 5
Didius, I. viii. 5; II. x. 1
Diodotus, III. xvi. 5
Dionysius, tyrant of Syracuse, I. viii. 11; III. iv. 3; III. iv. 4
Dis, another name for Pluto, god of the underworld, I. ii. 10
Domitian, the Emperor, I. i. 8; I. iii. 10; II. iii. 23; II. xi. 7; IV. iii. 14
Domitius Calvinus, III. ii. 1
Domitius Corbulo, a distinguished Roman general, II. ix. 5; IV. i. 21; IV. i. 28; IV. ii. 3; VI. vii. 2
Duellius, winner of the first naval victory won by the Romans, I. v. 6; II. iii. 24; III. ii. 2
Duillius, I. v. 17
Dyrrhachium, a seaport town of Illyria, II. vii. 13; III. xvii. 4; IV. i. 43

Elea, a city of Aeolis in Asia Minor, IV. v. 16
Eleusis, a city of Attica, famous for the Eleusinian mysteries, II. ix. 9
Enipeus, a river of Thessaly, II. iii. 22
Epaminondas, the famous Theban general, I. xi. 6; I. xi. 16; I. xi. 5; I. xii. 6; I. xii. 7; II. ii. 12; II. v. 26; III. ii. 7; III. xi. 5; III. xii. 3; IV. ii. 6; IV. iii. 6
Ephesii, the inhabitants of Ephesus, III. iii. 7; III. ix. 10
Ephesus, a city of Ionia in Asia Minor, III. iii. 7
Ephialtes, the betrayer of the Spartans at Thermopylae, II. ii. 13
Epicydes, III. iii. 2
Epidaurus, a town in Argolis, famed for the worship of Aesculapius, II. xi. 1
Epirotes, I. iv. 4; II. iii. 21; II. v. 9; III. iii. 1; III. vi. 3
Epirotica regio, II. v. 10
Epirus, a province of northern Greece, II. xiii. 8

Erythraei, the inhabitants of Erythraea in Boeotia, II. v. 15
Etrusca lingua, I. ii. 2
Etrusci, Etruscans, I. viii. 3; I. xi. 1; II. v. 2; II. vi. 7; II. vii. 11
Etruscum bellum, I. ii. 2; I. ii. 7
Eumenes: (1) the Cardian, IV. vii. 34
(2) king of Bithynia, I. xi. 15
Euphrates, the river, I. i. 6; III. vii 4; III. vii. 5
Europe, II. vi. 8
Eurymedon River, II. ix. 10

Fabius: (1) Q. Fabius Maximus Rullus, I. ii. 2; I. viii. 3; I. xi. 21; II. i. 8; II. iv. 2; II. v. 2; IV. i. 35; IV. i. 39
(2) Fabius Caeso, I. ii. 2
(3) Q. Fabius Maximus Servilianus, IV. i. 42
(4) Q. Fabius Maximus Cunctator, I. iii. 3; I. v. 28; I. viii. 2; I. xi. 4; II. v. 22; III. iv. 1; III. ix. 2; IV. v. 1; IV. vii. 36
(5) Q. Fabius Maximus, son of No. 4, III. ix. 2; IV. vi. 1
(6) M. Fabius Vibulanus, I. xi. 1; II. vi. 7; II. vii. 11
Fabricius, a hero of the Roman republic, IV. iii. 2; IV. iv. 2
Falisci, the inhabitants of Falerii in Etruria, II. iv. 18; II. v. 9; II. viii. 3; II. viii. 8; IV. iv. 1
Fidenae, a town in the Sabine territory, north of Rome, II. v. 1
Fidenates, II. iv. 19; II. viii. 9
Fimbria, a partisan of Marius and Cinna, III. xvii. 5
Flaccus, see Fulvius
Flaminius, a Roman general, defeated and slain by Hannibal, II. v. 24
Flamma, see Calpurnius
Fonteius Crassus, I. v. 12; IV. v. 8
Forum Gallorum, a town of Cisalpine Gaul, II. v. 39
Fulvius: (1) Cn. Fulvius, II. v. 9
(2) Cn. Fulvius Centumalus, I. viii. 3
(3) Q. Fulvius Flaccus, consul 212 B.C., IV. i. 44
(4) Cn. Fulvius Flaccus, consul 211 B.C., II. v. 27; IV. vii. 29
(5) Q. Fulvius Flaccus, censor 174 B.C., II. v. 8 (*cf.* note); IV. i. 32

473

INDEX OF PROPER NAMES

(6) Cn. Fulvius Flaccus, brother of No. 5, IV. i. 32
(7) Fulvius Nobilior, I. vi. 1; I. vi. 2; I. xi. 2
Furius, I. i. 11; I. v. 13
Furius Agrippa, II. viii. 2
M. Furius Camillus: (1) dictator 396 B.C., II. iv. 15; II. viii. 4; II. viii. 6; III. xiii. 1; III. xv. 1; IV. iv. 1; IV. vii. 40
(2) Son of No. 1, II. vi. 1

Gabii, a town in Latium, III. iii. 3
Gabini, the inhabitants of Gabii, I. i. 4; III. iii. 3
Galli, the Gauls, I. i. 6; I. viii. 3; I. xi. 15; II. i. 8; II. iii. 16; II. iii. 18; II. iv. 5; II. iv. 7; II. v. 20; II. v. 34; II. vi. 1; II. xiii. 1; III. xiii. 1; III. xv. 1; III. xvi. 2; III. xvi. 3; III. xvii. 6; III. xvii. 7; IV. v. 4
Gallia, I. i. 8; II. i. 16; II. xiii. 11; III. vii. 2; III. xvii. 6; IV. iii. 14
Gallic Way, II. vi. 1
Gallus, see Sulpicius
Gastron, II. iii. 13
Gelo, tyrant of Syracuse, I. xi. 18
Germani, the Germans, I. i. 8; I. iii. 10; I. xi. 3; II. i. 16; II. v. 20; II. v. 34; II. vi. 3; II. ix. 4; IV. v. 11; IV. vii. 8
Germanicum bellum, IV. iii. 14
Germanicus, see Domitian
Germany, IV. vii. 8
Gisgo, see Hasdrubal
Glabrio, see Acilius
Glaucia, IV. v. 1
Gracchus, see Sempronius
Graeca urbs, III. v. 3
Graeci, the Greeks, Pref. to Bk. I., II. ii. 14; II. iii. 6; II. iii. 13; III. ii. 6; IV. ii. 7; IV. ii. 8; IV. vii. 22
Graecinus, II. v. 31
Greece, I. iv. 6; II. ii. 14
Halys, a river in Asia Minor, I. v. 4
Hamilcar, the name of several Carthaginian generals: (1) III. xvi. 2
(2) Hamilcar Rhodinus, I. ii. 3
(3) Hamilcar Barca, II. iv. 17; III. x. 9
Hannibal, the name of several Carthaginian generals: (1) son of Gisgo, III. x. 3

(2) Hannibal, leader in First Punic War, IV. i. 19
(3) Hannibal, son of Hamilcar Barca, I. i. 9; I. ii. 9; I. iii. 3; I. iii. 8; I. v. 19; I. v. 28; I. vii. 2; II. viii. 2; I. viii. 7; I. viii. 10; II. ii. 6; II. ii. 7; II. iii. 7; II. iii. 9; II. iii. 16; II. v. 13; II. v. 21; II. v. 22; II. v. 23; II. v. 24; II. v. 25; II. v. 27; II. vi. 4; II. vii. 7; II. ix. 2; III. ii. 3; III. iii. 6; III. vi. 1; III. ix. 2; III. x. 3; III. x. 4; III. xiv. 2; III. xv. 3; III. xvi. 1; III. xvi. 4; III. xviii. 1; III. xviii. 2, III. xviii. 3; IV. iii. 7; IV. iii. 8; IV. v. 20; IV. vii. 10; IV. vii. 25; IV. vii. 26; IV. vii. 38
Hanno, the name of several Carthaginian generals: (1) I. v. 27
(2) Hanno, 261 B.C., III. xvi. 3
(3) Hanno, 218 B.C., II. iii. 1
Harrybas, king of the Molossians, a people in Epirus, II. v. 19
Hasdrubal, the name of several Carthaginian generals: (1) H., brother of Hannibal, I. i. 9; I. ii. 9; I. v. 19; II. iii. 8; II. ix. 2; II. vii. 15
(2) H., son of Gisgo, I. iii. 5; II. i. 1; II. iii. 4
(3) Others of this name : I. v. 12; II. v. 4; III. xvii. 1; III. xvii. 3; IV. iii. 8; IV. v. 8; IV. vii. 18
Hellespont, the modern Dardanelles, I. iv. 7; II. v. 44
Henna, a city of central Sicily, IV. vii. 22
Heraclea, III. x. 3
Hermocrates, a Syracusan general, II. ix. 6; II. ix. 7
Hernici, a people of Latium, II. viii. 2
Himera, a city of northern Sicily, III. x. 3
Himeraeans, III. iv. 4; III. x. 3
Himilco, a Carthaginian general, I. i. 2; III. x. 5
Hippias, tyrant of Athens, son of Pisistratus, II. ii. 9
Hirtius, consul 43 B.C., III. xiii. 7; III. xiii. 8; III. xiv. 3; III. xiv. 4
Hirtuleius, a lieutenant of Sertorius in Spain, I. v. 8; II. i. 2; II. iii. 5; II. vii. 5; IV. v. 19

474

Hispani, the Spaniards, II. iii. 1; II. iv. 17; II. v. 31; II. x. 1; IV. v. 19
Hispania, Spain, I. i. 1; I. i. 12; I. ii. 5; I.iii.5; I. v. 1; I. v. 8; I. v. 12; I. v. 19; II. i. 1; II. i. 2; II. i. 3; II.iii.1; II. ii. 4; II. iii. 5; II. iii. 11; II. v. 14; II. v. 31; II. v. 32; II. ix. 2; II. x. 1; II. xi. 5; II. xii. 2; II. xiii. 6; III. i. 2; III. vii. 3; III. xiv. 1; III. xviii. 1; IV.i.23; IV. iii. 4; IV. v. 8; IV. vii. 31; IV. vii. 42
Hispaniae, II. x. 2
Homericus versus, II. iii. 21
Horatius Cocles, the defender of the bridge against the Etruscans under Porsenna, II. xiii. 5
Hostilius (Tullus H.), third king of Rome, II. vii. 1
Hydaspes, a river in India, I. iv. 9
Hyllii, the Illyrians, II. iii. 2

Iapydes, a people of Illyria, II. v. 28
Ilerda, a town in Spain, II. v. 38; II. xiii. 6
Ilergetes, a people of Spain, IV. vii. 31
Illyrii, the Illyrians, a people on the eastern shore of the Adriatic, II. v. 10; II. v. 19; III. vi. 3
Indi, the people of India, I. iv. 9
Indibile, a town in Spain, II. iii. 1
Indus, the river in India, I. iv. 9a
Initia, a fortress in Armenia, IV. i. 21
Iphicrates, a famous Athenian general, I. iv. 7; I. v. 24; I. vi. 3; II. i. 5; II. i. 6; II. v. 42; II. xii. 4; III. xii. 2; IV. vii. 23
Isaura, a town in Asia Minor, III. vii. 1
Isthmos (of Corinth), II. v. 26
Italici, the Italians, II. iii. 16; II. iii. 17
Italy, I. iii. 2; I. iii. 8; I. iv. 11; I. v. 5; *et passim*

Juba, a Numidian king, II. v. 40
Judaei, the Jews, II. i. 17
Jugurtha, a Numidian king, I. viii. 8; II. i. 13; II. iv. 10
Jugurthinum bellum, III. ix. 3; IV. i. 2
Julius Caesar, I. i. 5; I. iii. 2; I. v. 5; I. v. 9; I. viii. 9; I. ix. 4; I. xi. 3; I. xii. 2; II. i. 11; II. i. 16; II. iii. 2; II. iii. 18; II. iii. 22; II. v. 20; II. v. 38; II. vi. 3; II. vii. 13; II. viii. 13; II. xiii. 6; II. xiii. 11; III. vii. 2; III. xiv. 1; III. xvii. 4; III. xvii. 6; IV. v. 2; IV. v. 11; IV. vii. 1; IV. vii. 32
Julius Civilis, IV. iii. 14
Junius: (1) D. Junius Brutus, consul, 138 B.C., IV. i. 20
(2) D. Junius Brutus, consul 43 B.C., III. xiii. 7; III. xiii. 8
(3) M. Junius Brutus, IV. ii. 1
(4) M. Junius Pera, II. v. 25
Jupiter, I. xii. 12

Kaeso, see Fabius

Laberius, I. v. 15; IV. v. 10
Labienus: (1) T. Labienus, a lieutenant of Caesar, II. v. 20; II. vii. 13
(2) T. Labienus, son of No. 1, II. v. 36
Lacedaemon, Sparta, I. i. 10; III. xi. 5; IV. vii. 13
Lacedaemonii, the Lacedaemonians, I. i. 10; I. iii. 7; I. iii. 9; I. iv. 2; I.iv.12; I. xi. 6; I. xi. 7; I. xi. 16; I. xii. 5; I. xii. 7; II. i. 6; II. i. 10; II. ii. 12; II. ii. 13; II. v. 26; II. v. 42; II. v. 47; II. x. 4; II. xi. 5; III. xii. 1; III. xv. 2; IV. ii. 6; IV. ii. 9; IV. v. 12; IV. vii.13
Lacetani, a people of Spain, III. x. 1
Laelii, IV. v. 14
Laelius: (1) C. Laelius, the friend of Scipio Africanus, the elder, I. i. 3; I. ii. 1; II. iii. 1
(2) D. Laelius, an adherent of Pompey, II. v. 31
(3) IV. v. 14
Laenas, see Popilius
Laevinus, see Valerius
Latin League, IV. vii. 25
Latina lingua, II. iv. 10
Latini, the people of Latium, I. xi. 8; II. viii. 4
Latinum nomen, IV. vii. 25
Lauron, a town in Spain, II. v. 31
Lentulus, see Cornelius
Leonidas, the hero of Thermopylae, IV. v. 13
Leotychidas, a Spartan admiral, I. xi. 7
Leptines, II. v. 15
Leucadia, a town and island in the Ionian Sea, III. iv. 5
Liburni, an Illyrian people, II. v. 43

475

INDEX OF PROPER NAMES

Licinius: (1) P. Licinius, proconsul, II. v. 28
  (2) P. Licinius Crassus Mucianus, IV. v. 16
  (3) P. Licinius Crassus, II. iv. 16; IV. vii. 41
  (4) M. Licinius Crassus, I. i. 13; I. v. 20; II. iv. 7; II. v. 34
  (5) Licinius Lucullus, III. x. 7
  (6) L. Licinius Lucullus, II. i. 14; II. ii. 4; II. v. 30; II. vii. 8; III. xiii. 6
Ligures, I. ii. 6; I. v. 16; I. v. 26; II. iii. 16; III. ii. 1; III. ix. 3; III. xvii. 2; IV. i. 46
Liguria, a district of Cisalpine Gaul, I. v. 16
Lilybaeum, a promontory of western Sicily, III. x. 9
Lingones, a people of Gaul, IV. iii. 14
Liparae Insulae, Liparian Islands, north of Sicily, IV. i. 31
Litana Forest, in Cisalpine Gaul, I. vi. 4
Livius: (1) Livy, the historian, II. v. 31; II. v. 34
  (2) M. Livius Macatus, III. iii. 6; III. xvii. 3
  (3) M. Livius Salinator, I. i. 9; I. ii. 9; II. iii. 8; IV. i. 45; IV. vii. 15
Locri, (Epizephyrii) a Greek settlement in southern Italy, IV. vii. 26
Longus, see Sempronius
Lucani, the Lucanians, a people of southern Italy, I. iv. 1; I. vi. 1; II. iii. 12; II. iii. 21; III. vi. 4
Luceria, a city in Apulia, IV. i. 29
Lucullus, see Licinius
Lueria, III. ii. 1
Lusitani, II. xiii. 4; III. v. 2
Lusitania, a country of western Spain, I. xi. 13
Lutatius Catulus, consul 102 B.C., I. v. 3
Lydia, a district of Asia Minor, I. viii. 12; IV. vii. 30
Lysander, a distinguished Spartan general, I. v. 7; II. i. 18; IV. i. 9
Lysimachus, one of Alexander's generals, I. v. 11; III. iii. 7

Macedones, I. iv. 6; I. iv. 13; II. iii. 2; II. iii. 16; II. iii. 17; II. iii. 20; II. vii. 8; III. iii. 7

Macedonia, IV. ii. 1; IV. vi. 3
Macedonicus habitus, III. ii. 11
Magnetes, the Magnesians, a people of Thessaly, III. viii. 2
Mago, the name of several Carthaginian generals: (1) brother of Hannibal, II. v. 23
  (2) another, III. vi. 5; IV. vii. 26
Maharbal, a Carthaginian, II. v. 12
Maleventum, IV. i. 14 (cf. note)
Mandro, III. iii. 7
Manlius: (1) A. Manlius, I. ix. 1
  (2) Cn. Manlius, I. xi. 1; II. vi. 7; II. vii. 11
  (3) Manlius Imperiosus, IV. i. 40
  (4) son of No. 3, IV. i. 41
  (5) A. Manlius Torquatus, III. v. 3
Mantinea, a city of Arcadia in the Peloponnesus, III. xi. 5
Marathon, a village in Attica, II. ix. 8
Marcellus, see Claudius
Marcius: (1) T. Marcius, II. vi. 2; II. x. 2
  (2) Marcius Coriolanus, I. viii. 1
  (3) Q. Marcius Rufus, II. iv. 7
Mariani muli, IV. i. 7
Marius, I. ii. 6; I. xi. 12; II. ii. 8; II. iv. 6; II. iv. 10; II. vii. 12; II. ix. 1; III. ix. 3; IV. i. 7; IV. ii. 2; IV. vii. 5
Marte aequo, II. iii. 13; II. iv. 1; II. v. 5
Masinissa, king of Numidia, II. iii. 16; III. vi. 1; IV. iii. 11
Massilienses, the inhabitants of Massilia, now Marseilles, I. vii. 4
Mauri, the Moors, II. iii. 16; III. xiv. 1
Maximus, see Fabius
Mediolanum, modern Milan, I. ix. 3
Megara, a city of Megaris in central Greece, IV. i. 8
Megarenses, II. ix. 9
Melanthus, II. v. 41
Memmius, IV. i. 1
Memnon, the Rhodian, II. v. 18; II. v. 46
Messana, a city of north-eastern Sicily, I. iv. 11; IV. i. 31
Messenii, the inhabitants of Messenia in the Peloponnesus, II. i. 10; III. ii. 4
Metellus, see Caecilius
Milesii, the inhabitants of Miletus in Asia Minor, III. ix. 7; III. xv. 6

IN THE STRATAGEMS

Milo, III. iii. 1
Miltiades, the celebrated Athenian general, II. ix. 8
Mindarus, II. v. 44
Minerva, the goddess, III. ii. 8; IV. vii. 13
Minucius: (1) M. Minucius Rufus, II. v. 22
(2) Minucius Rufus, II. iv. 3
(3) Q. Minucius Thermus, I. v. 16; I. viii. 10
Mithridates, king of Pontus, I. i. 7; I. v. 18; I. xi. 20; II. i. 12; II. i. 14; II. ii. 2; II. ii. 3; II. ii. 4; II. v. 30; II. v. 33; III. xiii. 6; III. xvii. 5; IV. v. 21
Mithridaticus exercitus, II. viii. 12
Molossi, a people of Epirus, II. v. 19
Mulucha, a river in northern Africa, III. ix. 3
Mummius, the general who destroyed Corinth, IV. vii. 3
Munda, a city in Spain, II. viii. 13
Munychia, one of the three Athenian harbours, I. v. 7
Mutina, a city of Cisalpine Gaul, I. vii. 5; III. xiii. 7
Mutinenses, III. xiv. 3
Myrina, a city of Aeolis in Asia Minor, IV. v. 16
Myronides, II. iv. 11; IV. vii. 21

Naevius, centurion, IV. vii. 29
Nasica, *see* Cornelius
Nautius, II. iv. 1
Nero, *see* Claudius
Nicostratus, king of the Aetolians, I. iv. 4
Nobilior, *see* Fulvius
Nolani, the inhabitants of Nola, a city in Campania, III. xvi. 1
Numantia, a city in Spain, II. viii. 7; IV. i. 1; IV. vii. 27
Numantini, III. xvii. 9; IV. v. 23
Numidae, Numidians, I. v. 16; II. ii. 11; II. iii. 16; II. v. 23; II. v. 27; II. v. 29; II. v. 40; III. vi. 1; IV. vii. 18
Numidia, a country of northern Africa, II. iv. 10; IV. vii. 18
Numistro, a city in Lucania, II. ii. 6

Oceanus, the British Channel, II. xiii. 11
Octavius Graecinus, II. v. 31

(Cn.) Octavius, IV. v. 7
Orchomenos, a city in Boeotia, I. xi. 5
Orestes, son of Agamemnon, I. ii. 8
Osaces, a Parthian general, II. v. 35
Otacilius Crassus, III. xvi. 3; IV. i. 19

Paches, an Athenian general, IV. vii. 17
Pacorus, a Parthian general, I. i. 6
Palaepharsalus, II. iii. 22
Pammenes, a Theban general, II. iii. 3
Pamphylia, a country of Asia Minor, II. ix. 10
Pannonii, the people of Pannonia, a country north of Illyricum, II. i. 15
Panormitani, III. xvii. 1
Panormus, the modern Palermo, II. v. 4
Pansa, II. v. 39
Papirius: (1) L. Papirius Cursor, dictator 325 B.C., IV. i. 39
(2) son of No. 1, II. iv. 1; III. iii. 1
Papus, *see* Aemilius
Parthi, the Parthians, an Asiatic people, I. i. 6; II. ii. 5; II. iii. 15; II. v. 35; II. v. 36; II. v. 37; II. xiii. 7; IV. ii. 3
Paulus, *see* Aemilius
Pelignus, a Pelignian, a man of Sabine descent, II. viii. 5
Pelopidas, a celebrated Theban general, I. v. 2; III. viii. 2; IV. vii. 28
Peloponnesii, the Peloponnesians, I. v. 10; II. xi. 4; III. ix. 9
Peloponnesus, the southern division of Greece, I. iii. 9; IV. v. 26
Perdiccas, a general of Alexander the Great, IV. vii. 20
Pericles, a great Athenian statesman, I. iii. 7; I. v. 10; I. xi. 10; I. xii. 10; III. ix. 5; III. ix. 9
Perperna, the murderer of Sertorius, II. v. 32
Persae, the Persians, I. xi. 17; II. ii. 13; II. iii. 3; II. iii. 13; II. ix. 8; II. ix. 10; IV. ii. 5; IV. v. 13; IV. vi. 3.
Perses, Perseus, king of Macedonia, II. iii. 20
Persicum bellum, II. viii. 5
Persicus habitus, III. viii. 3
Persis, Persia, II. iii. 6
Peticus, *see* Sulpicius

477

INDEX OF PROPER NAMES

Petilini, the inhabitants of Petelia, a town in southern Italy, IV. v. 18
Petilius, IV. i. 46
Petreius, one of Pompey's generals, I. viii. 9; II. i. 11
Phalaris, a tyrant of Agrigentum, III. iv. 6
Pharnabazus, a Persian satrap, II. vii. 6
Pharnaces, son of Mithridates, II. ii. 3
Pharnaeus, I. i. 6
Pharnastanes, a Parthian general, II. v. 37
Pharsalica pugna, II. vii. 13.
Philip: (1) of Macedon, father of Alexander the Great, I. iii. 4; I. iv. 13; I. iv. 13a; II. i. 9; II. iii. 2; II. viii. 14; III. iii. 5; IV. ii. 4; IV. v. 12; IV. vii. 37
(2) son of Demetrius, I. iv. 6; II. xiii. 8; III. viii. 1
(3) III. ix. 8; IV. i. 6
Phocenses, the inhabitants of Phocaea, a city of Ionia, III. xi. 2
Phormion, a celebrated Athenian general, III. xi. 1
Phrygia, a district in Asia Minor, I. iv. 2
Picentes, the inhabitants of Picenum, a district in central Italy, I. xii. 3
Pinarius, IV. vii. 22
Piraeus, the largest seaport of Athens, I. xi. 20
Pisidia, a country of Asia Minor, I. iv. 5
Pisistratus, a tyrant of Athens, II. ix. 9
Piso, see Calpurnius
Pius, see Metellus
Poeni, the Carthaginians, I. i. 9; I. iv. 11; I. iv. 12; I. viii. 6; I. xi. 4; I. xi. 18; I. xii. 9; II. i. 1; II. i. 4; II. ii. 11; II. iii. 4; II. iii. 16; II. iv. 14; II. v. 4; II. v. 29; II. vi. 2; II. ix. 2; II. x. 2; II. xiii. 9; II. xiii. 10; III. iii. 6; III. x. 3; III. x. 9; III. xiii. 2; IV. i. 44; IV. v. 18; IV. vii. 12; IV. vii. 22; IV. vii. 38
Pollux, brother of Castor, I. xi. 8; I. xi. 9
Pompeianae partes, Pompeiani, followers of Pompey, II. v. 31; II. vii. 13; III. xiv. 1; IV. vii. 32
Pompeius: (1) Cn. Pompeius Strabo, III. xvii. 8

(2) Cn. Pompeius Magnus, I. i. 7; I. iv. 8; I. v. 5; I. ix. 3; II. i. 3; II. i. 12; II. ii. 2; II. iii. 11; II. iii. 14; II. iii. 22; II. v. 31; II. v. 32; II. v. 33; II. xi. 2; III. xvii. 4; IV. v. 1
(3) Cn. Pompeius, son of No. 2, III. xiv. 1
Pomptinius, II. iv. 7
Pomptinus ager, a marshy district in Latium, II. vi. 1
Pontici, soldiers of Mithridates, II. i. 12
Pontius Cominius, III. xiii. 1
Popilius Laenas, III. xvii. 9
Porcius Cato: (1) the Censor, I. i. 1; I. i. 5; II. iv. 4; II. vii. 14; III. i. 2; III. x. 1; IV. i. 16; IV. i. 33; IV. iii. 1; IV. vii. 12; IV. vii. 31; IV. vii. 35
(2) son of No. 1, IV. v. 17
Porsenna, an Etruscan king, II. xiii. 5
Porus, a king of India, I. iv. 9
Postumius: (1) Aulus Postumius, consul 496 B.C., I. xi. 8
(2) L. Postumius Megellus, I. viii. 3
(3) L. Postumius, II. i. 4
(4) Postumius, consularis, IV. v. 3
Praeneste, a town in Latium, II. ix. 3
Prinassum, III. viii. 1
Priscus, see Servilius, Tarquitius
Prusias, king of Bithynia, IV. vii. 10
Ptolomaeus: (1) Ptolemy Soter, IV. vii. 20
(2) Ptolemy Ceraunus, III. ii. 11
Punica acies, II. iii. 1; classis, II. iii. 24; navis, I. iv. 12
Punicum bellum, I. iii. 5; I. iv. 11; I. viii. 6; II. ii. 11; II. iii. 8; IV. i. 25
Punicum vallum, III. x. 3
Pyrrhus, king of Epirus, II. ii. 1; II. iii. 21; II. iv. 9; II. iv. 13; II. vi. 9; II. vi. 10; III. vi. 3; IV. i. 3; IV. i. 14; IV. i. 18; IV. iv. 2
Pythias, IV. vii. 37

Quintius: (1) L. Quintius, II. v. 34
(2) T. Quintius Capitolinus, II. vii. 10; II. viii. 3 (cf. note); II. xii. 1; III. i. 1
(3) T. Quintius Crispinus, IV. vii. 38
(4) Quintius Crispinus, VI. vii. 26

IN THE STRATAGEMS

Regini, the inhabitants of Rhegium, III. iv. 3
Regulus, *see* Atilius
Rhegium, a Greek town in southern Italy, I. iv. 11; IV. i. 38
Rhodii, the inhabitants of the island Rhodes, I. v. 13a; I. vii. 4; III. ix. 10
Rhodnus, *see* Hamilcar
Rhyndacus, a river in Asia Minor, III. xvii. 5
Romani, Romans, *passim*: I. vi. 4; I. viii. 1; I. viii. 3; I. viii. 6; I. viii. 7; I. x. 1; II. i. 13; II. iii. 16; III. ii. 3; III. xiii. 1; III. xiii. 2; III. xv. 1; III. xv. 4; III. xvi. 2; III. xvii. 1; III. xviii. 1; III. xviii. 2; III. xviii. 3; IV. i. 14; IV. vi. 6
Romanum nomen, I. ii. 2
Rome, I. ii. 4; IV. i. 39
Romulus, founder of Rome, II. v. 1
Rufinus, *see* Cornelius
Rufus, *see* Aemilius, Marcius, Minucius
Rullus, *see* Fabius
Rutilius Rufus, a Roman statesman, IV. i. 12; IV. ii. 2

Sabboras, II. v. 40
Sabini, the Sabines, a people of central Italy, I. viii. 4; II. viii. 1; II. viii. 10; IV. iii. 12
Sabinus, *see* Titurius
Saepinum, a town of the Samnites, IV. i. 24
Saguntini, the inhabitants of Saguntum in Spain, III. x. 4
Salamis, an island off the west coast of Attica, II. ii. 14
Salapia, a city in Apulia, IV. vii. 38
Salinator, *see* Livius
Sallentini, a people in south-eastern Italy, II. iii. 21
Salvius, II. viii. 5
Samii, the inhabitants of Samos, an island in the Aegean Sea, I. iv. 14
Samnites, the people of Samnium, I. viii. 3; I. xi. 2; II. i. 8; II. iii. 21; II. iv. 1; II. iv. 2
Samniticum bellum, I. v. 14; II. viii. 11; IV. v. 9
Samnium, a country in south central Italy, I. vi. 1; II. iv. 2; IV. i. 29
Sanii, III. ii. 11; III. iii. 5
Sardes, Sardis, the capital of Lydia, III. viii. 3
Sardinia, III. ix. 4; III. x. 2

Saturni dies, II. i. 17
Scaurus, *see* Aemilius
Scipio, *see* Cornelius
Scipiones, II. vi. 2
Scordisci, a people on the borders of Illyria, II. iv. 3; III. x. 7
Scorylo, a Dacian chieftain, I. x. 4
Scribonius Curio: (1) consul 76 B.C., IV. i. 43
 (2) in Civil War, IV. v. 40
Scultenna River, in Cisalpine Gaul, III. xiii. 7; III. xiv. 3
Scythae, the people of Scythia, north of the Black and Caspian Seas, I. v. 25; II. iv. 20; II. v. 5; II. viii. 14
Segobrigenses, a people of Spain, III. x. 6; III xi. 4
Segovienses, a people of Spain, IV. v. 22
Semiramis, the famous Assyrian queen, III. vii. 5
Sempronius: (1) Ti. Sempronius Gracchus, IV. vii. 24
 (2) T. Sempronius Gracchus, I. xii. 3 (*cf.* note)
 (3) Ti. Sempronius Longus, II. v. 23
 (4) P. Sempronius Tuditanus, IV. v. 7
 (5) Ti. Sempronius Gracchus, II. v. 3; II. v. 14; III. v. 2; IV. vii. 33
Sentinas ager, the country near Sentinum in Umbria, I. viii. 3
Sertorius, a general of Marius, I. v. 1; I. v. 8; I. x. 1; I. x. 2; I. xi. 13; I. xii. 4; II. i. 3; II. iii. 11; II. v. 31; II. vii. 5; II. xii. 2; II. xiii. 3; II. xiii. 4; IV. vii. 6
Servilius: (1) Q. Servilius Priscus, II. viii. 8
 (2) P. Servilius Vatia, III. vii. 1; IV. v. 1
Servius Tullius, sixth king of Rome, II. viii. 1
Sicily, I. i. 2; I. iv. 11; I. viii. 11; II. i. 4; II. v. 4; II. vii. 4; III. iv. 6; III. vi. 6; III. x. 9; III. xvi. 3; IV. i. 25; IV. i. 30; IV. i. 44; IV. vii. 22
Siculum fretum, I. vii. 1
Sicyonii, the inhabitants of Sicyon, a city in the Peloponnesus, III. ii. 10; III. vii. 6; III. ix. 7
Siris, a river in southern Italy, IV. i. 24
Sosistratus, III. iii. 2

479

INDEX OF PROPER NAMES

Sparta, III. xi. 5
Spartacus, a Thracian gladiator, I. v. 20; I. v. 21; I. v. 22; I. vii. 6; II. v. 34
Spurius Nautius, II. iv. 1
Statilius, IV. vii. 36
Statorius, I. i. 3
στενά, the Hellespont, I. iv. 13
Sudines, a soothsayer, I. xi. 15
Suenda, a town in Cappadocia, III. ii. 9
Suessetani, a people of Spain, III. x. 1
Sulla, *see* Cornelius
Sulpicius: (1) C. Sulpicius Gallus, I. xii. 8
(2) C. Sulpicius Peticus, II. iv. 5
Superbus, *see* Tarquinius
Sutrini, the inhabitants of Sutrium in Etruria, II. v. 2
Syphax, a Numidian king, I. i. 3; I. ii. 1; II. v. 29; II. vii. 4
Syracusani, Syracusans, I. viii. 11; II. v. 11; II. ix. 7; III. iii. 2; III. vi. 6
Syracusanus portus, I. v. 6
Syracuse, a celebrated city in Sicily, I. iv. 12; I. viii. 11; I. xi. 18
Syria, I. xi. 12; II. v. 35; II. xiii. 2

Tarentini, the inhabitants of Tarentum, I. v. 1; II. iii. 21; II. iv. 13; III. iii. 1; III. iii. 6; III. xvii. 3
Tarentum, a city in southern Italy, III. iii. 1; III. iii. 6
Tarpeia Saxa, the Tarpeian rock on the Capitoline Hill, named from Tarpeia, who betrayed the Roman citadel to the Sabines, III. xiii. 1
Tarquinienses, the inhabitants of Tarquinii in Etruria, II. iv. 18
Tarquinius Superbus, the fifth king of Rome, I. i. 4; II. viii. 1; II. viii. 10; III. iii. 3
(S.) Tarquinius, the last king of Rome, I. i. 4; III. iii. 3
Tarquitius Priscus, II. v. 31
Taurus Mts., I. i. 6
Tegea, a city of Arcadia in the Peloponnesus, III. ii. 8
Terentius Varro, the Roman general defeated by Hannibal at Cannae, IV. i. 4; IV. v. 6
Teutoni, a people of Germany, II. ii. 8; II. iv. 6; II. vii. 12; II. ix. 1; IV. vii. 5

Teutonicum bellum, I. ii. 6
Thamyris, a Scythian queen, II. v. 5
Theagenes, a despot of Megara, IV. i. 8
Thebani, Thebans, I. iv. 3; I. x. 3; I. xi. 6; II. iv. 11; II. vi. 6; III. ii 10; IV. ii. 6; IV. vii. 19; IV. vii. 21
Thebes, a city of Boeotia, I. iv. 3; I. xi. 6
Themistocles, the famous Athenian statesman, I. i. 10; I. iii. 6; II. ii. 14; II. vi. 8
Thermopylae, a celebrated mountain pass, leading from Thessaly to Locris, I. iv. 6; II. ii. 13; II. iv. 4; IV. ii. 9
Thermus, *see* Minucius
Thessali, the inhabitants of Thessaly, a division of northern Greece, IV. vii. 28
Thessalicum bellum, I. v. 2
Thrace, a district north of Greece, I. iv. 13; I. v. 24; I. vi. 3; II. xi. 3
Thraces, Thracians, II. i. 4; III. v. 1; III. xv. 5; III. xvi. 5; IV. v. 16
Thrasybulus, tyrant of Miletus, III. ix. 7; III. xv. 6
Tiber, the river, II. vi. 1; III. xiii. 1
Tiberius, *see* Claudius
Tigranes, a king of Armenia, II. i. 14; II. ii. 4
Tigranocerta, a city in Armenia, II. i. 14; II. ii. 4; II. ix. 5
Timarchus, III. ii. 11
Timotheus, an Athenian general, I. xii. 11; II. v. 47
Tisamenus, son of Orestes, I. ii. 8
Tissaphernes, a Persian satrap, I. viii. 12
Titius, IV. i. 26
Titurius Sabinus, a lieutenant of Caesar, III. xvii. 6; III. xvii. 7
Torquatus, *see* Manlius
Trachinius, a native of Trachis in Thessaly, II. ii. 13
Trasumenus, a lake in Etruria, II. v. 24; II. vi. 4; IV. vii. 25
Trebia, a river of Cisalpine Gaul, II. v. 23
Triballi, a Thracian people, II. iv. 20
Tridentinus saltus, a pass of the Alps, IV. i. 13
Troezen, a town of Argolis, I. iii. 6
Troezenii, the inhabitants of Troezen, III. vi. 7

480

Tryphon, a king of Syria, II. xiii. 2
Tuditanus, see Sempronius
Servius Tullius, the sixth king of Rome, II. viii. 1
(Q) Tullius Cicero, brother of the orator, III. xvii. 6
Tullus Hostilius, see Hostilius
Tuscus habitus, I. ii. 2

Umbria, a district in north-eastern Italy, I. i. 9
Umbri, Umbrians, I. viii. 3

Vadandus, II. ix. 5
Valerius : (1) P. Valerius, II. xi. 1
 (2) P. Valerius Laevinus, II. iv. 9; IV. i. 24; IV. vii. 7
 (3) Valerius, IV. i. 30
Variana clades, III. xv. 4; IV. vii. 8
Varinius, I. v. 22
Varro, see Terentius
Veientes, inhabitants of Veii, II. iv. 19; II. vii. 1
Veii, an ancient city of Etruria, III. xiii. 1
Ventidius, a Roman general, famous for his victory over the Parthians, I. i. 6; II. ii. 5; II. v. 36; II. v. 37

Verginius, II. i. 7
Vespasian, the Emperor, II. i. 17; IV. vi. 4
Vesuvius, I. v. 21
Vetulonia, an ancient city of Etruria, I. ii. 7
Vibius Pansa, consul 43 B.C., II. v. 39
Viriathus, a celebrated Spanish general, II. v. 7; II. xiii. 4; III. x. 6; III. xi. 4; IV. v. 22
Voccaei, IV. vii. 33
Volsci, a people of Latium, II. i. 7; II. iv. 15; II. viii. 4; II. xii. 1; III. i. 1; IV. vii. 40
Volturnus, a river in Campania, II. ii. 7 (*cf.* note); III. xiv. 2

Xanthippus, the Spartan commander of Carthaginian troops, II. ii. 11; II. iii. 10
Xanthus, a Boeotian, II. v. 41
Xenophon, the historian, I. iv. 10; IV. ii. 8; IV. vi. 2
Xerxes, king of Persia, I. iii. 6; II. ii. 14; II. vi. 8; IV. ii. 9

Zeugma, a town in Syria, I. i. 6
Zopyrus, a Persian noble, III. iii. 4

INDEX OF PROPER NAMES IN THE AQUEDUCTS

(*The References are to Chapters.*)

Acilius Aviola, 102
Aelius Tubero, 2, 99, 100, 104, 106, 108, 125, 127
Agrippa, 3, 9, 10, 25, 98, 99, 104, 116
Albius Crispus, 102
Albudinus spring, 14
Alsietina, 4, 11, 18, 22, 71, 85
Alsietina, which is called Augusta, 4, 11
Alsietinus lake, 11, 71
Anio river, 90, 93
Anio Novus, 4, 13, 15, 18, 20, 21, 68, 86, 91, 93, 104, 105
Anio Vetus, 4, 6, 7, 9, 13, 18, 21, 66, 67, 80, 91, 92, 125
Anios, two aqueducts, 13, 90
Antistius Vetus, C., 102
Apollo, spring of, 4
Appia, 4, 5, 6, 7, 9, 18, 22, 65, 79, 125
Appian Way, 5
Appius Claudius Crassus (Caecus), **5**
Appius Claudius Pulcher, **7**
Aquila Julianus, 13, 102
Asinian Gardens, 21
Asinius Celer, 102
Asinius Pollio, C., 102
Ateius Capito, 97, 102
Augusta, 5, 12, 14, 65, 72
Augustus, 5, 9, 11, 12, 97, 98, 99, 100, 104, 108, 11**6**, 125
Aurelius Cotta, L., **7**
Aventine Mount, 20, 22, 76, 87

Caecilius, Q., **7**
Caelian Mount, 19, 20, 22, 76, 87
Caelius Rufus, 76
Caerulean spring, 13, 14, 72
Caesar, Caligula, 13
Caesar, emperor, 3, 23, 33, 78, 79, 80, 81, 82, 83, 84, 85, 86, 103, 105, 116, 118
Camenae, spring of, **4**
Campania, 24
Capena Gate, 5, 19
Capitolium, **7**, 8
Capua, 5
Careiae, 71
Carvilius, Sp., **6**
Cassius Longinus Ravilla, L., 8
Cejonius Commodus, 70
Circus Maximus, 97
Claudia, 4, 13, 14, 15, 18, 20, 69, 72, 76, 86, 87, 89, 91, 104, 105
Claudian Way, 11
Claudius, 13, 20, 76, 105, 116
Cocceius Nerva, M., 102
Collatian Way, 5, 10
Cominius, L., 99
Cornelius Cethegus, 102
Cossus, 102
Crabra brook, 9
Crassus Frugi, 102
Curius Dentatus, 6
Curtian spring, 13, 14, 72

Decius Mus, 5
Didius Gallus, 102
Domitian, 102, 118
Domitius Afer, 102

Epaphroditian Gardens, 5, 68
Esquiline Gate, 21

Fabius Maximus, 99, 100, 104, 106, 108, 125, 127
Fabius Persicus, 102
Fabius, 96
Fenestella, **7**

INDEX OF PROPER NAMES

Fonteius Agrippa, 102
Fulvius Flaccus, 6
Fulvius Flaccus, M., **8**

Galerius Trachalus, 102
Gemelli, 5, 65
Greeks, 16

Herculaneus rivus, 15, 19

Italia, 24
Italians, 18

Julia, 4, 9, 10, 18, 19, 68, 69, 76, 83, 125
Julius Caesar, 129
Juturna, spring of, 4

Labican Way, 21
Laecanius Bassus, 102
Laelius, 7
Latin Way, 8, 9, **19, 21**
Lentulus, 7
Lepidus, 7
Licinius, 96
Lucius Telesinus, 102
Lucretius, 10
Lucullan Gardens, 22
Lucullan Field, 5, 8, 10

Marcia, 4, 7, 9, 12, 13, 14, 18, 19, **67, 68, 72, 76, 81, 87, 89, 91, 92, 93**, 125
Marcius Rex, 7
Marius, P., 102
Martius, Campus, 22
Memmius Regulus, 102
Messala Corvinus, 99, 102

Naumachia, 11, 22
Nero, 7, 76, 102
Neronian Sublacensian villa, 93
Nerva, 1, 64, 87, 88, 102, 118
New Way, 21
Nonius Asprenas, 13, 102
Nonius Quintilianus, 102

Octavian channel, 21
Octavius Laenas, 102

Palatium, 20
Pallantian Gardens, **19, 20**, 69
Papirius Cursor, 6
Petronius Turpilianus, 102
Piso, 102

Plancus, 102
Plautius Hypsaeus, 8
Plautius Venox, 5
Pompeius Longus, 102
Pompeius Silvanus, 102
Porcius Cato, 102
Postumius Sulpicius, 99
Praenestine Way, 5
Publician Hill, 5, 22
Pyrrhus, 6

Quintius Crispinus, 129

Roman empire, 119
Roman people, 129
Romans, 4
Rome, 3, 7, 8, 9, 10, 91, 129

Sabatinus lake, 71
Saepta, 22
Salinae, 5
Sentius, C., 10
Sergian tribe, 129
Servilius Caepio, Q., 7
Servilius Caepio, Cn., 8
Sibylline books, 7
Silius, 102
Silius Italicus, 102
Simbruvium, 15
Spes Vetus, 5, 19, 20, 21, 65, **76**, 87
Sublacensian Way, 7, 14, 15
Sublacensian Neronian villa, 93
Suetonius Paulinus, 102
Sulla, 13
Sulpicius Galba, 7

Tampius Flavianus, 102
Tarius Rufus, 102
Tepula, 4, 8, 9, 18, 19, **67, 68, 69**, 82, 125
Tergemina Gate, 5
Tiber, 4
Tiberius, **13**
Tibur, 6
Tiburtines, 6, 66
Titianus, **13**
Titus, 102
Torquatian Gardens, 5
Trajan, 64, 87, 88, 93
Transtiberine ward, 11, 18, 20
Treba Augusta, 93
Tusculan Field, 8
Tusculanians, 9

INDEX OF PROPER NAMES

Valerian Way, 7
Valerius Maximus, 5
Valerius Messalinus, 102
Varro, Sex., 129
Veranius, 102
Verginius Rufus, 102
Vespasian, 102

Viminal Gate, 19
Viminal Hill, 19
Virgo, 4, 10, 18, 22, **70, 84**
Visellius Varro, 102
Vitellius, 102
Vitruvius, 25
Volcatius, 9

PORTA MAGGIORE

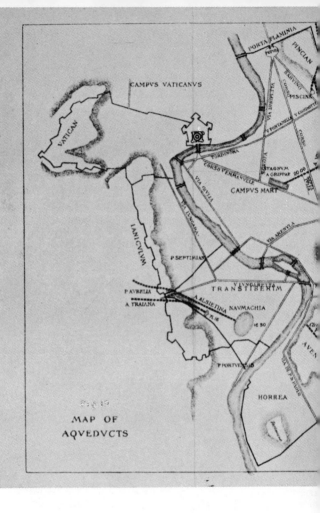

MAP OF AQVEDVCTS

REPRODUCED BY PERMISSION OF THE HOUGHTON MIFFLIN COMPANY, FROM LANCIANI'S "RUINS AND EXCAVATIONS OF ANCIENT ROME"

RUINS OF AQUA CLAUDIA

THE SEVEN AQUEDUCTS AT THE PORTA MAGGIORE

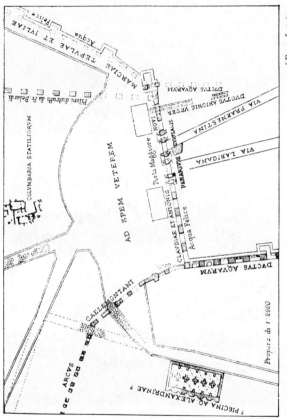

[From Lanciani

RUINS OF AQUA CLAUDIA NEAR THE APPIAN WAY

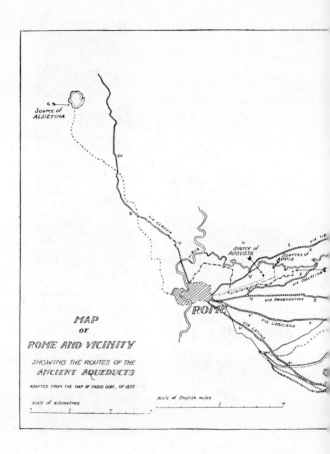

LEGEND

Roads: ············
Aqueducts:
1. APPIA
2. ANIO VETUS
3. MARCIA
4. TEPULA
5. JULIA
6. ALSIETINA
7. CLAUDIA
8. ANIO NOVUS
9. FELICE
10. VERGINE
11. AUGUSTA

493

THE WATER SUPPLY OF THE CITY OF ROME

TABLE I.

Names of the Aqueducts	Height of the water in Rome above the Tiber wharves	Wards within the City receiving Water	Taps in Actual Use	Outside the City			No. of De-livery Tanks	Within the City			
	Rondelet in ft. Eng.			Total	In the name of Caesar	By private parties		Total	In the name of Caesar	By private parties	For public uses
				Quinariae				*Quinariae*			
Appia	28	2, 8, 9, 11, 12, 13, 14	704	5	0	5	20	699	151	194	354
Anio Vetus	84	1, 3, 4, 5, 6, 7, 8, 9, 12, 14	1,610	508	104	404	35	1,102	60	490	552
Marcia	125	1, 3, 4, 5, 6, 7, 8, 9, 10, 14	1,935	837	269	568	51	1,098	116	543	439
Tepula	128	4, 5, 6, 7	445	114	58	56	14	331	34	247	50
Julia	133	2, 3, 5, 6, 8, 10, 12	803	206	85	121	17	597	18	196	383
Virgo	35	7, 9, 14	2,504	200	...	200	18	2,304	549	...	1,417
Alsietina	(?)	Outside the city	392	392	254	138		0	...	...	...
Claudia	158		5,625	1,801	217	439	92	3,824	779	1,839	1,206
Anio Novus	158	In all the 14 wards			731	414					
Totals			14,018	4,063	1,718	2,345	247	9,955	1,707	3,847	4,401

REMARK.—These figures are as adjusted by Poleni and others so as to conform to the rules of arithmetic.

TABLE II.

Names of the Aqueducts	For public uses within the city	For camps		For public structures		For ornamental fountains		For water basins	
	Quinariae	Number	*Quinariae*	Number	*Quinariae*	Number	*Quinariae*	Number	*Quinariae*
Appia	354	1	4	14	123	1	2	92	226
Anio Vetus	552	1	50	18	196	9	88	94	218
Marcia	439	4	41	15	41	12	104	113	253
Tepula	50	1	12	3	7	…	…	13	31
Julia	383	3	69	10	182	3	67	28	65
Virgo	1,417	…	…	16	1,330	2	26	25	61
Alsietina	…	…	…	…	…	…	…	…	…
Claudia	} 1,206	9	104	18	522	12	99	226	481
Anio Novus									
Totals	4,401	19	279	94	2,401	39	386	591	1,335

REMARK.—These figures are as adjusted by Poleni and others so as to conform to the rules of arithmetic.

TABLE III.

Names of the Aqueducts	Height of the water in Rome above the Tiber wharves Rondelet in ft. Eng	Wards within the city receiving water	Taps in actual use	Outside the City			No. of Delivery Tanks	Within the City			
				Total	In the name of Caesar	By private parties		Total	In the name of Caesar	By private parties	For public uses
				Quinariae				*Quinariae*			
Appia	28	2, 8, 9, 11, 12, 13, 14	704	5	0	5	20	699	151	194	354
Anio Vetus	84	1, 3, 4, 5, 6, 7, 8, 9, 12, 14	262 {1,348}	...	169	404	35	1,508.5	66.5	490	503
Marcia	125	1, 3, 4, 5, 6, 7, 8, 9, 10, 14	95 {1,840}	...	261.5	...	51	1,472	116	543	(?)
Tepula	128	4, 5, 6, 7	92 {190} 163	...	58	56	14	331	42	237	50
Julia	133	2, 3, 5, 6, 8, 10, 12	803	...	85	121	17	548	18	(?)	383
Virgo	35	7, 9, 14	2,504	200	...	...	18	2,304	509	338	1,167
Alsietina	(?)	Outside the city	392	392	354	138	...	0	...	...	...
Claudia	158	(1,750–162)	(1,588)	...	246	439	} 92	3,498	819	1,067	1,012
Anio Novus	158	In all the 14 Wards (4,200–163)	(4,037)	...	728	...					
Totals			14,018	4,063	1,718	2,345	247	9,955	1,707.5	3,847	4,401

REMARK.—These figures are as given in the original Monte Cassino codex. (?) means an omission due to a defect in the manuscript

TABLE IV.

Names of the Aqueducts	For public uses within the city Quinariae	For camps Number	For camps Quinariae	For public structures Number	For public structures Quinariae	For ornamental fountains Number	For ornamental fountains Quinariae	For water basins Number	For water basins Quinariae
Appia	354	1	4	14	123	1	2	92	226
Anio Vetus	503	1	50	19	196	9	88	94	218
Marcia	(?)	4	42.5	15	15	12	104	113	256
Tepula	50	1	12	3	7	...	...	13	32
Julia	383	(?)	69	(?)	181	3	67	28	65
Virgo	1,167	...	...	16	1,380	2	26	25	51
Alsietina	...	...	...	...	...	...	...	...	...
Claudia	} 1,012	9	149	18	374	12	107	226	481
Anio Novus									
Totals	4,401	(?)	279	75	2,301	39	386	591	1,335

Remark.—These figures are as given in the original Monte Cassino codex. (?) means an omission due to a defect in the manuscript.